I0056162

Encyclopedia of Zoology

Volume I

Encyclopedia of Zoology
Volume I

Edited by **Carlos Wyatt**

R CALLISTO REFERENCE

New York

Published by Callisto Reference,
106 Park Avenue, Suite 200,
New York, NY 10016, USA
www.callistoreference.com

Encyclopedia of Zoology: Volume I
Edited by Carlos Wyatt

© 2015 Callisto Reference

International Standard Book Number: 978-1-63239-307-4 (Hardback)

This book contains information obtained from authentic and highly regarded sources. Copyright for all individual chapters remain with the respective authors as indicated. A wide variety of references are listed. Permission and sources are indicated; for detailed attributions, please refer to the permissions page. Reasonable efforts have been made to publish reliable data and information, but the authors, editors and publisher cannot assume any responsibility for the validity of all materials or the consequences of their use.

The publisher's policy is to use permanent paper from mills that operate a sustainable forestry policy. Furthermore, the publisher ensures that the text paper and cover boards used have met acceptable environmental accreditation standards.

Trademark Notice: Registered trademark of products or corporate names are used only for explanation and identification without intent to infringe.

Printed in the United States of America.

Contents

Preface

Zoology is also known as animal biology. It is a wide branch of biological sciences, which is related to the studies and researches associated with the animal kingdom. If we turn the pages of history all over the world, zoology was initially associated with natural history and can be traced to the works of Aristotle and Galen in the ancient Greco-Roman world. With a renewed interest in empiricism, zoological thought was revolutionized in Europe in the 18th century.

This field of biology is mainly focussed on the structural aspects, embryology, evolution of living organisms, classifications, habits, and the distribution of organisms, both living and extinct. This field of study is very old but has evolved itself with time. In fact, today, the subject is referred to as modern zoology. This field has many areas of studies such as zoography (related to animal and their habitat), comparative anatomy, animal physiology, behavioural ecology, ethology, and taxonomically oriented disciplines such as mammalogy, herpetology, ornithology and entomology among others.

Instead of organizing the book into a pre-formatted table of contents with chapters and then asking authors to submit their respective chapters, the authors were encouraged by the publisher to submit chapters based on their area of expertise. The editor was then commissioned to examine the reading material and compile it together as a book.

I hope that this book can provide zoology professionals with a solid grounding in this subject so that they can actively involve and participate with the subject in future. I wish to thank all the authors for their efforts and time that they have given to this project. Without their dedication and timely submissions, this publication wouldn't have been possible. I must also acknowledge the editorial team at the publishing house, who have done a tremendous job with this book in terms of collecting the most relevant data. Last but not the least, I wish to thank my family and friends, who have supported me in my life through everything.

Editor

From Citizen Science to Policy Development on the Coral Reefs of Jamaica

M. James C. Crabbe

Institute of Biomedical and Environmental Science Technology and Faculty of Creative Arts, Technologies and Science, University of Bedfordshire, Park Square, Luton LU1 3JU, UK

Correspondence should be addressed to M. James C. Crabbe, james.crabbe@beds.ac.uk

Academic Editor: Richard Stafford

This paper explores the application of citizen science to help generation of scientific data and capacity-building, and so underpin scientific ideas and policy development in the area of coral reef management, on the coral reefs of Jamaica. From 2000 to 2008, ninety Earthwatch volunteers were trained in coral reef data acquisition and analysis and made over 6,000 measurements on fringing reef sites along the north coast of Jamaica. Their work showed that while recruitment of small corals is returning after the major bleaching event of 2005, larger corals are not necessarily so resilient and so need careful management if the reefs are to survive such major extreme events. These findings were used in the development of an action plan for Jamaican coral reefs, presented to the Jamaican National Environmental Protection Agency. It was agreed that a number of themes and tactics need to be implemented in order to facilitate coral reef conservation in the Caribbean. The use of volunteers and citizen scientists from both developed and developing countries can help in forging links which can assist in data collection and analysis and, ultimately, in ecosystem management and policy development.

1. Introduction

Coral reefs throughout the world are under severe challenges from a variety of anthropogenic and environmental factors including overfishing, destructive fishing practices, coral bleaching, ocean acidification, sea-level rise, algal blooms, agricultural run-off, coastal and resort development, marine pollution, increasing coral diseases, invasive species, and hurricane/cyclone damage [1–3]. It is the application of citizen science to help generation of scientific data and capacity-building, and so underpin scientific ideas and policy development in the area of coral reef management, that are explored in this paper, concentrating on Jamaican coral reefs.

The "compulsive" appetite for increasing mobility [4] allied to a social desire for extraordinary "peak experiences" [5] has led to the modern "ethical consumer" for tourism services [4, 6] derived from the "experiential" and "existential" tourist of the 1970s [7]. Several organisations have taken the concept of ecotourism further to embracing tourism with citizen science, whereby the tourist gets to work on research projects under the supervision of recognised researchers.

Several organisations worldwide have developed citizen science programmes. The drivers behind these activities vary significantly between scientific studies, education, and/or getting the public more engaged and raising awareness of the natural environment. The overall driver cannot only determine the type, quality, and quantity of data required but also the level of volunteer expertise needed. Three organisations that have developed citizen science with tourism are the Earthwatch Institute, (http://www.earthwatch.org/), Operation Wallacea (http://www.opwall.com/), and Coral Cay Conservation (http://www.coralcay.org/). All are international environmental charities, working with a wide range of partners, from individuals who work as conservation volunteers on research teams through to corporate partners (such as HSBC with Earthwatch), governments, and institutions. Research volunteers work with scientists and social scientists around the world to help gather data needed to address environmental and social issues. It is the long-term strategy of these organisations combined with their citizen science funding models that underpins their successes; they are "in for the long haul" and can effect conservation in a different way to a standard 3-year research grant. Key elements are developing projects that can be used by volunteers and verifying the scientific information in a statistically significant way. This paper shows how an Earthwatch programme using volunteers on coral reefs generated scientific information which was used to inform management strategies in Jamaica.

2. Materials and Methods

2.1. Training of Volunteers. All volunteers were SCUBA divers of at least PADI Open Water standard. Training took place at the Discovery Bay Marine Laboratory, Jamaica, and consisted of lectures and interactive discussions covering scleractinian coral biology and taxonomy, coral recognition, data measurements and analysis, and health and safety. Volunteers all had to accomplish open water diving tests, and coral recognition tests in the field, after studying coral taxonomy books and passing land-based tests.

2.2. Reef Sites. Four randomly located transects, each 15 m long and separated by at least 5 m, were laid at 5–8.5 m depth at each of five sites on the North coast of Jamaica near Discovery Bay: Rio Bueno (18° 28.805′ N; 77° 21.625′ W), M1 (18° 28.337′ N; 77° 24.525′ W), Dancing Ladies (18° 28.369′ N; 77° 24.802′ W), Dairy Bull (18° 28.083′ N; 77° 23.302′ W), and Pear Tree Bottom (18° 27.829′ N; 77° 21.403′ W). These sites were chosen as being workable by volunteers, as they were with 20 min boat ride from the Discovery Bay Marine Laboratory, where all volunteers stayed. Sites had been studied before over a number of years by marine scientists from many countries. GPS coordinates were determined using a hand-held GPS receiver (Garmin Ltd., UK).

2.3. Citizen Science Data Collection. Corals 2 m either side of the transect lines were photographed for archive information, and surface areas were measured with flexible tape as described previously using SCUBA [8–10]. Depth of samples was between 5 and 8.5 m, to minimise variation in growth rates due to depth [11]. To increase accuracy, surface areas rather than diameters of live nonbranching corals were measured [8, 9]. Sampling was over as wide a range of sizes as possible. Colonies that were close together (<50 mm) or touching were avoided to minimise age discontinuities through fission and altered growth rates [12–14]. In this study *Montastrea annularis* colonies were ignored, because their surface area does not reflect their age [12], and because hurricanes can increase their asexual reproduction through physical damage [13]. Overall, over 6,000 measurements were made on over 1,000 coral colonies, equally distributed between the sites for species and numbers of colonies.

This work was conducted at Discovery Bay during July 15–31 and December 19–30 in 2000, March 26–April 19 in 2002, March 18–April 10 in 2003, July 23–August 21 in 2004, July 18–August 13 in 2005, April 11–18 in 2006, December 30 in 2006–January 6 in 2007, and July 30–August 16 in 2008. Surveys were made at the same locations at the same sites

each year. Data from ninety volunteers was used over this period.

2.4. Storm Severity. Data on storm severity as it impacted the island was obtained from UNISYS (http://weather.unisys.com/hurricane/atlantic/), the NOAA hurricane site (http://www.nhc.noaa.gov/pastall.shtml). Information on bleaching was obtained from the NOAA coral reef watch site (http://coralreefwatch.noaa.gov/satellite/current/sst_series_24reefs.html).

2.5. Data Analysis. Data analysis on corals was using ANOVA. Skewness (sk, [15]) was used to estimate the distribution of small and large colonies in the coral populations around Discovery Bay in Jamaica. In a normal distribution, approximately 68% of the values lie within one standard deviation of the mean. If there are extreme values towards the positive end of a distribution, the distribution is positively skewed, where the mean is greater than the mode (the mode is the value that occurs the most frequently in a data set) (right tail is longer). The opposite is true for a negatively skewed distribution, where the mean is less than the mode (left tail is longer). With regard to coral populations, negative skewness implies more large colonies than small colonies, while positive skewness implies more small colonies than large colonies.

3. Results

3.1. Coral Sizes and Growth. All the Jamaican sites showed some similarities in distribution of the size classes for the species studied between 2002 and 2008. However, there were differences between the different sites and between the different species studied at the sites. Skewness values (sk) were used to compare the distribution of the data between 2002 and 2008. For *S. siderea*, all sk values were positive, with more small colonies than in a normal distribution for 2002 and 2008, with little change between the dates (all sk values between 0.5 and 1.6). With *D. labyrinthiformis* colonies, there was a change from negative skewness in 2002 at Dairy Bull and Pear Tree Bottom, with more large colonies than in a normal distribution (sk values −0.25 and −0.006, resp.) to smaller colonies than in a normal distribution in 2008 (sk values of 0.20 and 0.97, resp.). There were no significant changes from 2002 to 2008 at the other sites, with positive sk values from 0.1 to 0.89. *M. meandrites* colonies at Rio Bueno and Dairy Bull showed a relative decrease in the distribution of larger colonies from 2002 to 2008, with changes in sk values from −0.03 in 02 to 0.78 in 08, and from −0.05 to 0.03, respectively; the other sites all exhibited slightly positive sk values in both years from 0.1 to 0.5. For *Agaricia* species, there was very little change between the years at all the sites, with sk values from 0.4 to 1.6. For *P. astreoides*, all values were positive for both years, with an increase in skewness at Rio Bueno from 0.2 to 2.6, showing a marked change in distribution towards the smaller colony sizes. At the other sites there were only small increases in sk values from 2002 to 2008, with Pear Tree Bottom showing a decrease in skewness from 0.9 to 0.6. *D. strigosa* colonies showed similar results to *P. astreoides*, all sk values being positive for 2002 and 2008, with an increase at Rio Bueno from 0.2 to 2.2 and at Pear Tree Bottom from 0.4 to 2.4; other sites showed similar sk values for 2002 and 2008 from 0.6 to 1.6. *C. natans* skewness changed from -0.07 to 0.68 at Rio Bueno from 2002 to 2008 (a decrease in larger colonies relative to a normal distribution) and at Dancing Ladies from -0.31 to 0.38. Other sites showed similar skewness in 2002 and 2008 (sk values between 0.5 and 0.6), except Pear Tree Bottom, which exhibited near normal distribution of colonies about the mean for both 2002 and 2008 (sk values <0.01). Interestingly, in 2005, the year after hurricane Ivan, the most severe storm to impact the reef sites over the study period, there was a slight reduction in the numbers of the smallest size classes, particularly notable at Dairy Bull.

In addition, our volunteer studies showed that radial growth rates (mm/yr) of non-branching corals calculated on an annual basis from 2000 to 2008 showed few significant differences either spatially or temporally along the North coast, although growth rates tended to be higher on reefs of higher rugosity and lower macroalgal cover [16].

3.2. Extreme Climate Events. The only extreme climate event that significantly impacted the Jamaican reef sites during the study period was the mass Caribbean bleaching event of 2005 [17]. Analysis of satellite data showed that there were 6 degree heating weeks (dhw) for sea surface temperatures in September and October 2005 near Discovery Bay, data which was mirrored by data loggers on the reefs.

3.3. Development of Coral Reef Action Plan. The coral size and growth data collected by the citizen scientists show that corals of above average size for their species at the sites studied lack resilience, particularly after the major bleaching event of 2005. Because of this, there is a need for different zones to have different levels of protection. To this end, the data was used in the development of an action plan for Jamaican coral reefs, presented to the Jamaican National Environmental Protection Agency, and described in Table 1.

4. Discussion

4.1. Citizen Science and Use of Volunteer Data. Citizen science and use of data measured by volunteers has been very helpful in a number of zoological areas, including amphibian population and biodiversity studies [18, 19], reporting invasive species [20], environmental monitoring [21], evolutionary change [22], marine species abundance and monitoring [23–25], dryland mapping [26], and conservation planning [27, 28].

This study used self-selected as "Earthwatch volunteers", and all were SCUBA divers. Motivation was high in all the volunteers, as was the validity of the data presented by volunteers. A key element in citizen science is good training of volunteers. In the area of coral reef research described in this study, training was given in species recognition, quantitative measurement techniques and validation, and

TABLE 1: Seven-point action plan for Jamaican coral reefs.

(1) The reefs around Jamaica could be designated as the Jamaican Coral Reef Marine Park. This could include all the fringing reefs, seagrass beds, and mangroves from Negril to all along the north coast to the eastern tip of the island. On the south coast it could include Port Royal and Portland Bight. The advantage of this is that one can then consider protection of the Jamaican reefs as a whole. Another advantage is that climate change effects can be considered in a more holistic way

(2) There could be a single body, possibly the National Environment Protection Agency (NEPA), or a subset of NEPA, given authority to manage the Park

(3) There could be a statement drawn up on "protection and wise use" of the Park. Drawing up that statement should include all stakeholders, from fishermen through Industry and tourism to policy makers

(4) The Park could be managed using a "zoning" system. This has been valuable in a number of areas, not least the Great Barrier Reef. This will allow some areas to have greater restrictions (e.g., fishing, resort pollution, ship pollution) than others. Such zoning should help avoid the "tragedy of the commons". Zoning Plans define what activities can occur in which locations, both to protect the marine environment and to separate potentially conflicting activities

(5) Divisions into zones could be
General Use,
Conservation Park,
Habitat Protection,
Marine National Park,
Another zone might be a Buffer Zone, next to a Marine National Park

(6) Each zone should have at least one of the following: (i) Community Partnerships, (ii) Local Marine Advisory Committees, and (iii) Reef Advisory Committees. These bodies should be responsible for regulating their own area and should be responsible to the overall Marine Park Management body. They would also be responsible for community involvement and information

(7) Permissions within the zones (e.g., for tourism, fishing, etc.) would be given by the Jamaican Government, through NEPA

data analysis. Independent validation of volunteer data, once training had been given, was consistent with previous findings by other groups [29]. The validation of the data produced by the volunteers indicated that with appropriate training, data collection by citizen scientists is appropriate for scientific applications in marine biology.

4.2. Coral Health and Resilience. What is apparent from our studies is that despite the chronic and acute disturbances between 2002 and 2008, demographic studies indicate good levels of coral resilience on the fringing reefs around Discovery Bay in Jamaica (see also [30]). The bleaching event of 2005 resulted in mass bleaching but relatively low levels of mortality unlike corals in the US Virgin islands and Tobago where there was extensive mortality [17, 31], probably because of their greater degree heating week values.

This data shows that while recruitment of small corals is returning after the major bleaching event of 2005 [32], larger corals are not necessarily so resilient and so need careful management if the reefs are to survive such major extreme events.

4.3. From Information to Policy Development: Themes and Tactics. Marine reserves are an important tool in the sustainable management of many coral reefs [33]. However, it is important that the reef ecosystems share regulatory guidelines, enforcement practices and resources, and conservation initiatives and management, underpinned by scientific research. An example of a single marine reserve is the Great Barrier Reef in Australia operated and managed solely by the Great Barrier Reef Marine Park Authority (GBRMPA). In contrast, the second largest barrier reef in the world, the MesoAmerican Barrier Reef, is bounded by four countries (Mexico, Belize, Guatemala, and Honduras), each with its own laws and policies. Here, a number of single and separated marine reserves exist along the barrier reef. In Belize we have successfully transferred scientific expertise in Belize to local volunteers to generate scientific evidence to underpin future management and conservation decisions, as judged, for example, by scientific findings on the impact of hurricanes on reefs in Belize, which showed that hurricanes and severe storms limited the recruitment and survival of nonbranching corals of the Mesoamerican barrier reef [10].

For Jamaica, the Action Plan developed (Table 1) was well received by managers of the National Environment Protection Agency (NEPA). It was felt by managers that this approach could link together the environment with tourism and business, so that environmental issues are seen as part of the way forward, not part of the problem, as has been all too evident in the past. Even if smaller Marine Protected Areas (MPAs) were developed around the island, the adoption of shared ownership of reef ecosystems was felt to be useful way to proceed.

In order to take this forward, it was felt necessary to develop a number of themes and tactics. In a separate capacity building exercise [34], for the MesoAmerican Barrier Reef in Sothern Belize, one officer from the Belize Fisheries Department, three senior officers from NGOs involved in managing Belize MPAs (TIDE, the Toledo Institute for Development and Environment; TASTE, the Toledo Association for Sustainable Tourism and Empowerment; and Friends of Nature), and a Facilitator (the author) from the UK developed six-month Personal/Professional Action Plans which involved

(a) tactics for leading, educating, and supporting issues regarding sustainable development of coral reefs;

(b) tactics for collaboration with other stakeholders to collectively influence policy decisions for coral reef conservation.

Discussion among the participants and facilitator resulted in the generation of a series of generic tactics to be adopted around a number of themes. These are enumerated in Table 2. Such themes and tactics may be

TABLE 2: Themes and tactics to facilitate conservation of coral reefs.

Organisation and Management

Tactic number 1: establish a key leader in the Organization/Department to effectively manage the Marine reserves on a day-to-day basis

Tactic number 2: have a selected key leader provide general Terms of Reference of what is expected of staff and immediate/major stakeholders in order to easily facilitate the process of decision making

Education

Tactic number 1: financial resources need to be allocated for an education program. The program should focus on both broad and specific issues that may create friction among stakeholders in the process

Tactic number 2: a group consisting of community leaders and key/immediate stakeholders should be established to create ways and methods of educating different levels of stakeholders in the effectiveness of sustainable development in the marine parks

Tactic number 3: surveys need to be conducted to evaluate level of success and failure. Too often programmes have been formed and implemented but end results have not been evaluated. Surveys should be carried back to stakeholders for a presentation to establish further steps

Support

Tactic number 1: a well-put together presentation needs to be developed and be presented to the key authority that will have over-all say in the marine park(s). This will stress on the support needed to accomplish both the mission and vision statements and will have positive effects in sustainable development

Tactic number 2: nonmonetary incentives need to be established in order to have full support of stakeholders who would otherwise deter progress in sustainable development

Policies

Tactic number 1: establish a set of policies that is considered necessary for proper management of the marine reserves. Such policies will be established by all stakeholders involved

Tactic number 2: create an influencing program for stakeholders to adhere to such policies through an education/retreat program

Tactic number 3: establish exchanges with other organizations in capacity building in policy creation and effective implementation

useful in development of coral reef policies in the Caribbean and elsewhere.

5. Conclusion

The use of volunteers and citizen scientists from both developed and developing countries can help in forging links which can assist in data collection and analysis and, ultimately, in ecosystem management and policy development. There is much progress internationally in involving organisations to utilize citizen science effectively and efficiently (e.g., [35]).

A number of questions remain for the future, for example, assessing how citizen science could be used to better effect, for example, identifying the potential for citizen science to fill known data gaps, for example, gaps in marine and terrestrial taxonomies. In addition, we need greater understanding of where and how technology (software, statistics) can transform the quality and quantity of data from nonexperts, and how scientists can make best use of technology, for example, in using smart phone apps to identify and/or record species and measurements.

Acknowledgments

The author thanks the Earthwatch Institute, the Royal Society, and the Oak Foundation (USA) for funding, Mr. Anthony Downes, Mr. Peter Gayle, and the staff of the Discovery Bay Marine Laboratory for their invaluable help and assistance, to the two anonymous referees for their valuable comments which improved the manuscript, to E. Martinez, C. Garcia, J. Chub, L. Castro, J. Guy, and Earthwatch colleagues in Belize, and to many volunteers for their considerable help underwater during this project.

References

[1] T. A. Gardner, I. M. Côté, J. A. Gill, A. Grant, and A. R. Watkinson, "Long-term region-wide declines in Caribbean corals," *Science*, vol. 301, no. 5635, pp. 958–960, 2003.

[2] D. R. Bellwood, T. P. Hughes, C. Folke, and M. Nyström, "Confronting the coral reef crisis," *Nature*, vol. 429, no. 6994, pp. 827–833, 2004.

[3] M. J. C. Crabbe, E. E. L. Walker, and D. B. Stephenson, "The impact of weather and climate extremes on coral growth," in *Climate Extremes and Society*, H. F. Diaz and R. J. Murnane, Eds., pp. 165–188, Cambridge University Press, New York, NY, USA, 2008.

[4] P. Burns, "Tribal tourism-cannibal tours: tribal tourism to hidden places," in *Niche Tourism: Contemporary Issues, Trends, Cases*, pp. 101–110, Elsevier, Oxford, UK, 2005.

[5] A. Holden, *Environment and Tourism*, Routledge, London, UK, 2nd edition, 2008.

[6] D. Fennell, *Ethical Tourism*, Routledge, London, UK, 2006.

[7] E. Cohen, "A phenomenology of tourist experiences," *Sociology*, vol. 13, pp. 179–201, 1979.

[8] M. J. C. Crabbe, J. M. Mendes, and G. F. Warner, "Lack of recruitment of non-branching corals in Discovery Bay is linked to severe storms," *Bulletin of Marine Science*, vol. 70, no. 3, pp. 939–945, 2002.

[9] M. J. C. Crabbe and D. J. Smith, "Sediment impacts on growth rates of Acropora and Porites corals from fringing reefs of Sulawesi, Indonesia," *Coral Reefs*, vol. 24, no. 3, pp. 437–441, 2005.

[10] M. J. C. Crabbe, E. Martinez, C. Garcia, J. Chub, L. Castro, and J. Guy, "Growth modelling indicates hurricanes and severe storms are linked to low coral recruitment in the Caribbean," *Marine Environmental Research*, vol. 65, no. 4, pp. 364–368, 2008.

[11] M. Huston, "Variation in coral growth rates with depth at Discovery Bay, Jamaica," *Coral Reefs*, vol. 4, no. 1, pp. 19–25, 1985.

[12] T. P. Hughes and J. B. C. Jackson, "Do corals lie about their age? Some demographic consequences of partial mortality, fission, and fusion," *Science*, vol. 209, no. 4457, pp. 713–715, 1980.

[13] N. L. Foster, I. B. Baums, and P. J. Mumby, "Sexual vs. asexual reproduction in an ecosystem engineer: the massive coral Montastraea annularis," *Journal of Animal Ecology*, vol. 76, no. 2, pp. 384–391, 2007.

[14] R. Elahi and P. J. Edmunds, "Consequences of fission in the coral Siderastrea siderea: growth rates of small colonies and clonal input to population structure," *Coral Reefs*, vol. 26, no. 2, pp. 271–276, 2007.

[15] J. H. Zar, *Biostatistical Analysis*, Prentice-Hall, Upper Saddle River, NJ, USA, 4th edition, 1999.

[16] M. J. C. Crabbe, "Climate change and tropical marine agriculture," *Journal of Experimental Botany*, vol. 60, no. 10, pp. 2839–2844, 2009.

[17] C. M. Eakin, J. A. Morgan, S. F. Heron et al., "Caribbean corals in crisis: record thermal stress, bleaching, and mortality in 2005," *PLoS One*, vol. 5, no. 11, article e13969, 2010.

[18] A. Bonardi, R. Manenti, A. Corbetta et al., "Usefulness of volunteer data to measure the large scale decline of "common" toad populations," *Biological Conservation*, vol. 144, no. 9, pp. 2328–2334, 2011.

[19] D. Sewell, T. J. C. Beebee, and R. A. Griffiths, "Optimising biodiversity assessments by volunteers: the application of occupancy modelling to large-scale amphibian surveys," *Biological Conservation*, vol. 143, no. 9, pp. 2102–2110, 2010.

[20] T. Gallo and D. Waitt, "Creating a successful citizen science model to detect and report invasive species," *BioScience*, vol. 61, no. 6, pp. 459–465, 2011.

[21] C. C. Conrad and K. G. Hilchey, "A review of citizen science and community-based environmental monitoring: issues and opportunities," *Environmental Monitoring and Assessment*, vol. 176, pp. 273–291, 2011.

[22] J. Silvertown, L. Cook, R. Cameron et al., "Citizen science reveals unexpected continental-scale evolutionary change in a model organism," *PLoS One*, vol. 6, no. 4, article e18927, 2011.

[23] C. A. Ward-Paige, C. Pattengill-Semmens, R. A. Myers, and H. K. Lotze, "Spatial and temporal trends in yellow stingray abundance: evidence from diver surveys," *Environmental Biology of Fishes*, vol. 90, pp. 263–276, 2010.

[24] S. Goffredo, F. Pensa, P. Neri et al., "Unite research with what citizens do for fun: recreational monitoring of marine biodiversity," *Ecological Applications*, vol. 20, no. 8, pp. 2170–2187, 2010.

[25] P. G. Finn, N. S. Udy, S. J. Baltais, K. Price, and L. Coles, "Assessing the quality of seagrass data collected by community volunteers in Moreton Bay Marine Park, Australia," *Environmental Conservation*, vol. 37, no. 1, pp. 83–89, 2010.

[26] D. S. Turner and H. E. Richter, "Wet/dry mapping: using citizen scientists to monitor the extent of perennial surface flow in dryland regions," *Environmental Management*, vol. 47, no. 3, pp. 497–505, 2011.

[27] P. De Ornellas, E. J. Milner-Gulland, and E. Nicholson, "The impact of data realities on conservation planning," *Biological Conservation*, vol. 144, no. 7, pp. 1980–1988, 2011.

[28] D. B. Oscarson and A. J. K. Calhoun, "Developing vernal pool conservation plans at the local level using citizen-scientists," *Wetlands*, vol. 27, no. 1, pp. 80–95, 2007.

[29] P. J. Mumby, A. R. Harborne, P. S. Raines, and J. M. Ridley, "A critical assessment of data derived from coral cay conservation volunteers," *Bulletin of Marine Science*, vol. 56, no. 3, pp. 737–751, 1995.

[30] M. James and C. Crabbe, "Coral resilience on the reefs of Jamaica," *Underwater Technology*, vol. 30, no. 2, pp. 65–70, 2011.

[31] J. Mallela and M. J. C. Crabbe, "Hurricanes and coral bleaching linked to changes in coral recruitment in Tobago," *Marine Environmental Research*, vol. 68, no. 4, pp. 158–162, 2009.

[32] M. J. C. Crabbe, "Environmental effects on coral growth and recruitment in the Caribbean," *Journal of the Marine Biological Association of the UK*. In press.

[33] L. Cho, "Marine protected areas: a tool for integrated coastal management in Belize," *Ocean and Coastal Management*, vol. 48, no. 11-12, pp. 932–947, 2005.

[34] M. J. C. Crabbe, E. Martinez, C. Garcia, J. Chub, L. Castro, and J. Guy, "Is capacity building important in policy development for sustainability? A case study using action plans for sustainable marine protected areas in Belize," *Society and Natural Resources*, vol. 23, no. 2, pp. 181–190, 2010.

[35] "UK—Environmental Observation Framework, Workshop Report," pp. 45, 2011, http://www.ukeof.org.uk/documents/. ukeof-citizen-science-workshop-report.pdf.

Identifying Large- and Small-Scale Habitat Characteristics of Monarch Butterfly Migratory Roost Sites with Citizen Science Observations

Andrew K. Davis,[1] Nathan P. Nibbelink,[2] and Elizabeth Howard[3]

[1] *Odum School of Ecology, The University of Georgia, Athens, GA 30602, USA*
[2] *D.B. Warnell School of Forestry and Natural Resources, The University of Georgia, Athens, GA 30602, USA*
[3] *Journey North, 1321 Bragg Hill Road, Norwich, VT 05055, USA*

Correspondence should be addressed to Andrew K. Davis, akdavis@uga.edu

Academic Editor: Anne Goodenough

Monarch butterflies (*Danaus plexippus*) in eastern North America must make frequent stops to rest and refuel during their annual migration. During these stopovers, monarchs form communal roosts, which are often observed by laypersons. Journey North is a citizen science program that compiles roost observations, and we examined these data in an attempt to identify habitat characteristics of roosts. From each observation we extracted information on the type of vegetation used, and we used GIS and a national landcover data set to determine land cover characteristics within a 10 km radius of the roost. Ninety-seven percent of roosts were reported on trees; most were in pines and conifers, maples, oaks, pecans and willows. Conifers and maples were used most often in northern flyway regions, while pecans and oaks were more-frequently used in southern regions. No one landcover type was directly associated with roost sites, although there was more open water near roost sites than around random sites. Roosts in southern Texas were associated primarily with grasslands, but this was not the case elsewhere. Considering the large variety of tree types used and the diversity of landcover types around roost sites, monarchs appear highly-adaptable in terms of roost site selection.

1. Introduction

Research on one of the world's most famous insects, the monarch butterfly (*Danaus plexippus*, Figure 1), has benefitted greatly from numerous citizen science programs in North America devoted to tracking this species at various life stages. The attention given to this insect no doubt stems from its large size, easily identifiable orange and black colors (Figure 1), and its well-known and spectacular migrations, which are unique among butterflies. All of these factors make this butterfly extremely charismatic, and this helps to promote public participation in various citizen science programs. For example, the larval stages of this insect are monitored each summer by volunteers of the Monarch Larval Monitoring Project (http://www.mlmp.org/), and these data have been used to document geographic and temporal variation in population recruitment [1, 2]. In the western North American population, volunteers count numbers of adult monarchs that overwinter in clusters along the California coast (Western Monarch Thanksgiving Count), and a recent analysis of these data showed the importance of climatic conditions at the natal sites for predicting overwintering numbers [3]. There is another citizen science program whereby volunteers submit samples of a monarch-specific protozoan parasite (MonarchHealth; http://www.monarchparasites.org/), which has led to the identification of trends in disease prevalence during the summer and fall [4]. Finally, numerous scientific investigations have made use of data from a citizen science program called Journey North (http://www.learner.org/jnorth/), which asks volunteers in North America to report sightings of adult monarchs during the winter [5], during the spring migration [6–8], and during the fall migration when monarchs from

FIGURE 1: Photograph of an adult monarch butterfly (*Danaus plexippus*), nectaring on milkweed (*Asclepias* sp.). Photo taken by Pat Davis in New York City, NY.

the eastern population are travelling to their Mexican overwintering site [9]. The primary fall sightings are of nocturnal roosts, which monarchs form during their southward migration [10], and that are easily recognized by laypersons, since they often consist of hundreds or thousands of monarchs (Figure 2).

Monarch roosts can be considered stopover sites, which are essentially places where migratory animals pause during their journey to rest and/or refuel. Like most migratory organisms, monarchs utilize stopover sites to feed and deposit fat reserves [11] and to rest at night. Moreover for monarchs, depositing fat reserves during the migration not only provides fuel for the flight, but is essential to their overwintering survival [12]. As such, determining where stopover sites are for monarchs is an important issue in conserving their migration [9]. Further, while there is a wealth of research into stopover ecology of migrating birds (e.g., [13–18]), there are comparatively few studies examining the nature of stopover behavior in monarchs [19–21]. Moreover, there are no published studies where monarch stopover habitat is documented, other than anecdotal observations of roost trees [10]. In fact, it is not known even if monarchs select specific large- or small-scale habitat features at all when they stop or if roost site selection is completely random. Prior examination of roost observations indicated that few locations are utilized by monarchs for roosting year after year [9], which argues for the latter scenario, although more thorough investigation on this idea is warranted. Furthermore, like most migratory animals, monarchs must face continually changing landscapes throughout the entire flyway, including prairies and farmland in the American Midwest, deciduous forests in the eastern seaboard, and dry scrublands in Texas and northern Mexico. Given these changing landscapes they encounter, how then would their stopover habitat preferences (if there are any) change as they progress southward?

The Journey North roost observation database is uniquely positioned to offer insights into this question. When volunteers observe a migratory roost, they not only report the location and date, but are also encouraged to

submit general observations on the roost, such as the type of tree or vegetation in which the roost was observed. In this study, we screened four years of Journey North's migratory roost sightings (from eastern North America only) and, from these records, we recorded the type of tree (or other vegetation) in which the roost was observed. We also examined the landscape-level features of the roost site using a GIS approach; here we compared the land use surrounding each roost location to those of randomly selected locations at similar stages of the migration, which we arbitrarily divided into five flyway regions. Our goals for this study were to (1) document the large- and small-scale habitat preferences of monarchs at roosting sites and (2) determine if monarchs display a uniform preference for specific stopover habitats throughout the migration flyway or does their preference change as the migration progresses. Results from this study will not only further scientific understanding of monarch butterfly migration, but should also be relevant to the science of animal migration in general. In fact, to our knowledge this study is the first to examine how stopover habitat preferences of a single migratory animal vary throughout an entire migration flyway.

2. Methods

2.1. Roost Observations. We examined roost observations from the Journey North program between 2005 and 2008 (Figure 3), which are accessible online in the archives section of the program (http://www.learner.org/jnorth/maps/archives.html). For the purposes of this study, we focused on the primary flyway only (the central flyway) and did not consider observations from the Atlantic flyway [9], since very few tagged monarchs from that region are ever recovered in Mexico [20, 22, 23]. Each roost observation in the database is associated with a date (of the first night of observation), and latitude and longitude (of the zip code of the observer's mailing address, see below). While all roost observations have at least these components, observers are also encouraged to record notes about the roost and even take pictures, which are also archived with the sightings. For this study we screened these written notes and recorded what the monarchs were reported roosting on (i.e., tree, shrub, etc.). Moreover, since the aim of this study was to compare roost characteristics along the migratory flyway, we arbitrarily created 5 "flyway regions" of 4° latitude blocks that encompassed the majority of the flyway and roost observations in Canada and the United States (Figure 3). We then categorized the roost observation data (type of vegetation, etc.) into these regions based on the latitude of the observation.

2.2. Landscape Features of Roost Sites. The latitude and longitude associated with roost observations were imported into ArcGIS for analyses of land use surrounding roost sites. We point out that the coordinates of roosts in the Journey North database are not necessarily for the roost tree itself; when new participants sign up, they are asked to report their home address, and from this information, coordinates

Identifying Large- and Small-Scale Habitat Characteristics of Monarch Butterfly Migratory Roost Sites with Citizen Science Observations

9

(a)

(b)

(c)

(d)

FIGURE 2: Photographs of monarch migration roosts on various tree types submitted by Journey North citizen scientists. Photograph credits: (a) Iris Tower, Youngstown, NY; (c) Emily McCormick, Mount Cory, OH; (b) Ron & Bobbie Streible, Fort Morgan, AL; (d) Bruce Morrison, Hartley, IA.

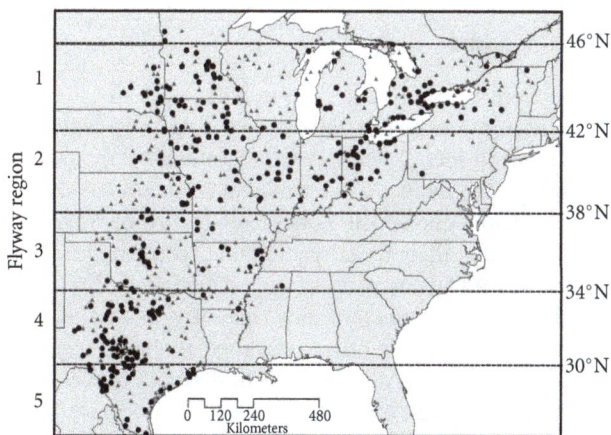

FIGURE 3: Map of roost observations (circles) reported to Journey North from 2005–2008 ($n = 310$) with arbitrarily created flyway regions used in this study indicated. Triangles indicate randomly selected locations ($n = 352$) for land use analysis.

are generated by Journey North personnel using a database of coordinates for North American zip (postal) codes. This practice was started for ease of overlaying points on an online, continent-scale map and since most observers do not know their latitude and longitude. For our purposes this means that the coordinates for any given observation could be centered on a point several kilometers distant from the roost (the center of the zip code). However, we attempted to minimize this problem by creating a buffer around each point with a 10 km radius ($314 \, km^2$), and evaluating the land cover within this area, which should be large enough to encompass the roost itself. The average area of zip codes in the United States is $222.7 \, km^2$ [24], and for urban areas that have multiple zip codes in the same city this number is likely to be much smaller, which only improves the chance that the buffer encompasses the roost. To minimize spatial autocorrelation, we eliminated all duplicate coordinates of roosts that were reported in the same city (which happens when two separate observers reported the same roost, from a roost being spotted in multiple years, or from two roosts sighted near one another). This left 310 spatially independent roost observations for analysis. In addition, we randomly selected a series of points ($n = 352$) throughout each flyway region for comparison to the monarch-selected locations. For this we generated a minimum convex polygon around the entire flyway and, within that area, randomly generated points within ArcGIS, preventing points from occurring within 20 km of one another.

To evaluate land cover characteristics of both the monarch-selected and random locations we overlaid a national land cover map [25] where land cover had been digitally classified into 19 categories (though for the purposes of this project we only considered 7 of the largest categories—deciduous forest, coniferous forest, cropland, grassland, urban, open water, and wetland). Then, we calculated the percent cover of each category within the buffered area surrounding each point.

2.3. Data Analyses. There were 217 observations where the type of vegetation was specified. Using these data we compared the frequency of the most commonly reported tree species across flyway regions using chi-square statistics. Using the large-scale land cover data set containing both monarch-selected sites (n = 310) and randomly selected locations (n = 352), we used logistic regression to simultaneously examine the effects of each land use category and flyway region (predictor variables) on whether a location was monarch-selected or random (response variable). We also included two-way interaction terms between each land use category and flyway region to determine if habitat preferences vary throughout the flyway. The full model with all main and interaction effects was simplified using likelihood ratio tests (Δ deviance) to evaluate the importance of nonsignificant terms following Crawley [26]. Nested models without interaction terms were compared against the full model prior to the removal of any main effects. Significance of terms remaining in the final model are reported based on Wald χ^2. All analyses were conducted using Statistica 6.1 software [27].

3. Results

3.1. Small-Scale Roost Habitat Characteristics. Of all roost observations where the type of vegetation was specified (n = 217), 97.7% of the roosts were reported on trees, with the remainder being on herbaceous vegetation, including two observations of monarchs roosting on seaside goldenrod (*Solidago sempervirens*), and one each of common groundsel (*Senecio vulgaris*), beach grass (*Ammophila* sp.), and golden crownbeard (*Verbesina encelioides*). There was no clear preference for one tree type; there were a total of 38 tree species reported overall (as hosting roosts) in the four years examined. The 10 most common tree species reported are listed in Table 1, broken down by flyway region. The most frequently reported trees included pines or other conifers (21.8%), maple species (20.7%), followed by oaks (15.6%), pecans (14.5%), and willows (7.8%). Collectively, these made up 80.4% of the observations (where the tree type was specified). The frequency of these 5 tree types (i.e., their use as roost sites) appeared to vary across the flyway regions (Table 1); a 5 × 5 contingency table analysis based on these top five rows revealed that these frequencies differed significantly (df = 16, χ^2 = 108, P < 0.001). In general, pines/conifers and maples were used most often in the northern areas of the flyway, while pecans and oaks were more frequently used in the southern regions.

3.2. Landscape Characteristics of Roost Sites. Of the 7 land use categories evaluated in both monarch-selected locations and random ones, the majority (~50%) of the landscapes across most flyway regions were composed of crops (Figure 4), which makes sense given that much of the central flyway traverses the agricultural region of the American Midwest (Figure 3). Following that were the broadleaf (deciduous) forest and grasslands categories. Visual comparison of the breakdown of all land use categories at monarch-selected sites (Figure 4(a)) versus random sites in the same region (Figure 4(b)) gives the impression that land use at random sites is fairly uniform throughout the flyway while that of actual roost sites varies to some degree. In particular, there appeared to be a distinct shift in the relative proportions of land use in the two southernmost regions (northern and southern Texas). In region 4, most of the land around selected roost sites was cropland (67%), while in the last region, 61% of the land around roost sites was grassland, compared to 15% around random locations in that region.

In the logistic regression model examining large-scale land use at monarch-selected sites versus random ones the results were complex. The probability of a monarch roost appeared to depend on the amount of deciduous forest, urban area, open water area, and wetland area around the site (all significant main effects; Table 2). In direct comparison of land use between roost sites and random sites, it appears that monarch-selected sites had less overall deciduous forest cover than random sites, more urban area, a higher percentage of open water nearby, and less wetland cover than random locations (Figure 5). Further, there were significant interaction effects (i.e., meaning that the strength of the main effect depended on the flyway region) in the percent deciduous forest, grassland, and urban area (Table 2).

4. Discussion

Places where monarch butterflies stop during their migration represent important links between breeding and overwintering areas, and identifying habitat requirements of roosting monarchs is therefore a key component to our understanding of this phenomenon. The ephemeral nature of migratory roosts [9], plus their broad geographic scope, makes them difficult to study using conventional scientific methodology. However, by using observations made by this nationwide network of citizen scientists, we hope to have made the first steps in addressing this question. For example, while monarch roosts were nearly exclusively on trees, we found no overwhelming preference for a tree species or type (i.e., conifer versus deciduous), other than a general tendency for maples and conifers in the north and pecans and oaks in the south (Table 1). A tendency to use males was also casually noted by F. A. and N. R. Urquhart [32] who were located in the northernmost region. When one considers the entire flyway however, given the diverse branch and leaf morphology of the various trees reported as used, it appears that monarchs are highly adaptable in terms of their roost tree use. This is also evidenced by the pictures submitted

Identifying Large- and Small-Scale Habitat Characteristics of Monarch Butterfly Migratory Roost Sites with Citizen Science Observations

11

TABLE 1: Summary of 10 most commonly reported tree types used for monarch roosts from 2005 to 2008, in all 5 flyway zones. Only observations where the roost tree type was specified are included.

Tree species	Flyway region					Total (%)
	1	2	3	4	5	
Pine/conifer (multiple species)	12	21	0	5	1	39 (21.8)
Maple (*Acer* sp.)	26	10	1	0	0	37 (20.7)
Oak (*Quercus* sp.)	4	4	1	9	10	28 (15.6)
Pecan (*Carya illinoinensis*)	0	0	4	17	5	26 (14.5)
Willow (*Salix* sp.)	4	2	1	5	2	14 (7.8)
Walnut (*Juglans* sp.)	5	3	1	0	0	9 (5.0)
Ash (*Fraxinus* sp.)	3	4	0	0	0	7 (3.9)
Elm (*Ulmus* sp.)	2	0	1	3	0	6 (3.4)
Hackberry (*Celtis occidentalis*)	0	1	0	4	0	5 (2.8)
Palm (type not specified)	0	0	0	0	4	4 (2.2)

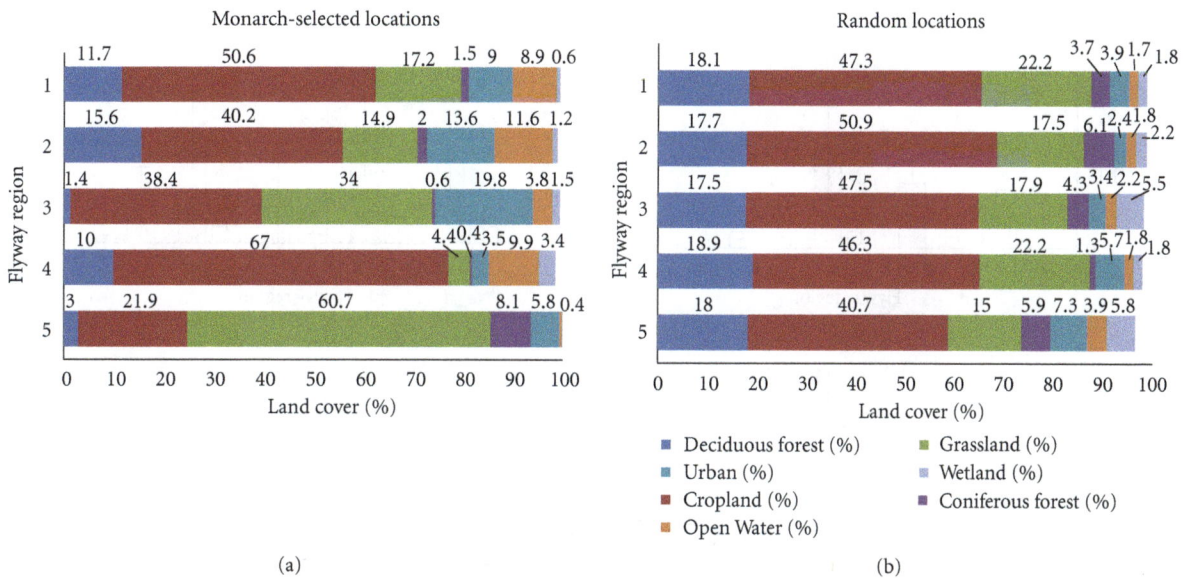

FIGURE 4: Relative proportions of all land-use categories in areas where roosts were observed ("monarch-selected", (a)) and in randomly selected locations (b) across all flyway regions in this study.

by Journey North participants (Figure 2); one can see that monarchs are capable of settling on a wide variety of branch structures (from needles to small-leafed trees to large-leafed trees).

Similar to the small-scale patterns obtained, from a large-scale habitat perspective, there was no one land cover type that best predicted the location of monarch roosts throughout the flyway; there was statistically significant difference in the proportion of multiple land cover types between monarch-selected and random sites (Table 2), and there were multiple interactions with flyway region. These complex results make interpreting these data difficult. One clear pattern in the land cover analyses was that, in nearly all flyway regions, there was a greater proportion of open water in the ($314 \, km^2$) roost area than around random locations (Figures 4 and 5). This land cover category would include large rivers, ponds, and lakes. It may be that monarchs use these land features as beacons while searching from the

air for potential roost sites, as these areas would tend to be lush with vegetation and possibly support a variety of nectaring plants. Conversely, monarchs roosted in areas that had significantly less wetland land cover than random sites did (Figure 5). Here we can only speculate as to the reason for this dichotomy; it may be that such areas are not as visible from the air as are open water bodies.

There were certain land cover patterns uncovered that may have resulted from inherent biases in observer distribution; most Journey North participants tend to live in urban areas (E. Howard, *unpublished data*). For example, monarch roosts had a higher proportion of urban land use around them than did random sites (Figure 5). This was probably an artifact of the tendency for most observers to live in or near cities (i.e., fewer observers in rural areas means fewer roost sightings). Similarly, in nearly all flyway regions, roost sites tended to have less deciduous forest area around them than did random sites (Figure 5), which could indicate either

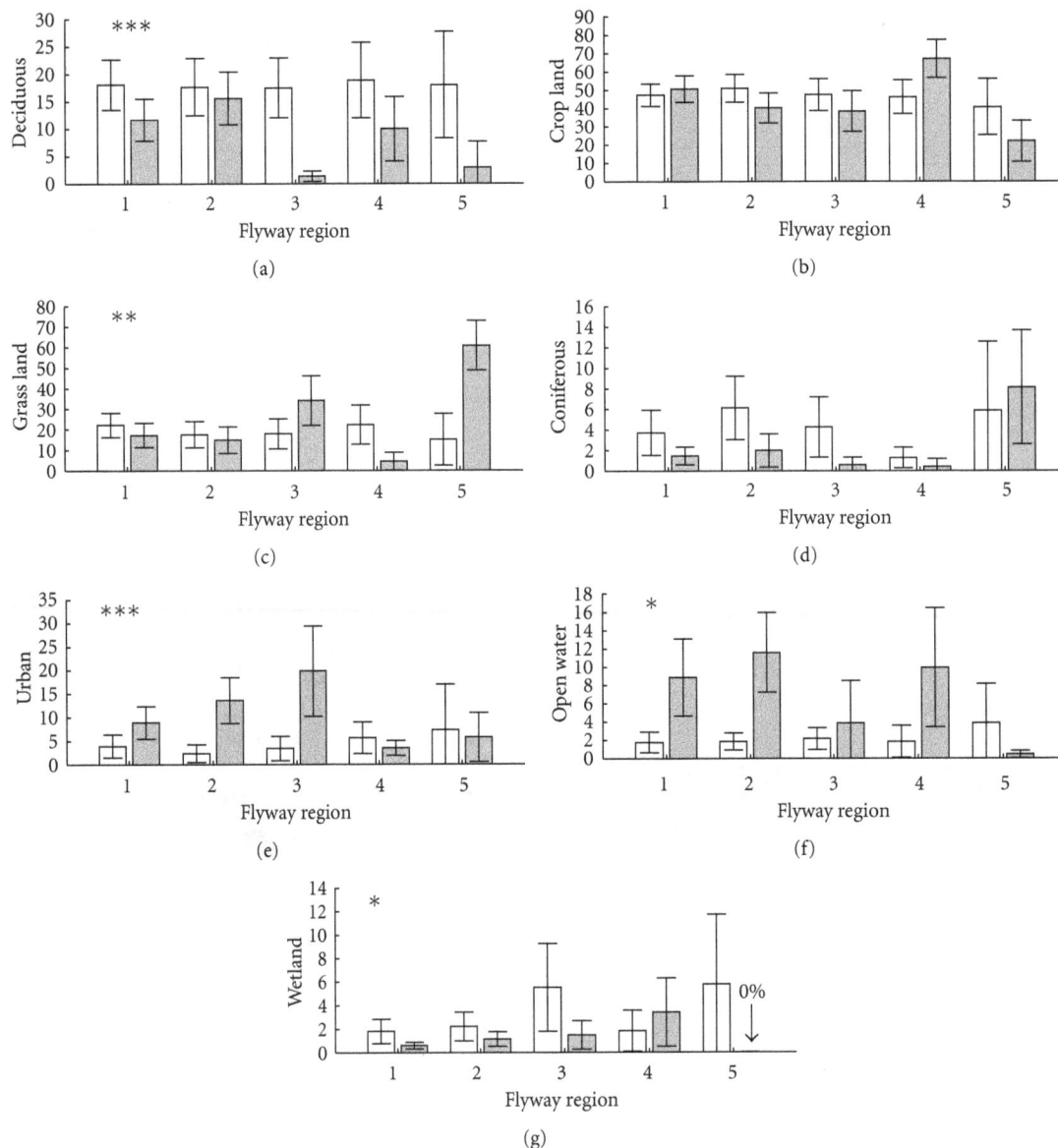

FIGURE 5: Percentage of land use categories in all monarch-selected (grey bars) and randomly selected (open bars) roost locations in all flyway regions. Mean percentages shown with 95% confidence intervals. Asterisks indicate significance of the main effect (*), the interaction with flyway region (**), or both (***) in the logistic regression model (Table 2).

a general avoidance of these land types for roosting purposes, or (more likely) that roosts in these habitats are not often spotted by laypeople. A roost of several hundred monarchs in a forest would not stand out as readily as one in a lone tree in the middle of a cornfield.

In addition to the significant main effects of land cover types, the logistic regression model revealed several significant interaction effects with flyway region and certain land cover types (Table 2, Figure 5), which indicates the effect of the land cover in question varied depending on the flyway region. In other words, there was some degree of change to the land cover preference (or avoidance) throughout the flyway. For example, the tendency for roosts to be associated with greater urban area was stronger in the northern regions than the southern (Figure 5), and the avoidance of deciduous

forest was most pronounced in the southern regions. Further, there was a statistical effect of grassland, but it depended on the flyway region; in the southernmost region in particular, monarch roosts were in areas with 61% grassland, but this habitat was not widespread throughout that region (random sites had 15% grassland; Figure 4). In this region (southern Texas), monarchs appear to be drawn to this type of large-scale habitat.

The collective results from both the small- and large-scale analyses in this study should have conservation implications, but not in the manner we anticipated. Conserving migratory habitats is an important issue for all migrant species [28]. With this issue in mind, we had attempted to ascertain if there were certain types of small- or large-scale habitat features that could be identified as being important to

Identifying Large- and Small-Scale Habitat Characteristics of Monarch Butterfly Migratory Roost Sites with
Citizen Science Observations

13

TABLE 2: Summary of logistic regression model examining effects of land use categories and flyway regions (predictors) on whether a location is monarch-selected or random (response). Full model with all main effects and interactions was simplified using likelihood ratio tests (Δ deviance) to evaluate the importance of nonsignificant terms following Crawley [26]. Nested models without interaction terms were compared against the full model prior to the removal of any main effects. Significance of terms remaining in the final model are reported based on Wald χ^2. The effects of cropland and coniferous forest were not significant and removed from the final model.

Variable	df	Wald	P
Flyway region	4	17.88	0.0013
Deciduous forest	1	16.04	0.0001
Grassland	1	0.027	0.8690
Urban	1	11.89	0.0006
Open water	1	21.45	0.0000
Wetland	1	10.05	0.0015
Flyway region * grassland	4	23.45	0.0001
Flyway region * urban	4	12.97	0.0114
Flyway region * deciduous	4	13.18	0.0104

monarchs for at least one part of the stopover, their overnight roosting. However, the data we gathered in this effort did not point to a select few landscape features or roost tree types, but instead indicate that monarchs are capable of using a wide variety of habitats for roosting. Knowing this, it then becomes challenging, from a management perspective, to pinpoint what could be done to conserve important stopover areas. It may be that conservation efforts need to be targeted at the areas where it is most clear what (primary) habitats monarchs are using, such as in southern Texas, and less so in areas where there are less "habitat" preferences that can be identified, such as in the northern flyway regions. Texas is also an area of high importance for monarch migration, since, here, monarchs deposit considerable amounts of fat [29], which they will use to sustain themselves over the winter [12].

Throughout this paper we have stressed that we would not have been able to study this phenomenon of migratory roosting by any other means than by using Journey North's nationwide citizen science network. We recognize, however, that this data set was not ideal for answering our original questions regarding habitat characteristics of roosts and that there are areas where this program could be strengthened. Moreover, by highlighting these limitations (below), managers of other citizen science projects may be able to learn from these problematic issues, which may be common to many programs. Perhaps the largest drawback of the Journey North program is the fact that the coordinates of all sightings, including those of "roosts," are of the geographic center of the zip code from the observer's address, which could be several kilometers away from the actual roost. There is no remedy for this problem, unless observers take GPS readings of roost trees, which would certainly be difficult

to implement into the Journey North protocol. It would also have been helpful if the protocol for reporting roosts included providing information about the surrounding habitat, such as how many trees are nearby (and not being used by monarchs) and what species they are. This would have allowed for more direct evaluation of roost tree "preference" at the sites where monarchs stop over (i.e., by comparing trees that were used to those that were not). Furthermore, in addition to habitat data, one area where Journey North could strengthen its protocol is in the reporting of the size of the roosts (i.e., number of monarchs). Many people state in their notes that they saw "hundreds" or (very often) "thousands" of monarchs in the roost observed. Estimating numbers of clustering monarchs is notoriously difficult, even for trained scientists (e.g., [30]), so this would also be difficult to implement in the current protocol. However, if actual numbers were associated with roost observations (and if we were confident in their accuracy), it would theoretically allow for annual estimates of the size of the entire migratory generation. Estimating long-term trends in abundance for this population is something that has been attempted with other data sets, but with inconsistent results [31, 32]. With the vast number of observers in the Journey North program, such data would undoubtedly be of value in this regard.

Finally, this study may well represent the first-ever examination of habitat requirements of a single migratory organism across an entire migration flyway. Such an approach allows us to identify any changes in habitat requirements at different stages of the migration. And indeed, although the monarchs' "habitat preferences" during migration appear to be broad, we did see certain changes in large- and small-scale habitat preferences as the migration advanced southward. The next step may be to assess the availability of nectaring sources at all stages of the migration, especially in the latter stages of the migration where monarchs are accumulating the most fat [29]. Additional questions regarding roosting or stopover behavior could also be addressed in the future, using citizen science observational data or direct study at specific stopover sites [20, 21]. Thanks to the efforts of hundreds of dedicated and observant people in North America who participate in citizen science programs, the answers to these and other questions are now within reach.

Acknowledgments

This project could not have been completed without the contributions of the thousands of Journey North participants who watch the skies each fall and faithfully submit roost observations. Lincoln Brower has provided expert advice to the Journey North program over the years. Funding for Journey North was provided by the Annenberg Foundation. The authors thank the members of the MonarchNet working group (Karen Oberhauser, Sonia Altizer, Leslie Ries, Dennis Frey, Becky Bartel, Elise Zipkin, James Battin, and Rebecca Batalden) for helpful discussion about the project, as well as two anonymous reviewers for suggestions for improvement on the paper.

References

[1] M. D. Prysby and K. Oberhauser, "Temporal and geographical variation in monarch densities: citizen scientists document monarch population patterns," in *The Monarch Butterfly: Biology & Conservation*, K. S. Oberhauser and M. J. Solensky, Eds., pp. 9–20, Cornell University Press, Ithaca, NY, USA, 2004.

[2] E. Lindsey, M. Mehta, V. Dhulipala, K. Oberhauser, and S. Altizer, "Crowding and disease: effects of host density on response to infection in a butterfly-parasite interaction," *Ecological Entomology*, vol. 34, no. 5, pp. 551–561, 2009.

[3] S. R. Stevens and D. Frey, "Host plant pattern and variation in climate predict the location of natal grounds for migratory monarch butterflies in western North America," *Journal of Insect Conservation*, vol. 14, no. 6, pp. 731–744, 2010.

[4] R. A. Bartel, K. S. Oberhauser, J. C. de Roode, and S. M. Altizer, "Monarch butterfly migration and parasite transmission in eastern North America," *Ecology*, vol. 92, no. 2, pp. 342–351, 2011.

[5] E. Howard, H. Aschen, and A. K. Davis, "Citizen science observations of monarch butterfly overwintering in the southern United States," *Psyche*, vol. 2010, Article ID 689301, 6 pages, 2010.

[6] E. Howard and A. K. Davis, "Documenting the spring movements of monarch butterflies with Journey North, a citizen science program," in *The Monarch Butterfly: Biology & Conservation*, K. Oberhauser and M. Solensky, Eds., pp. 105–114, Cornell University Press, Ithaca, NY, USA, 2004.

[7] A. K. Davis and E. Howard, "Spring recolonization rate of monarch butterflies in eastern North America: new estimates from citizen-science data," *Journal of the Lepidopterists' Society*, vol. 59, no. 1, pp. 1–5, 2005.

[8] E. Howard and A. K. Davis, "A simple numerical index for assessing the spring migration of monarch butterflies using data from Journey North, a citizen-science program," *Journal of the Lepidopterists' Society*, vol. 65, pp. 267–270, 2011.

[9] E. Howard and A. K. Davis, "The fall migration flyways of monarch butterflies in eastern North America revealed by citizen scientists," *Journal of Insect Conservation*, vol. 13, no. 3, pp. 279–286, 2009.

[10] F. A. Urquhart and N. R. Urquhart, "Breeding areas and overnight roosting locations in the northern range of the monarch butterfly (*Danaus plexippus plexippus*) with a summary of associated migratory routes," *Canadian Field Naturalist*, vol. 93, pp. 41–47, 1979.

[11] L. P. Brower, L. S. Fink, and P. Walford, "Fueling the fall migration of the monarch butterfly," *Integrative and Comparative Biology*, vol. 46, no. 6, pp. 1123–1142, 2006.

[12] A. Alonso-Mejía, E. Rendon-Salinas, E. Montesinos-Patiño, and L. P. Brower, "Use of lipid reserves by monarch butterflies overwintering in Mexico: implications for conservation," *Ecological Applications*, vol. 7, no. 3, pp. 934–947, 1997.

[13] S. R. Morris, M. E. Richmond, and D. W. Holmes, "Patterns of stopover by warblers during spring and fall migration on Appledore Island, Maine," *Wilson Bulletin*, vol. 106, no. 4, pp. 703–718, 1994.

[14] K. Winker, "Autumn stopover on the isthmus of tehuantepec by woodland nearctic-neotropic migrants," *Auk*, vol. 112, no. 3, pp. 690–700, 1995.

[15] C. E. Gellin and S. R. Morris, "Patterns of movement during passerine migration on an island stopover site," *Northeastern Naturalist*, vol. 8, no. 3, pp. 253–266, 2001.

[16] J. Jones, C. M. Francis, M. Drew, S. Fuller, and M. W. S. Ng., "Age-related differences in body mass and rates of mass gain of passerines during autumn migratory stopover," *Condor*, vol. 104, no. 1, pp. 49–58, 2002.

[17] J. D. Carlisle, G. S. Kaltenecker, and D. L. Swanson, "Stopover ecology of autumn landbird migrants in the boise foothills of southwestern Idaho," *Condor*, vol. 107, no. 2, pp. 244–258, 2005.

[18] D. W. Mehlman, S. E. Mabey, D. N. Ewert et al., "Conserving stopover sites for forest-dwelling migratory landbirds," *Auk*, vol. 122, no. 4, pp. 1281–1290, 2005.

[19] A. K. Davis and M. S. Garland, "Stopover ecology of monarchs in coastal Virginia: using ornithological methods to study monarch migration," in *The Monarch Butterfly: Biology & Conservation*, K. Oberhauser and M. Solensky, Eds., pp. 89–96, Cornell University Press, Ithaca, NY, USA, 2004.

[20] J. W. McCord and A. K. Davis, "Biological observations of monarch butterfly behavior at a migratory stopover site: results from a long-term tagging study in coastal South Carolina," *Journal of Insect Behavior*, vol. 23, no. 6, pp. 405–418, 2010.

[21] J. W. McCord and A. K. Davis, "Characteristics of monarch butterflies (*Danaus plexippus*) that stopover at a site in coastal South Carolina during fall migration," *Journal of Research on the Lepidoptera*, vol. 45, pp. 1–8, 2012.

[22] M. S. Garland and A. K. Davis, "An examination of monarch butterfly (*Danaus plexippus*) autumn migration in coastal Virginia," *American Midland Naturalist*, vol. 147, no. 1, pp. 170–174, 2002.

[23] L. Brindza, L. P. Brower, A. K. Davis, and T. Van Hook, "Comparative success of monarch butterfly migration to overwintering sites in Mexico from inland and coastal sites in Virginia," *Journal of the Lepidopterists' Society*, vol. 62, no. 4, pp. 189–200, 2008.

[24] Geography Division, United States Census Bureau, and U.S. Gazetteer Files, 2010, http://www.census.gov/geo/www/gazetteer/gazetteer2010.html.

[25] United States Geological Survey, Land Cover of North America at 250m, and Sioux Falls SD, 2005, http://www.cec.org/naatlas/.

[26] M. J. Crawley, *Statistical Computing: An Introduction to Data Analysis Using S-Plus*, John Wiley & Sons, Chichester, UK, 2002.

[27] "Statistica," Statistica version 6.1, Statsoft Inc., 2003.

[28] F. R. Moore, J. Sidney, A. Gauthreaux, P. Kerlinger, and T. R. Simons, "Habitat requirements during migration: important link in conservation," in *Ecology and Management of Neotropical Migratory Birds*, T. E. Martin and D. M. Finch, Eds., pp. 121–144, Oxford University Press, New York, NY, USA, 1995.

[29] L. P. Brower, "New perspectives on the migration biology of the monarch butterfly, *Danaus plexippus*, L.," in *Migration: Mechanisms and Adaptive Significance*, M. A. Rankin, Ed., pp. 748–785, The University of Texas, Austin, Tex, USA, 1985.

[30] W. H. Calvert, "Two methods of estimating overwintering monarch population size in Mexico," in *The Monarch Butterfly: Biology & Conservation*, K. Oberhauser and M. Solensky, Eds., pp. 121–127, Cornell University Press, Ithaca, NY, USA, 2004.

[31] L. P. Brower, O. R. Taylor, E. H. Williams, D. A. Slayback, R. R. Zubieta, and M. I. Ramirez, "Decline of monarch butterflies overwintering in Mexico: is the migratory phenomenon at

Identifying Large- and Small-Scale Habitat Characteristics of Monarch Butterfly Migratory Roost Sites with Citizen Science Observations

15

risk?" *Insect Conservation and Diversity*, vol. 5, no. 2, pp. 95–100, 2012.

[32] A. K. Davis, "Are migratory monarchs really declining in eastern North America? Examining evidence from two fall census programs," *Insect Conservation and Diversity*, vol. 5, no. 2, pp. 101–105, 2012.

The Behaviour of Stallions in a Semiferal Herd in Iceland: Time Budgets, Home Ranges, and Interactions

Hrefna Sigurjonsdottir,[1] **Anna G. Thorhallsdottir,**[2,3] **Helga M. Hafthorsdottir,**[3] **and Sandra M. Granquist**[4]

[1] *School of Education, University of Iceland, Stakkahlíð, 105 Reykjavík, Iceland*
[2] *Bioforsk Ost, Heggenes, 2940 Volbu, Norway*
[3] *Division of Environmental Sciences, The Agricultural University of Iceland, Hvanneyri, 311 Borgarnes, Iceland*
[4] *Institute of Freshwater Fisheries and The Icelandic Seal Center, Brekkugata 2, 530 Hvammstangi, Iceland*

Correspondence should be addressed to Hrefna Sigurjonsdottir, hrefnas@hi.is

Academic Editor: Randy J. Nelson

A permanent herd of Icelandic horses with four stallions and their harems was studied for a total of 316 hours in a large pasture (215 ha) in May 2007 in Iceland. Interactions between stallions of different harems and other aspects of the horses' behaviour were studied. One stallion and nine horses were introduced into the pasture prior to the study to examine the reactions of the resident stallions to a newcomer. The stallions spent significantly less time grazing than other horses and were more vigilant. Home ranges overlapped, but harems never mixed. The stallions prevented interactions between members of different harems indirectly by herding. Generally, interactions between resident stallions were nonviolent. However, encounters with the introduced stallion were more aggressive and more frequent than between the other stallions. Here, we show that four harems can share the same enclosure peacefully. The social network seems to keep aggression at a low level both within the harems and the herd as a whole. We encourage horse owners to consider the feasibility of keeping their horses in large groups because of low aggression and because such a strategy gives the young horses good opportunities to develop normally, both physically and socially.

1. Introduction

Wild and feral horses form herds composed of harems and bachelor groups. Harems are defended by one or more stallions against other males [1], while bachelor groups consist of young stallions that have not yet managed to form their own harems. Usually one to three adult females and their offsprings make up the harem [2–4], while more numerous harems have been observed in some cases; the highest number recorded is 26 [1]. The stallion does not defend a territory but keeps his harem within a certain home range [5]. Both female and male offspring disperse from their natal group when sexually mature [6, 7]. Adult mares have been seen to change harems in feral populations (see [8]). but many stay in the same harem all their lives such that breeding group membership is very stable [9]. Changing harems will often mean some fitness loss for the mares since drifting

mares are harassed and herded more often by the stallions [10]. This mating system, that is, female defence polygyny, is remarkably stable in different habitats where domestic or feral horses live [1].

Traditionally predation risk and distribution of resources are considered to be the fundamental factors behind the evolution of this mating system [11]. Thus, when females form groups because of increased survival, the males can use that to their advantage and can defend the females [12]. Linklater et al. [13] and Rubenstein and Nunez [10] argue on the other hand that social bonding is more likely to explain the prevalence of the mating system because social bonds between unrelated females both increase birth rate and survival [9] and decrease male harassment.

Home ranges vary in size with season and between populations. Their sizes correlate with availability of resources, but not with group size [11]. Minimum spacing between the harems is maintained by the stallions. When harems get close to each other, the stallions approach each other and display. The interactions escalate to mild fights on some occasions. This behaviour has been documented in populations living in a variety of environments. Interactions between horses from different harems, other than stallions, seem to be very rare or absent [8, 14, 15].

Environmental conditions, time of year, level of competition for resources, and age composition of the groups have been found to be important variables deciding time budgets of feral horses. Grazing has been reported to take 56–86% of the time, resting standing and lying down 9–35%, standing alert 6–20%, moving 3–13%, allogrooming and playing 2-3%, and other behaviour 2–4% [2, 16].

Here we report on a study of a semiferal herd in Iceland. The only breed found on the island is the Icelandic horse. The breed has been isolated since the time of settlement (870 A.D.) and shows adaptations to the harsh climate and the rough terrain [17, 18]. No horses are allowed to be imported to Iceland because of reduced immunity of the breed [19, 20]. Many horses are kept outside all year around, but for winter, supplementary feeding is now required [20]. Often, the horses are kept under conditions with little human interference. Adult mares, foals, subadults, and gelding are in some places kept in large herds consisting of hundreds of individuals over the summertime in extensive grazing areas. Thus the young horses have good opportunity to develop normal equine social skills [21, 22]. However, free roaming of stallions is prohibited and only in exceptional cases do harem keeping stallions share the same enclosure.

The aim of this research was primarily to study the behaviour of stallions in a semiferal herd. Interactions between the stallions were analysed, their home ranges were mapped, and their time budgets were compared to the other horses. We also wanted to see how residency of the stallions affected their behaviour. This was done to deepen our understanding of the social structure of horse groups and to get information about the feasibility and possible advantages of keeping horses in large groups.

2. Methods

2.1. Study Area and Animals.
The study was conducted at the farm Sel in East-Landeyjar in the south of Iceland. Horses have been kept there in a 215 ha enclosure with very limited human contact for approximately 30 years, giving a herd with close to natural social organization. Occasionally old stallions have been replaced with new ones. In the autumn, male foals and most of the fillies have been removed (some fillies had been left in the pasture to serve as replacements for the ageing mares). Replacement foals were given anthelmintics once in their first year, but were otherwise not handled. Foals born late in the summer stayed over the winter in the herd and were removed the following autumn. In addition, the horses were gathered weekly during August and September, for sampling blood from some of the breeding mares. The blood sampling, which was done for commercial purposes, was done by veterinarians.

Table 1: Composition, size of the harems, and age and residency of the four stallions.

Harems	Total number	Adult mares	Immature horses (female/male)	Age and residency (in years) of the stallions
H1	20	16	3 (1/2)	13–2
H2	31	21	9 (8/1)	17–8
H3	30	23	6 (3/3)	20–15
H4	12	8	3 (3/0)	13–8

The land in the enclosure is productive, native grassland, flat with small tussocks. The dominant species are the grasses *Agrostis capillaris*, *Agrostis vinealis*, *Poa pratensis*, and *Anthoxanthum odoratum* and the sedge *Juncus arcticus*. A large ditch (3 m wide, 1500 m long) runs along part of the west border and in the middle there is a manmade shelter (turf). In the wintertime the horses are fed supplementary hay.

In the enclosure, four resident stallions (St 1–4) had divided the herd into four harems (H1–H4, see Table 1). The total herd consisted of 93 adult mares and subadults (3yrs and younger). In addition, 42 foals were born shortly before and during our observations.

One day before regular observations started, an additional group, a young stallion with 6 subadult mares, three adult mares, and a foal, were put into the enclosure. Six of the mares and the foal joined H2 within 30 minutes, one subadult joined H3 within one day and these eight horses became members of the resident harems. These horses are included in the respective harems in Table 1. Two adult mares and the young stallion were removed from the pasture one week later.

2.2. Observations and Mapping.
The four harems (H1–H4) were observed for 81 (H1), 77 (H2), 77 (H3), and 81 (H4) hours, respectively, on 18 days between the 9th and the 31st of May 2007. This was done during 5-hour shifts (with exceptions) during daylight hours organized in a balanced way between 04:00 and 00:00. Data was recorded in field books and binoculars were used when needed. Interactions between stallions were sometimes videotaped. Two persons observed H1 and H2 and two other persons observed H3 and H4. Interobserver reliability was used to assess the degree to which different observers recorded. Two persons were on guard each shift, observing one harem each. The observers approached the horses quietly and sat on the ground, taking care not to disturb or interact with the horses. All the horses were easily recognized by the observers, either by colour, special markings, or freeze brand marks on their backs.

For time budget estimates different behaviour (see Table 2) was scored by *instantaneous scanning* [23]. For the stallions it was done every 10 minutes during all the shifts, ($N = 486, 462, 462,$ and 486 for St 1–4, resp.) but less frequently for the other horses ($N = 296, 268, 319,$ and 259 for H 1–4, resp.). The part of the pasture, that each harem had stayed on during every hour was marked on a map. This was done by hand by rough estimation since no GPS

TABLE 2: Behavioural classes (for detailed descriptions of these behavioural acts see [25]).

(1) Time budget analyses	(2) Interactions between stallions
Grazing, standing resting, standing alert, lying down, walking, herding*, sexual behavior, allogrooming (other behaviours were pooled)	(A) Aggressive and violent interactions: fighting (neck wrestling, boxing, and dancing where both are upright), attack where one lunges on the other, aggressive bites, aggressive chase, foreleg strike, stamping
	(B) Nonviolent interactions: approach with arched necks, parallel walk, posturing, marking

* Applies to stallions only.

measurements were taken. Changes in group compositions were also noted based on individual recognition of the horses.

The method *all occurrences of some behaviour* [23] was used to collect data on interactions between stallions. The ethogram is listed in Table 2. Other interactions between horses of different harems were very few and are described more generally in the results. Nature and frequencies of interactions between members within harems are published elsewhere [24].

2.3. Analyses. Chi square tests were applied to the time budget data to test if proportions of behaviours differed between stallions and the other horses. We compared the behaviour of the stallions in the same way when analysing their interactions with respect to residency in the pasture and location of their home ranges.

The home ranges were estimated by giving each map 4 points on an ArcGis map. The likelihood of finding the harem within the home range is then shown by fading colours from red (10x or more) to yellow (seen once) and the core area defined as the most red part of the home range. Statistics and computations were done with the help of Excel.

3. Results

3.1. Time Budget. An overall average time budget for the horses in the pasture is shown in Figure 1.

The horses spent most of their time grazing (77%). The four stallions spent proportionally less time grazing ($\chi^2 = 135.4$, $P < 0.001$) and more time standing than the other horses ($\chi^2 = 601.9$, $P < 0.001$) (Figure 2). This was due to proportionally much higher frequency of standing alert while standing resting as not different between the stallion and the other horses ($\chi^2 = 2.93$ ns).

3.2. Home Range, Herding Rate, and Changes in Group Compositions. Although the home ranges overlap by ca. 20%, the harems never used the same area at the same time and the horses in one harem hardly ever mixed with horses from another harem. The sizes of the home ranges estimated in this way were 93, 54, 65, and 49 ha for H1, H2, H3, and H4 respectively (see Figure 3).

The herding rates differed between the stallions. Size of the home range, number of mares and experience clearly matters. The stallion which had the largest home range, St1, herded his band 84 times (1.04 per hour), while St4,

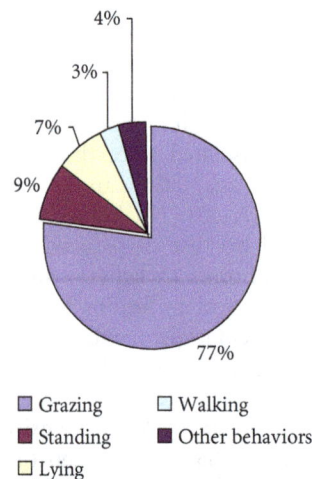

FIGURE 1: Time budget of all the horses in the herd.

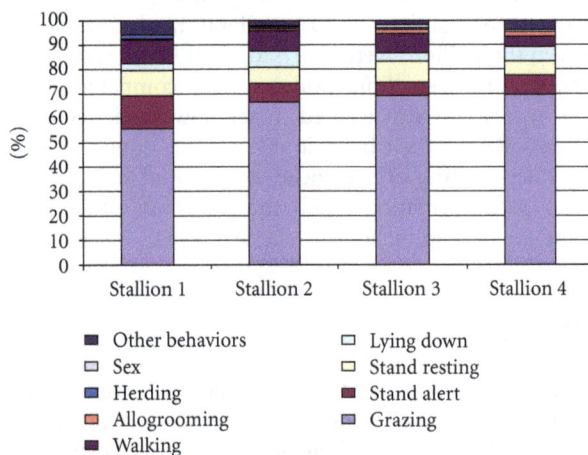

FIGURE 2: Time budget of the four stallions.

who had the smallest home range, was only seen herding 5 times in total (0.06 per hour). The other stallions were seen herding 56 and 50 times, respectively (see also [24]). When the numbers of mares in the harems are considered, the difference between the stallions is less but nevertheless of interest. Although St1 herded his mares 5.2 times on average compared to the average rate of 2.6, 2.2, and 0.6 time for St2, St3, and St4, respectively, he mounted his mares less frequently (0.8 times on average per mare compared to 1.3,

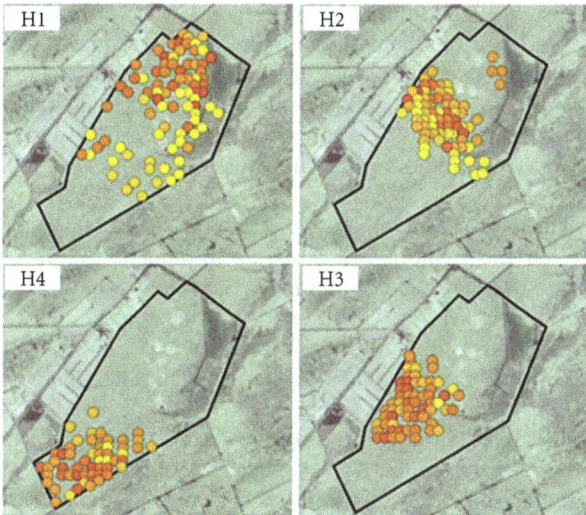

FIGURE 3: Location of the four harems in the pasture. The horses are most often found where the reddest spots are shown.

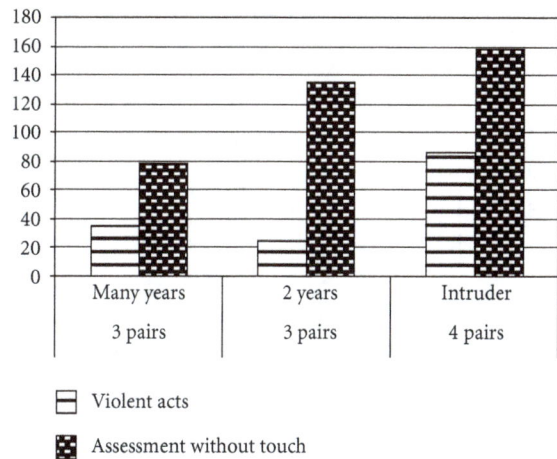

Violent acts

Assessment without touch

FIGURE 4: Number of behavioural acts shown during encounters between pairs of stallions that have shared the same pasture for different length of time, split into violent and nonviolent acts.

1.9, and 1.4 for St2, St3, and St4, respectively, see [24]). This is probably due to his lack of experience and large home range.

Only three mares changed harems. One adult mare moved without being chased or threatened on the 25th of May from H1 to H2. One young mare was driven from her natal group, H4, on the 19th of May by the stallion (St4) and was accepted the following day into H1. One young mare was excluded from H2 on the 23rd of May by St2 and was on her own positioned between H2 and H3 until the end of the observations (when checked a month later, she had been accepted into H3).

3.3. Interactions between Members of Different Harems. A total of 112 encounters between individuals from different harems were observed during the study. Most, or 81.5%, occurred between two stallions (including the newcomer), 8.5% involved aggressive behaviours of stallions against mares and subadults from other bands, and 10% of cases involved interactions between subadults mares from two bands.

3.3.1. Interactions between Stallions. The average rate of close encounters (distance less than one horse length) between the resident stallions was 0.30 events per hour. St2 had the highest frequency (0.44/hour) and St4 the lowest (0.14/hour). St2 showed most often the initiative to approach other stallions (25 cases of 34 encounters, which is a significantly higher proportion than for the other stallions ($\chi^2 = 70.17$, d.f. = 1, $P < 0.001$), while St1 was relatively the least initiative (7/31). The oldest stallion who had been in the pasture for the longest time, St3, and St4 who had the smallest band were less active (18 and 11 encounters, resp.) than the others. As expected, the resident stallions interacted proportionally more with their neighbours than with other stallions ($\chi^2 = 88.7$, d.f. = 1, $P < 0.001$).

When the encounters are split into different behavioural classes (see Table 2) the data shows that the frequency of nonviolent acts (374), that is, assessment without touch, was much more frequent than violent acts (152). In Figure 4 the behavioural acts are shown with respect to how long each pair has been sharing the same pasture.

Not surprisingly, the newcomer was frequently, or in total 36 times, challenged by St1 and St2, who had home ranges located to the area where he spent most of his time while in the pasture. He only had three encounters in total with the other two stallions. On average he had an encounter with St1 or St2 once per hour during the time he was in the pasture which is much higher than between neighbouring resident stallions (St4/St3: 0.07, St3/St2: 0.12, and St2/St1: 0.35 per hour). Also, encounters between the newcomer and the resident stallions were significantly of the more aggressive nature than between the resident stallions (Figure 4, $\chi^2 = 161.7$, d.f. = 1, $P < 0.001$). Residency is clearly an important variable as can be seen in Figure 5 which shows the behaviour of the most active stallion, St2, against both his resident neighbouring stallions and the newcomer (intruder). His interactions were most violent against the newcomer, but he also interacted much more with his neighbour St1 than St3 which is probably due to shorter residence time of St1 in the pasture.

4. Discussion

The results presented here show many similarities with results from studies on feral horses [2] although no bachelor groups were present. Thus, three of the four stallions herded the harems/bands every hour or every other hour preventing the adult mares and the subadults to interact with members of other harems and keeping them within the home range. During our observation time we witnessed two stallions expel young mares from their harems with fierce aggression and later that summer we witnessed two more cases within the herd. Clearly, the behaviour of the males is typical of

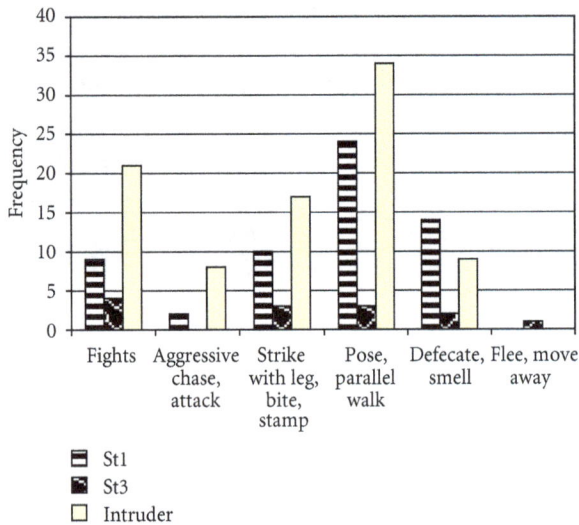

FIGURE 5: Frequencies of different behavioural acts between St2 and three stallions (his neighbours and the newcomer/intruder).

what has been described for feral horses when natal dispersal occurs [6]. The hierarchies of the harems were less linear and rigid than in nonstallion groups [21, 24] just as Feist and McCullough [15] reported in their study on 44 bands of feral horses in USA.

Time budgets for the studied herd showed that the horses spent on average 77% of their time grazing which is similar to the grazing time measured in the late summer and fall for Icelandic horses but less than in the middle of the summer in Iceland [16]. At the time of the study, in the spring, some of the horses were rather thin after the winter and almost all the mares were already lactating or close to having a foal. Acquiring enough nutrients was therefore a high priority for them. Similar grazing time has been shown in other areas during harsh times [8, 14, 26]. The stallions spent significantly less time grazing and more time standing alert than the average harem member. This is expected since they need to keep the harem within the home range and be aware of the time when the mares become sexually receptive after foaling.

Home ranges overlapped considerably and all harems spent more time close to the water than in other parts of the home ranges. The same behaviour has been described for other populations [27]. The pasture was adequate for keeping the horses in good condition from spring till late in the autumn and during the observation time there were no signs of competition for grazing areas. Thus, the home ranges seemed to satisfy the nutrient requirements of the horses during the growing season.

Interactions between members of different bands were very few, except between stallions. The low number of interactions between other horses is mainly due to the herding behaviour by the stallions preventing the harems to come too close to each other. The herding rate was low compared to some other studies [8] but similar to Feh's study [28] on Camargue horses. However, comparison is difficult

because of high female/male ratio and lack of bachelors in this study. In this herd the pressure to defend mares from other stallions is less than in feral herds because of the absence of bachelors and secondary males (in multistallion groups). The stallions communicate to great extent by scent through urine and dung but dung piles were numerous in the pasture.

The low sex ratio can also explain why direct interactions between the stallions are relatively infrequent in spite of having overlapping home ranges. In Franke-Stevens [8] study on feral horses (60 horses in 12 harems on 172 ha pasture), the resident stallions interacted 0.88 times/hour in the summer time and 0.48 in the wintertime compared to 0.30 on average in this study. In the study by Linklater et al. [13] the stallions in the single stallion harems interacted 0.50 times per hour. As expected, neighbours interacted more than others in this study. Length of time the stallions have been together in the pasture seems to influence their interaction rates (Figure 4). Thus, encounters with the newcomer were significantly more frequent than with the other stallions, and the stallion that had been in the pasture for a shorter time (St1) than the other resident stallions was approached by his neighbour (St2) relatively frequently (Figure 5). For the stallions that have been together for the longest time, it is probably often sufficient to leave dung and urine markings instead of engaging in direct interactions. Berger [14] argued that marking is a significant factor in maintaining the organizational systems of feral horses. Rubenstein [29] has shown that the scent is an honest signal and serves to communicate information about the quality and presence of males.

When the interactions between stallions are analysed in more detail it is clear that most behavioural acts which the stallions exhibit during interactions were nonviolent. Aggressive chases and fights were relatively rare. This is in contrast with what Berger [14] found but in agreement with most other studies [2]. For instance, Rubenstein and Hack [30] analysed 231 encounters and of those only 19% resulted in violent contests. As expected, interactions with the newcomer were more aggressive. The horses' behaviour is a good example of assessment strategies, where assessment of opponents is a part of the behavioural sequence and withdrawal of the one with a smaller RHP (resource holding potential) is the rule [29, 31].

5. Conclusion

Our study shows that the behaviour of semiferal horses in a herd where stallions with harems are kept in a relatively large and productive pasture (ca. 2 ha/horse) is similar to that found in feral and semiferal herds living in larger areas. Also, when the stallions have shared the same pasture for many years they clearly recognize each other [32, 33] and rarely interact. They avoid the cores of the home ranges of other stallions and seldom fight. This is in accordance with findings of Cozzi et al. [34] which showed that horses show highly developed skills in dealing with conflict by reconciling and appeasing in a similar way as has been described for apes

[35], dogs [36], and birds [37, 38]. Thus, horses seem to realize their position in the hierarchy and the social network which characterizes their group. Keeping horses in large herds where they have the opportunity to mix with horses of all ages and both sexes allows them to develop their social skills when young. More research and deeper understanding of the consequences of not giving horses such opportunities is needed [22, 39, 40]. In this herd the stallions kept their harems separate from others and their interactions were in general nonviolent. It is concluded that it can be a good management strategy to keep horses in large herds with harems, provided good tempered stallions are chosen and the horses have good access to water, shelter, and sufficient space.

Acknowledgments

The project was supported by The Icelandic Centre for Research (Rannís) and University of Iceland Research Fund. Thanks to Margrét Björk Sigurðardóttir for help in the field, Sigmundur Helgi Brink for help with the maps, and the farmers at Sel in East Landeyjar for allowing the authors to work in their fields and for all their help.

References

[1] W. L. Linklater, "Adaptive explanation in socio-ecology: lessons from the Equidae," *Biological Review*, vol. 75, no. 1, pp. 1–20, 2000.

[2] G. H. Waring, *Horse Behavior*, Noyes Publication/Willam Andrew Publishing, Norwich, NY, USA, 2nd edition, 2003.

[3] S. R. B. King and J. Gurnell, "Habitat use and spatial dynamics of takhi introduced to Hustai National Park, Mongolia," *Biological Conservation*, vol. 124, no. 2, pp. 277–290, 2005.

[4] J. E. Roelle, F. J. Singer, L. C. Zeigenfun, J. I. Ranson, L. Coates- Markle, and K. A. Schoenecker, "USGS demography of the pryer mountain wild horses 1993-2007," U.S. Geological Scientific Investigations Report 2010, USGS Publications Warehouse, USA, 2010.

[5] K. Krueger, "Social ecology of horses," in *Ecology of Social Evolution*, J. Korb and J. Heinze, Eds., pp. 195–206, Springer, Berlin, Germany, 2008.

[6] A. M. Monard and P. Duncan, "Consequences of natal dispersal in female horses," *Animal Behaviour*, vol. 52, no. 3, pp. 565–579, 1996.

[7] W. L. Linklater and E. Z. Cameron, "Social dispersal but with philopatry reveals incest avoidance in a polygynous ungulate," *Animal Behaviour*, vol. 77, no. 5, pp. 1085–1093, 2009.

[8] E. F. Stevens, "Instability of harems of feral horses in relation to season and presence of subordinate stallions," *Behaviour*, vol. 112, no. 3-4, pp. 149–161, 1990.

[9] E. Z. Cameron, T. H. Setsaas, and W. L. Linklater, "Social bonds between unrelated females increase reproductive success in feral horses," *Proceedings of the National Academy of Sciences of the United States of America*, vol. 106, no. 33, pp. 13850–13853, 2009.

[10] D. I. Rubenstein and C. Nunez, "Sociality and reproductive skew in horses and zebras," in *Reproductive Skew in Vertebrates; Proximate and Ultimate Causes*, R. Hager and C. Jones, Eds., pp. 196–226, Cambridge University Press, Cambridge, UK, 2009.

[11] G. H. Waring, *Horse Behaviour: The Behavioural Traits and Adaptations of Domestic and Wild Horses, Including Ponies*, Noyes, Park Ridge, NJ, USA, 2nd edition, 1983.

[12] N. B. Davies, J. R. Krebs, and S. A. West, *An Introduction to Behavioural Ecology*, Wiley-Blackwell, Chichester, UK, 2012.

[13] W. L. Linklater, E. Z. Cameron, E. O. Minot, and K. J. Stafford, "Stallion harassment and the mating system of horses," *Animal Behaviour*, vol. 58, no. 2, pp. 295–306, 1999.

[14] J. Berger, "Organizational systems and dominance in feral horses in the grand canyon," *Behavioral Ecology and Sociobiology*, vol. 2, no. 2, pp. 131–146, 1977.

[15] J. D. Feist and D. R. McCullough, "Behavior patterns and communication in feral horses," *Zeitschrift fur Tierpsychologie*, vol. 41, no. 4, pp. 337–371, 1976.

[16] A. G. Thórhallsdóttir, G. Ágústsson, and J. Magnússon, "Grazing behaviour of the Icelandic horse," in *Proceedings of the Havemeyer Foundation Workshop, Horse Behavior and Welfare*, Holar, Iceland, June 2002.

[17] B. Hendricks, *International Encyclopedia of Horse Breeds*, University of Oklahoma Press, Norman, Okla, USA, 1995.

[18] G. B. Björnsson and H. J. Sveinsson, *The Icelandic Horse*, Forlagid, Reykjavik, Icelands, 2006.

[19] V. Fridriksdottir, E. Gunnarsson, S. Sigurdarson, and K. B. Gudmundsdottir, "Paratuberculosis in Iceland: epidemiology and control measures, past and present," *Veterinary Microbiology*, vol. 77, no. 3-4, pp. 263–267, 2000.

[20] "Mast-Icelandic food and veterinary authority," http://www2.mast.is/index.aspx?GroupId=1057&TabId=1064.

[21] H. Sigurjónsdóttir, M. C. Van Dierendonck, S. Snorrason, and A. G. Thórhallsdóttir, "Social relationships in a group of horses without a mature stallion," *Behaviour*, vol. 140, no. 6, pp. 783–804, 2003.

[22] M. Van Dierendonck and D. Goodwin, "Social contact in horses: implications for human—horse interactions," in *The Human—Animal Relationship: Forever and a Day*, F. de Jonge and R. van den Bos, Eds., pp. 65–82, Royal van Gorcum, Assen, The Netherlands, 2005.

[23] P. N. Lehner, *Handbook of Ethological Methods*, Cambridge University Press, Cambridge, UK, 2nd edition, 1996.

[24] S. M. Granquist, A. G. Thorhallsdottir, and H. Sigurjonsdottir, "The effect of stallions on social interactions in domestic and semi feral harems," *Applied Animal Behavior Science*, vol. 141, no. 1, pp. 49–56, 2012.

[25] S. McDonnell, *A Practical Field Guide to Horse Behaviour—The Equid Ethogram*, The Bloodhorse, Hong Kong, 2003.

[26] P. Duncan, "Time budgets of Camargue horses. II. time-budgets of adult horses and weaned sub-adults," *Behaviour*, vol. 72, no. 1-2, pp. 26–49, 1980.

[27] L. Boyd and R. Keiper, "Behavioural ecology of feral horses," in *The Domestic Horse. The Origins, Development and Management of Its Behaviour*, D. Mills and S. McDonnell, Eds., pp. 55–82, Cambridge University Press, Cambridge, UK, 2005.

[28] C. Feh, "Alliances and reproductive success in Camargue stallions," *Animal Behaviour*, vol. 57, no. 3, pp. 705–713, 1999.

[29] D. I. Rubenstein, "Behavioral ecology of island feral horses," *Equine Veterinary Journal*, vol. 13, pp. 27–34, 1981.

[30] D. I. Rubenstein and M. A. Hack, "Horse signals: the sounds and scents of fury," *Evolutionary Ecology*, vol. 6, no. 3, pp. 254–260, 1992.

[31] H. Sigurjónsdóttir and G. A. Parker, "Dung fly struggles: evidence for assessment strategy," *Behavioral Ecology and Sociobiology*, vol. 8, no. 3, pp. 219–230, 1981.

[32] L. Proops, K. McComb, and D. Reby, "Cross-modal individual recognition in domestic horses (*Equus caballus*)," *Proceedings*

of the National Academy of Sciences of the United States of America, vol. 106, no. 3, pp. 947–951, 2009.

[33] K. Krueger and B. Flauger, "Olfactory recognition of individual competitors by means of faeces in horse (*Equus caballus*)," *Animal Cognition*, vol. 14, no. 2, pp. 245–257, 2011.

[34] A. Cozzi, C. Sighieri, A. Gazzano, C. J. Nicol, and P. Baragli, "Post-conflict friendly reunion in a permanent group of horses (*Equus caballus*)," *Behavioural Processes*, vol. 85, no. 2, pp. 185–190, 2010.

[35] F. Aureli and F. B. M. de Waal, *Natural Conflict Resolution*, University of California Press, Berkeley, Calif, USA, 2000.

[36] G. Cordoni and E. Palagi, "Reconciliation in wolves (*Canis lupus*): new evidence for a comparative perspective," *Ethology*, vol. 114, no. 3, pp. 298–308, 2008.

[37] A. M. Seed, N. S. Clayton, and N. J. Emery, "Postconflict third-party affiliation in rooks, *Corvus frugilegus*," *Current Biology*, vol. 17, no. 2, pp. 152–158, 2007.

[38] O. N. Fraser and T. Bugnyar, "Do ravens show consolation? responses to distressed others," *PLoS ONE*, vol. 5, no. 5, Article ID e10605, 2010.

[39] M. Bourjade, M. Moulinot, S. Henry, M. A. Richard-Yris, and M. Hausberger, "Could adults be used to improve social skills of young horses, *Equus caballus*?" *Developmental Psychobiology*, vol. 50, no. 4, pp. 408–417, 2008.

[40] M. Bourjade, A. de Boyer des Roches, and M. Hausberger, "Adult-young ratio, a major factor regulating social behaviour of young: a horse study," *PLoS ONE*, vol. 4, no. 3, article e4888, 2009.

4

Influence of Local Wind Conditions on the Flight Speed of the Great Cormorant *Phalacrocorax carbo*

Ken Yoda,[1] Tadashi Tajima,[1] Sachiho Sasaki,[1] Katsufumi Sato,[2] and Yasuaki Niizuma[3]

[1] *Graduate School of Environmental Studies, Nagoya University, Furo-cho, Chikusa-ku, Nagoya 464-8601, Japan*
[2] *International Coastal Research Center, Atmosphere and Ocean Research Institute, University of Tokyo,*
 5-1-5 Kashiwanoha, Kashiwa, Chiba 277-8564, Japan
[3] *Faculty of Agriculture, Meijo University, 1-501 Shiogamaguchi, Tenpaku-ku, Nagoya 468-9502, Japan*

Correspondence should be addressed to Ken Yoda, yoda.ken@nagoya-u.jp

Academic Editor: Inma Estevez

In seabirds, the relationship between flight speed and wind direction/speed is thought to be particularly important for studying energy-saving strategy and foraging habitat selection. In this study, we examined whether the ground and calculated air speeds of four great cormorants (*Phalacrocorax carbo*) were affected by wind conditions using high-resolution GPS data loggers. The birds increased their ground flight speed in tailwinds, decreased it in headwinds, and changed their air speed in relation to wind components. However, they did not change their foraging sites according to the wind conditions. They were likely to respond to moderate wind conditions by adjusting their air speed without changing their foraging sites.

1. Introduction

Understanding how environmental factors affect animal movements is of central importance to movement ecology [1]. In seabirds, the relationship between flight speed and wind direction/speed is thought to be particularly important for studying energy-saving strategy and foraging habitat selection and has been well examined [2, 3]. For example, the family Procellariiformes is likely to favour side or tailwinds at large scales (thousands of kilometres) [4, 5], which can reduce their energy expenditure [4] or lead sexually dimorphic species to be segregated in foraging areas [6]. Thus, the effects of winds are expected to exert strong selection pressures on morphology, behaviour, and the life histories of birds [7, 8]. However, several factors can bias the relationship between wind and flight speed in seabirds, especially at fine spatial scales (several to tens of kilometres). First, the flight paths of seabirds are often convoluted, making it difficult to relate flight paths with wind conditions. Second, because wind conditions change dramatically during short time periods, large-scale meteorology such as satellite-derived wind data is insufficient for detecting the effects of winds on bird flight at fine scales. Third, conventional analyses of air

speed have been used in many bird studies (see papers cited in [9]), but this leads to erroneous interpretations of wind effects on bird flight [10]; statistically rigorous approaches need to be used [9].

In this study, we deployed fine-scaled GPS data loggers on free-ranging great cormorants (*Phalacrocorax carbo*) to evaluate the effects of winds on their ground speed (i.e., the speed of the bird with respect to the ground) and calculated air speed (i.e., the speed relative to the air in which the bird is flying). Cormorants often commute in a straight line to memorised foraging areas [11–13] and use powered flight with uninterrupted flapping without gliding; therefore, they are suitable for examining wind effects on seabird movements and expected to follow the theory of powered flight (e.g., [14]). In addition, because the maximum foraging range of the great cormorant is relatively small (less than 15 km; [15]), we could use local wind measurements that would be nearly identical to the winds that the birds encountered. We examined whether the ground and calculated air speeds of great cormorants were affected by wind conditions and we determined if they changed foraging sites according to wind conditions.

2. Materials and Methods

This study was conducted in June 2010 at the Unoyama breeding colony (34°48′N, 136°53′E) and the Ishigakiike colony (34°51′N, 136°36′E) in the Tokai region of Japan. Great cormorants nest on trees in these regions [16]. Because they temporally leave their nests when humans approach, we used alpha-chloralose (Tokyo Chemical Industry Co., Ltd., Tokyo, Japan) to capture specimens. Alpha-chloralose is an anaesthetic and it allowed us to capture cormorants safely. We inserted 35–40 mg of alpha-chloralose powder into the mouth of a fish (e.g., ayu *Plecoglossus altivelis*) and placed it on a nest that contained chicks using a fishing rod. After a bird ate the fish and was immobilised (within 1-2 h), we captured it. In total, we captured eight birds in this way. No birds were injured during the procedure. The procedures used in this study were approved by the Ministry of the Environment, Japan.

We deployed GPS loggers on four breeding adults, we deployed acceleration data loggers on two other birds for different research purposes, and we did not deploy any loggers on the remaining two birds. A small plastic base was attached to the bird's back feathers with adhesive tape (Tesa, Hamburg, Germany) and glue (Loctite 401). A data logger was attached using a cable tie that could be cut remotely (RC-150-150T, Little Leonard, Tokyo, Japan), which entered beneath the feathers that were glued to the base. To avoid having to recapture of birds using alpha-chloralose, we used a remote release system that could send a signal to cut the cable tie (RX-100N, Little Leonard). After several days of attachment, we recovered the data loggers when the birds were on their nests.

A GPS data logger consisted of a GPS receiver with an antenna (GiPSy, Technosmart, Rome, Italy) and it was powered by a Li-SOCl$_2$ battery (LS14500, SAFT, Paris, France). The GPS loggers were programmed to take positional fixes every 9 or 20 s. The overall mass was 59–84 g, which corresponded to <5% of each bird's body mass. The birds did not appear to be negatively affected by the loggers or from being handled by the researchers.

We calculated trip duration and trip range, which was defined as the maximum distance from the colony. The cormorants commuted in a straight line without stopovers (Figure 1); therefore, we could easily define outward and homing flights using the sudden initial decrease and the final increase in ground flight speed, which marked the first flight and the final flight, respectively, during the trip (Figure 2). We defined foraging time as the period between the outward and homing flights; the foraging area was identified as the area where the bird remained during that period. For each outward and homing flight, flight speed and direction were averaged.

Wind direction and speed were recorded at two climatological stations operated by the Japan Meteorological Agency (JMA). We used 10-min wind direction and speed measurements for each outward and homing flight. For each flight, several wind measurements were averaged. The weather stations were within 15 km of each colony (Figure 1; Minamichita station: 34°44′N, 136°56′E, 6.5 m asl; Tsu

station: 34°44′N, 136°31′E, 39.6 m asl). For the outward and homing flights of each foraging trip, the absolute difference in the angle between flight direction and wind direction (i.e., 0–180°) was calculated. Wind speed in the direction of flight was calculated as the product of wind speed and the cosine of the angle. We calculated air speed using ground speed, wind speed, and the angle between the ground and wind vectors.

To examine the relationship between ground flight speed and wind speed in the direction of flight, a linear mixed model that treated individual birds as a random effect was run using the lme4 package in the statistical software R [17]. The significance of the fixed effects, as well as their 95% confidence intervals, was obtained from 100,000 Markov chain Monte Carlo (MCMC) simulations performed using the pvals.fnc function in R [18].

A two-dimensional generalised additive model (GAM) was used to analyse the relationship between air speed and wind [9]. The GAM was used instead of the conventional method that tests the relationship between air speed as a function of ground speed and air speed (see papers cited in [9]) because the conventional analysis can produce erroneous correlations [9, 10]. We divided the wind variable into two components (i.e., zonal and meridional components); the eastern direction was represented by positive values of x_w and the northern direction by positive values of y_w. The two variables were implemented in a GAM by first transforming them via a LOESS smoother (a locally weighted regression) with a maximum span of 80% and 2 degrees of freedom, following [9]. To fit the GAM, we used the gam package in R [19].

To examine the degree of overlap in foraging areas between successive trips, we calculated the proportion of the foraging area of one trip that was covered by the foraging area of another trip, that is HR[i, j] = A[i, j]/A[i], where A[i, j] is the area of the intersection between the two foraging areas and A[i] is the foraging area of trip i [20]. These were calculated for two birds that conducted six successive trips. This analysis was performed using the adehabitat package in R [21]. We excluded one trip from this analysis because the bird's position was not obtained consistently in the foraging area (Trip no. 4 in Table 1).

Data were analysed using Matlab version 7.1 (Math-Works, Inc., Natick, MA, USA) and R version 2.14.1 [22].

3. Results

We obtained GPS data from four birds with body masses ranging from 1570 to 2140 g (mean = 1905 g). The birds were named A, B (Unoyama colony), C, and D (Ishigakiike colony). The GPS loggers recorded 15 trips that consisted of 12 complete trips and 3 trips with truncated data due to exhausted batteries. Mean trip duration and trip range were 5.1 ± 4.8 h (0.7–19.0 h, n = 12) and 13.1 ± 3.8 km (8.6–18.7 km, n = 12), respectively, for complete trips. For the truncated data, we only used the outward flights in our analysis. Finally, we obtained 27 outward and homing paths.

Average flight duration was 11 min (4–23 min, n = 27). Average wind speed on each path was 2.7 ± 1.4 m/s (0.8–5.3,

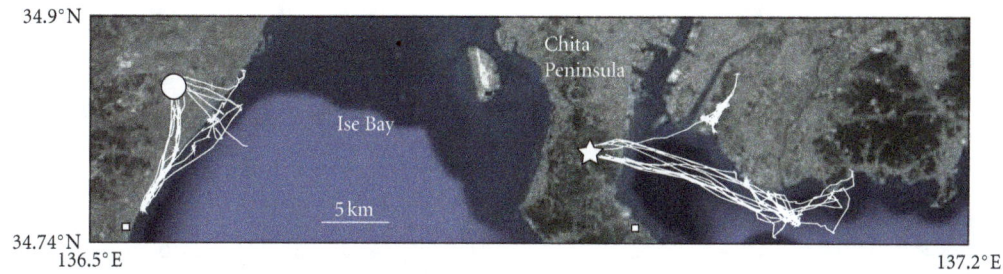

FIGURE 1: GPS tracks for four great cormorants during 15 foraging trips. The star and circle indicate the Unoyama and Ishigakiike breeding colonies, respectively. The two squares indicate the closest meteorological stations to each colony from which we derived wind information.

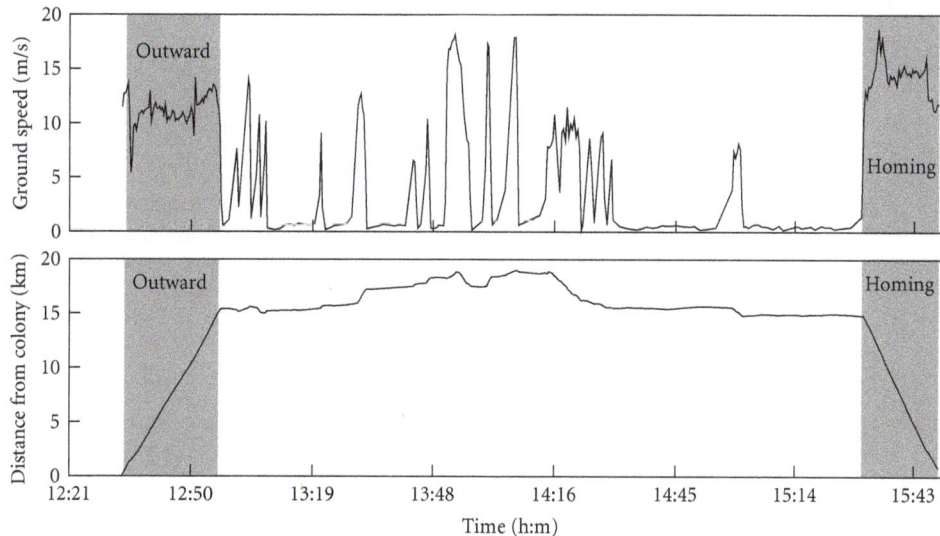

FIGURE 2: Example of time-series data of ground flight speed and distance from the colony. The shadows indicate outward and homing flights.

$n = 27$). Mean ground and air speeds on the paths were 14.4 ± 1.7 m/s (11.2–18.5, $n = 27$) and 14.6 ± 1.3 m/s (11.2–17.5, $n = 27$; Figure 4(a)), respectively.

Ground flight speed increased significantly with wind strength relative to the birds (Figure 4(b)). A linear mixed model estimated $S = 0.70$ (± 0.10, 0.48–0.91; SD, 95% CI)· $W + 14.4$ (± 0.31, 13.4–15.1), where S and W were ground flight speed and wind speed, respectively, in the direction of flight ($P < 0.001$). Air speed was significantly related to x_w ($F = 5.2$, $P < 0.05$), but not to y_w ($F = 1.7$, $P = 0.2$). Thus, air speed was only influenced by wind that was blowing east-west (Figure 5).

Successive trips by two birds qualitatively showed high foraging-site fidelity and the foraging areas overlapped (Table 1; Figure 6).

4. Discussion

Great cormorants encountered various wind speeds and directions (Figure 3). The birds sometimes encountered contrasting wind conditions between outward and homing trips. Although sample size was small (four individuals), our data clearly demonstrated that the ground speed of great

cormorants can be predicted by wind speed in the direction of flight (Figure 4(b)). Their ground flight speed increased in tailwinds and decreased in headwinds. Qualitatively, this matches very well with observations from visual surveys in which ground flight speeds of great cormorants in tailwinds were higher than in headwinds [23]. In addition, [23] reported a mean ground speed of 18.8 m/s in a strong tailwind (6.5 m/s), which agrees very well with our mixed-model prediction of 19.0 m/s. On the other hand, [23] also reported a ground speed of 12.4 m/s in an extreme headwind (8.0 m/s), which was higher than our predicted value of 8.8 m/s in a similar wind. Great cormorants prefer to fly close to the ground (i.e., ground effect; weak wind speeds closer to the ground ~0.3 m) when flying into strong headwinds [23]; therefore, our predictive equation seems to underestimate their ground speed in such extreme headwinds, that is, outside our measured range (>5 m/s of wind speed).

The calculated air speeds of our birds (14.6 m/s) ranged between previously reported minimum power speeds (V_{mp}) for this species (13.5 [24] to 16.5 m/s [23]), which is the speed at which the least mechanical power is required from the flight muscles to maintain the bird flying at a constant air speed [25]. Therefore, great cormorants might adopt V_{mp} as

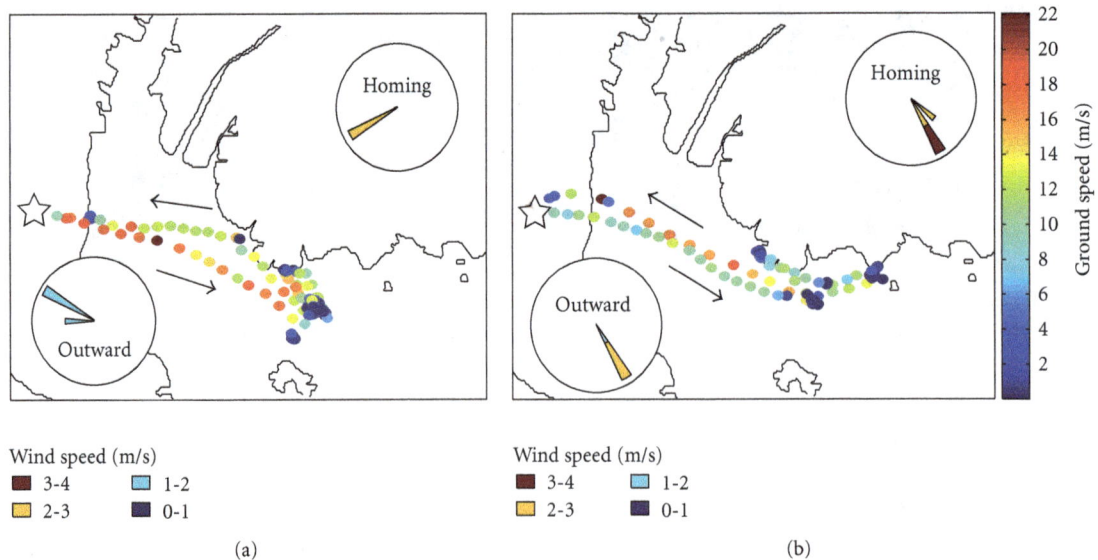

FIGURE 3: Two examples of wind conditions and GPS positions. The Bird A encountered different wind conditions between outward and homing trips in (a) Trip no. 2 and (b) Trip no. 6, respectively. The star indicates the Unoyama breeding colony. The dots show the GPS positions that were interpolated at 1 min to produce a clearer display and the colour indicates the ground speed of the bird. Arrows indicate travel direction. The two rose plots on each panel show the frequency of occurrence of wind direction, which is the direction from which the wind was blowing (not where it was blowing to), and wind speed (colour) during outward and homing flights.

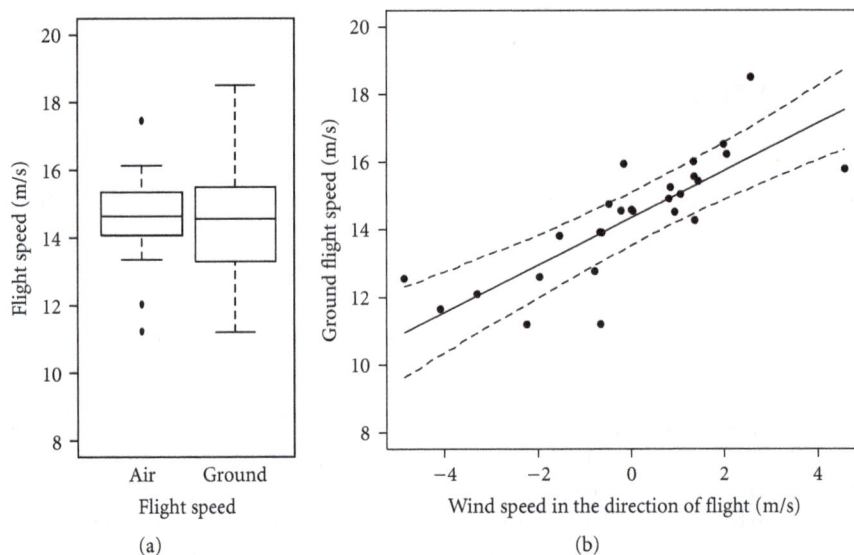

FIGURE 4: (a) Measured ground speeds and calculated air speeds for great cormorants and (b) the relationship between ground speed and wind speed in the direction of flight (tailwind component along the track direction). The solid line is the best fit that was estimated by a linear mixed model: the dashed lines indicate the 95% confidence intervals.

was indicated for Kerguelen shag *P. verrucosus* [26]. However, the great cormorants showed changes in calculated air speed in relation to wind components (Figure 5). Optimal flight theory predicts that V_{mp} will be unaffected by winds [7]; therefore, air speed optimization [27] might explain changes in calculated air speed in relation to wind conditions, as found in this study. Because air and ground flight speed are approximately equivalent in calm wind conditions, the actual air speed in this study might show the same relationship to the wind as the ground speed (Figure 4(b)). If air speed

decreases in tailwinds and increases in headwinds, this means that the birds adopted maximum range speed (V_{mr}), which is the air speed where energy expenditure per distance travelled is minimal [27], and our data might support the optimality theory in relation to wind condition. Further study that incorporates morphological and actual air-speed measurements would be needed to clarify this issue.

Notably, the great cormorants showed strong foraging-site fidelity (Table 1; Figure 6), which is common in cormorants [12, 28, 29], irrespective of wind conditions. This

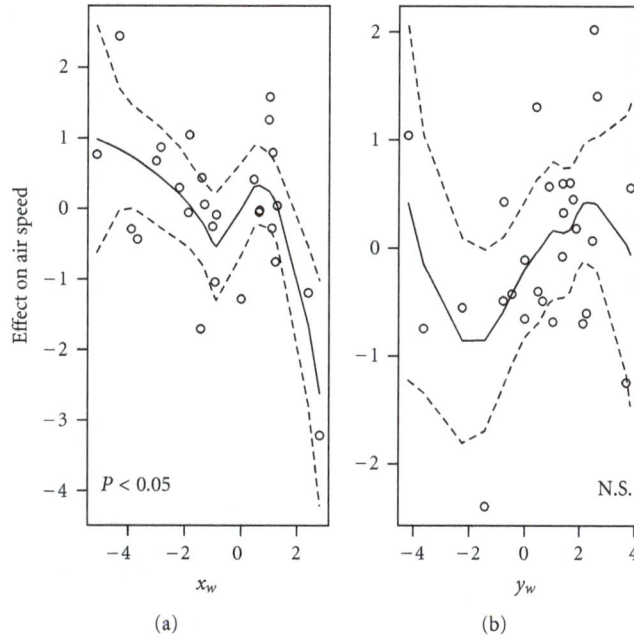

(a)   (b)

FIGURE 5: The predicted relative influence of the wind components (x_w and y_w) on the air speed of great cormorants. The eastern direction was represented by positive values of x_w and the northern direction by positive values of y_w. Calculated air speed was significantly related to x_w, but not to y_w. The solid lines are the fitted functional response and the broken lines represent standard error curves.

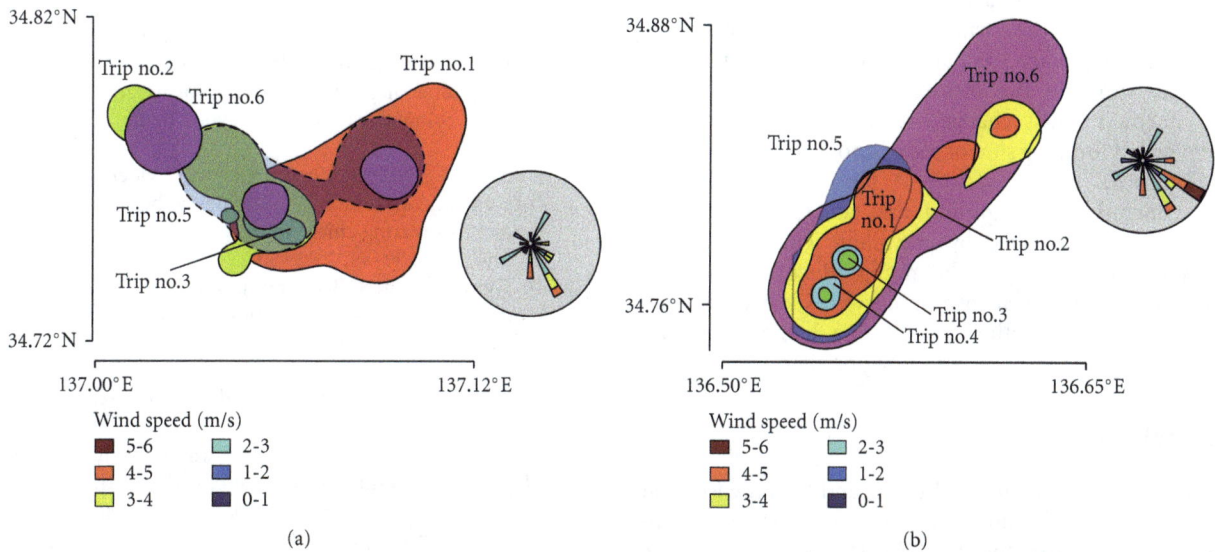

FIGURE 6: Foraging area overlap in two great cormorants that conducted multiple foraging trips. The 95% contours of the kernel density estimates of the foraging areas are shown for (a) Bird A and (b) Bird C (see Table 1). The two rose plots show the frequency of occurrence of wind direction, which is the direction from which the wind was blowing (not where it was blowing to), and wind speed (colour) during outward and homing flights.

might be related to low predictability of wind conditions in this study area, and therefore, foraging site selection based on wind would not be efficient in contrast to some seabirds that live in "predictable" local wind conditions at large spatial scales [3, 5, 29]. In addition, because the cormorants using pure powered flight could not efficiently extract energy compared to gliders such as shearwaters [5], albatrosses [4], boobies, and gannets [3, 30], they did not select foraging sites

based on winds. The significant levels of foraging-site fidelity in successive foraging trips also suggest that their prey may be predictable in space within a period of several days and that the cormorants are able to remember specific foraging sites [12, 31, 32]. In general, seabirds with small foraging ranges and corresponding short trip durations, as was the case in this study (5 h in mean trip duration), are likely to have higher foraging site fidelity because the low probability of

TABLE 1: Foraging area overlap in two great cormorants, the proportion of the foraging area of one trip that was covered by the foraging area of another trip, that is $HR[i, j] = A[i, j]/A[i]$, where $A[i, j]$ is the area of the intersection between the two foraging areas and $A[i]$ is the foraging area of trip i. The trip no. 4 of Bird A was not used because the bird's position was not obtained consistently in the foraging area.

i	j					
	1	2	3	4	5	6
Bird A						
1	—	0.18	0.06	—	0.46	0.14
2	0.39	—	0.12	—	0.72	0.26
3	0.94	0.90	—	—	0.99	0.35
5	0.70	0.50	0.09	—	—	0.24
6	0.47	0.39	0.07	—	0.51	—
Bird C						
1	—	0.87	0.04	0.13	0.72	0.99
2	0.54	—	0.02	0.08	0.62	0.98
3	1.00	1.00	—	1.00	1.00	1.00
4	1.00	1.00	0.29	—	1.00	1.00
5	0.59	0.82	0.03	0.11	—	0.89
6	0.27	0.43	0.01	0.04	0.30	—

extreme changes in food availability [32]. Additionally, one of the main prey of the cormorants in this study area is the mullet (*Mugil cephalus*) [33], an herbivorous bottom feeder. Therefore, although the distribution might be relatively stable, it could slowly change over time, reflecting subtle movements of foraging sites of the cormorants between their trips (Figure 6).

In conclusion, the ground and air speeds of great cormorants were affected by prevailing wind conditions. However, wind conditions did not produce variations in foraging sites. Thus, great cormorants were likely to respond to moderate wind conditions by adjusting their air speed but without changing the location of foraging sites.

Acknowledgments

The authors thank H. Fujii, H. Kuroki, and Y. Inoue for their help during data collection. This work was financially supported by Grants from the Japan Society for the Promotion of Science to K. Yoda (20519002 and 24681006) and by the program Bio-Logging Science of the University of Tokyo.

References

[1] R. Nathan, W. M. Getz, E. Revilla et al., "A movement ecology paradigm for unifying organismal movement research," *Proceedings of the National Academy of Sciences of the United States of America*, vol. 105, no. 49, pp. 19052–19059, 2008.

[2] M. Mateos-Rodríguez and B. Bruderer, "Flight speeds of migrating seabirds in the Strait of Gibraltar and their relation to wind," *Journal of Ornithology*, vol. 153, no. 3, pp. 881–889, 2012.

[3] H. Weimerskirch, M. Le Corre, Y. Ropert-Coudert, A. Kato, and F. Marsac, "The three-dimensional flight of red-footed boobies: adaptations to foraging in a tropical environment?" *Proceedings of the Royal Society B*, vol. 272, no. 1558, pp. 53–61, 2005.

[4] H. Weimerskirch, T. Guionnet, J. Martin, S. A. Shaffer, and D. P. Costa, "Fast and fuel efficient? Optimal use of wind by flying albatrosses," *Proceedings of the Royal Society B*, vol. 267, no. 1455, pp. 1869–1874, 2000.

[5] V. H. Paiva, T. Guilford, J. Meade, P. Geraldes, J. A. Ramos, and S. Garthe, "Flight dynamics of Cory's shearwater foraging in a coastal environment," *Zoology*, vol. 113, no. 1, pp. 47–56, 2010.

[6] E. D. Wakefield, R. A. Phillips, M. Jason et al., "Wind field and sex constrain the flight speeds of central-place foraging albatrosses," *Ecological Monographs*, vol. 79, no. 4, pp. 663–679, 2009.

[7] A. Hedenström, T. Alerstam, M. Green, and G. A. Gudmundsson, "Adaptive variation of airspeed in relation to wind, altitude and climb rate by migrating birds in the Arctic," *Behavioral Ecology and Sociobiology*, vol. 52, no. 4, pp. 308–317, 2002.

[8] H. Weimerskirch, M. Louzao, S. de Grissac, and K. Delord, "Changes in wind pattern alter albatross distribution and life-history traits," *Science*, vol. 335, no. 6065, pp. 211–214, 2012.

[9] J. Shamoun-Baranes, E. Van Loon, F. Liechti, and W. Bouten, "Analyzing the effect of wind on flight: pitfalls and solutions," *Journal of Experimental Biology*, vol. 210, no. 1, pp. 82–90, 2007.

[10] F. Liechti, "Birds: blowin' by the wind?" *Journal of Ornithology*, vol. 147, no. 2, pp. 202–211, 2006.

[11] F. Quintana, R. Wilson, P. Dell'Arciprete, E. Shepard, and A. G. Laich, "Women from Venus, men from Mars: inter-sex foraging differences in the imperial cormorant *Phalacrocorax atriceps* a colonial seabird," *Oikos*, vol. 120, no. 3, pp. 350–358, 2011.

[12] J. Kotzerka, S. A. Hatch, and S. Garthe, "Evidence for foraging-site fidelity and individual foraging behavior of pelagic cormorants rearing chicks in the gulf of Alaska," *Condor*, vol. 113, no. 1, pp. 80–88, 2011.

[13] H. Weimerskirch, S. Bertrand, J. Silva, J. C. Marques, and E. Goya, "Use of social information in seabirds: compass rafts indicate the heading of food patches," *PLoS one*, vol. 5, no. 3, p. e9928, 2010.

[14] C. J. Pennycuick, *Modelling the Flying Bird*, Academic Press, London, UK, 2008.

[15] T. Hino and A. Ishida, "Home ranges and seasonal movements of great cormorants *Phalacrocorax carbo* in the Tokai area, based on GPS-Argos tracking," *Japanese Journal of Ornithology*, vol. 61, no. 1, pp. 17–28, 2012.

[16] Y. Inoue, K. Yoda, H. Fujii, H. Kuroki, and Y. Niizuma, "Nest intrusion and infanticidal attack on nestlings in great cormorants *Phalacrocorax carbo*: why do adults attack conspecific chicks?" *Journal of Ethology*, vol. 28, no. 2, pp. 221–230, 2010.

[17] D. Bates and M. Maechler, "lme4: linear mixed-effects models using S4 classes," R package version 0.999375-42, http://cran.r-project.org/web/packages/lme4/, 2011.

[18] R. H. Baayen, "languageR: data sets and functions with 'Analyzing Linguistic Data: a practical introduction to statistics," R package version 1.4, http://cran.r-project.org/web/packages/languageR/, 2011.

[19] T. Hastie, "Gam: generalized additive models," R package version 1.06.2, http://cran.r-project.org/web/packages/gam/, 2012.

[20] J. Fieberg and C. O. Kochanny, "Research and management viewpoint-quantifying home-range overlap: the importance

of the utilization distribution," *Journal of Wildlife Management*, vol. 69, no. 4, pp. 1346–1359, 2005.

[21] C. Calenge, "Adehabitat: analysis of habitat selection by animals," R package version 1.8.10, http://cran.r-project.org/web/packages/adehabitat/, 2012.

[22] R Development Core Team, *R: A Language and Environment For Statistical Computing*, R Foundation for Statistical Computing, Vienna, Austria, 2011.

[23] J. Finn, J. Carlsson, T. Kelly, and J. Davenport, "Avoidance of headwinds or exploitation of ground effect—why do birds fly low?" *Journal of Field Ornithology*, vol. 83, no. 2, pp. 192–202, 2012.

[24] B. Bruderer and A. Boldt, "Flight characteristics of birds: I. Radar measurements of speeds," *Ibis*, vol. 143, no. 2, pp. 178–204, 2001.

[25] C. Pennycuick, P. L. F. Fast, N. Ballerstädt, and N. Rattenborg, "The effect of an external transmitter on the drag coefficient of a bird's body, and hence on migration range, and energy reserves after migration," *Journal of Ornithology*, vol. 153, no. 3, pp. 633–644, 2012.

[26] Y. Y. Watanabe, A. Takahashi, K. Sato, M. Viviant, and C. A. Bost, "Poor flight performance in deep-diving cormorants," *Journal of Experimental Biology*, vol. 214, no. 3, pp. 412–421, 2011.

[27] C. J. Pennycuick, "Fifteen testable predictions about bird flight," *Oikos*, vol. 30, no. 2, pp. 165–176, 1978.

[28] D. Grémillet, R. P. Wilson, S. Storch, and Y. Gary, "Three-dimensional space utilization by a marine predator," *Marine Ecology Progress Series*, vol. 183, pp. 263–273, 1999.

[29] J. T. H. Coleman, M. E. Richmond, L. G. Rudstam, and P. M. Mattison, "Foraging location and site fidelity of the double-crested Cormorant on Oneida Lake, New York," *Waterbirds*, vol. 28, no. 4, pp. 498–510, 2005.

[30] S. Garthe, W. A. Montevecchi, and G. K. Davoren, "Flight destinations and foraging behaviour of northern gannets (*Sula bassana*) preying on a small forage fish in a low-Arctic ecosystem," *Deep-Sea Research*, vol. 54, no. 3-4, pp. 311–320, 2007.

[31] D. B. Irons, "Foraging area fidelity of individual seabirds in relation to tidal cycles and flock feeding," *Ecology*, vol. 79, no. 2, pp. 647–655, 1998.

[32] H. Weimerskirch, "Are seabirds foraging for unpredictable resources?" *Deep-Sea Research*, vol. 54, no. 3-4, pp. 211–223, 2007.

[33] M. Sato, Y. Inoue, M. Ishigaki, R. Yamawaki, Y. Nakagawa, and Y. Niizuma, "Food habit of great cormorants at two locations in Aichi Prefecture," *Japanese Journal of Ornithology*, vol. 58, no. 2, pp. 196–200, 2009.

5

Distribution and Structure of Purkinje Fibers in the Heart of Ostrich (*Struthio camelus*) with the Special References on the Ultrastructure

Paria Parto,[1] **Mina Tadjalli,**[2] **S. Reza Ghazi,**[2] **and Mohammad Ali Salamat**[3]

[1] Biological Department, Faculty of Science, Razi University, Kermansha 6714967346, Iran
[2] School of Veterinary Medicine, Shiraz University, Shiraz 1731-71345, Iran
[3] Biological Department, Faculty of Science, Razi University, Kermansha 6714967346, Iran

Correspondence should be addressed to Paria Parto; pariaparto@gmail.com

Academic Editor: Greg Demas

Purkinje fibers or Purkinje cardiomyocytes are part of the whole complex of the cardiac conduction system, which is today classified as specific heart muscle tissue responsible for the generation of the heart impulses. From the point of view of their distribution, structure and ultrastructural composition of the cardiac conduction system in the ostrich heart were studied by light and electron microscopy. These cells were distributed in cardiac conducting system including SA node, AV node, His bundle and branches as well as endocardium, pericardium, myocardium around the coronary arteries, moderator bands, white fibrous sheet in right atrium, and left septal attachment of AV valve. The great part of the Purkinje fiber is composed of clear, structure less sarcoplasm, and the myofibrils tend to be confined to a thin ring around the periphery of the cells. They have one or more large nuclei centrally located within the fiber. Ultrastructurally, they are easily distinguished. The main distinction feature is the lack of electron density and having a light appearance, due to the absence of organized myofibrils. P-cells usually have two nuclei with a mass of short, delicate microfilaments scattered randomly in the cytoplasm; they contain short sarcomeres and myofibrillar insertion plaque. They do not have T-tubules.

1. Introduction

Jan Evangelista Purkinje was born on 18 December, 1787, in Libochovice (Bohemia). Between the ages of 35 and 63, he made his most significant discoveries. This very age generally displays the physiological measure of the biggest creative waves of the human being. From 1850 until his death in 1869, Purkinje worked in the Institute of Physiology in Prague, which he had based. The acquisition of the big Plösl microscope in 1835 represented an important turning point for Purkinje's histological and embryological research. Purkinje published his discovery of a part of the heart conduction system, nowadays called Purkinje fibers or Purkinje cardiomyocytes [1].

Tawara (1906) was able to follow proximally from the Purkinje fibers (P-fiber) to the besides bundles, which, he organize, were connected to his bundle. Afterwards, he found that his bundle was connected proximally to a well-set plexus of fibers (which he called a node, in his book).

This is the arterioventricular node. In addition, he was able to display interdigitating connections between the Purkinje fibers and ventricular muscles, as well as between the node and arterial muscles [2]. Electrical activity of the specialized conducting system of canine hearts has been recorded in situ through electrodes attached to the endocardium during total cardiopulmonary bypass [3]. Armiger et al. [4] stated that the connective tissue of the trabeculae got from the puppies and the young dogs had little elastic fibers, but this element was well extended in the connective tissue of the adult dogs. The trabeculae of older dogs also showed distributed foci of extracellular fat droplets, and their junctional areas nearest to the ventricular wall were often heavily full with fat. The Purkinje cells were uniform in each group but differed from one group to another [4].

Forsgren et al. [5] identified that in the bovine fetal heart subendocardial bundles of cells could be prominent from the main myocardial mass. Their morphological characteristics

Distribution and Structure of Purkinje Fibers in the Heart of Ostrich (Struthio camelus) with the Special References
on the Ultrastructure

31

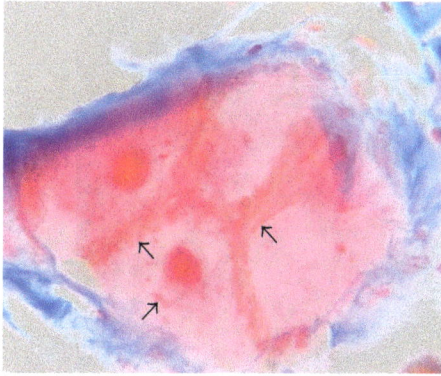

FIGURE 1: Photomicrograph from the microscopic location of the Purkinje cell in the endocardial layer of the sinus venosus of left sinus—arterial valve in the male ostrich heart. Arrow—myofibril cells.

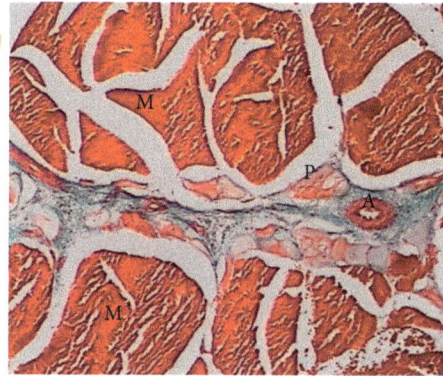

FIGURE 3: Photomicrograph from the microscopic location of the Purkinje fibers (P) around coronary artery (A), between the heart muscle bundles, (M) Green Masson's Trichrome.

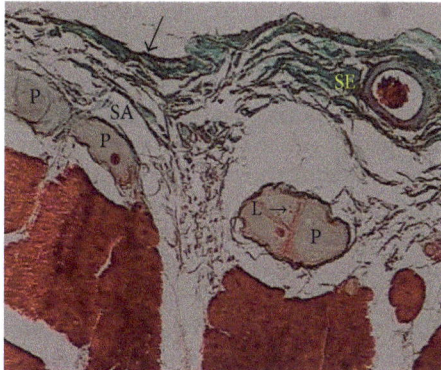

FIGURE 2: Photomicrograph from the microscopic location of the Purkinje fibers in endocardium, that showing the layers, Green Masson's Trichrome. Arrow—endothelium cells, SE—subendothelium layer, SA—subendocardial, P—Purkinje cell, *—connective tissue sheath, M—myocardial cells, L—lipofuscin pigment.

express that they represent bundles of Purkinje fibers. A severe fluorescence after incubation in antisera up to the intermediate filament protein skeleton is also backing this suggestion [5].

A comparative ultrastructural study of bovine Purkinje fibers and common myocytes during fetal development has been undertaken by Forsgren and Thornell [6].

Differences between the two cell types with respectability to the accommodated disc, amount of myofibrils, sequence of mitochondria, amount of glycogen, and formation of T-tubules became apparent gradually. In all stages studied, a redundancy of intermediate filaments was typical for the Purkinje fibers. Myofibrillar M-bands developed an earlier stage in Purkinje fibres than in ordinary myocytes. Myofilament-polyribosome complexes typical of adult cow Purkinje fibers were not seen in the fetal hearts [6].

The ostrich heart has some different features from the other birds. In the ostrich, fibrous pericardium as sternopericardial ligament attaches along the thoracic surface of the sternum. The central edge of muscular valve hangs down into the right ventricle and gives attachment to its rough parietal wall by a thick muscular stalk. The left and right pulmonary veins enter the left atrium independently, and their openings were completely separated from each other by a septum. In the heart of the ostrich, the moderator bands were found in both the right and left ventricles in different locations. The right ventricle presents one tendinous moderator band near the base of the ventricle that extends from septum to the muscular valve. Also the moderator bands as tendinous thread like or flat sheet are usually present at about apex of the right ventricle that extends from septum to the parietal wall. In the left ventricle, there were some tendinous moderator bands close to the apex that extends from septum to the parietal wall and between trabeculae carneae of the parietal wall [7].

Cardiovascular diseases in human and animals are one of the main causes of death. Thus, correct interpretation of heart function or physiology needs full understanding of anatomy, histology, and cardiac conduction system. Macroscopic and microscopic anatomy of heart and its conduction system has been studied in several animal species, but there is no comprehensive research on the ostrich (*Struthio camelus*) heart. The results of this survey in future will be used as basic knowledge.

2. Materials and Methods

2.1. Light Microscopy. Five hearts from healthy male ostriches were used. The hearts were collected at the slaughter house immediately after slaughter. The average weight of the hearts was 1054.33 ± 172.34 g; the length of the long axis was 19.33 ± 1.05 cm; and the circumferential length at the coronary groove was 35.66 ± 1.04 cm. After removal of pericardium, the hearts were flushed with normal saline and subsequently immersed in 10% buffer neutral formalin for 72 hrs (ventricular apex was cut off to permit the penetration of formalin into the lumen). The right and left atrium, right muscular valve, and interarterial septum were separated and divided into several segments. Each segment was dehydrated, cleared, and embedded in paraffin. Serial sections of 6–8 Lm

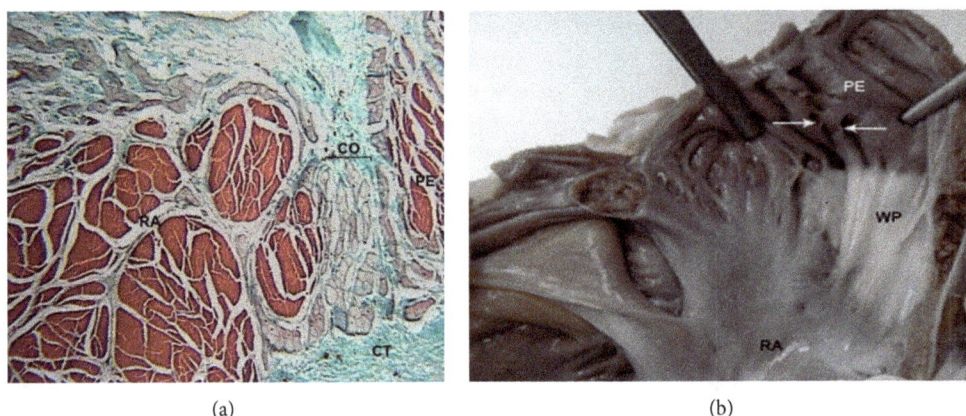

(a)   (b)

FIGURE 4: (a) Photomicrograph from the microscopic location of the fibrous white sheet in between the pectinate muscles (PE), in the right atrium (RA) in the male ostrich heart. Green Masson's Trichrome: CT—connective tissue mass, CO—columns of Purkinje cells. (b) Photomicrograph from the microscopic location of the fibrous white sheet (WP) and related clauses (arrow) in right atrium the male ostrich heart: RA—right atrium bottom, PE—shoulder muscles.

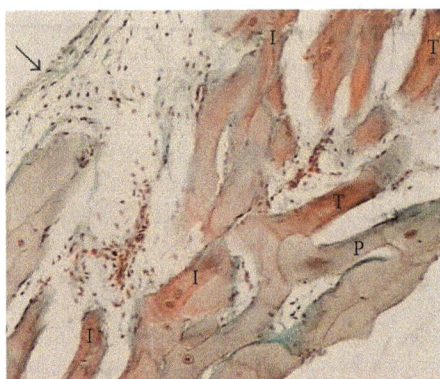

FIGURE 5: Photomicrograph from the microscopic location of the sinoatrial node (SAN), showing P cells (P), transitional cells (T), and intermediate (I). Arrow—endothelial cells in surface sinuatrial valve of the sinus venosus, Green Masson's Trichome.

thickness were cut, mounted, and stained with H&E and Green Masson's Trichrome [8].

2.2. Electron Microscopy. Three other hearts were removed just after slaughter and quickly immersed in Karnovsky's solution for fixation. Cubes of tissue (about 1 mm) were post-fixed in 1% osmium tetroxide with 0.1 m phosphate buffer. Specimens were dehydrated in ethanol and embedded in epoxy resin. Thin sections (0.5–1 Lm) were stained with toluidine blue to identify the location of AV node. Ultra-thin sections (600A°) were mounted on the copper grids and stained with uranyl acetate and lead citrate and were examined with a Philips CM-10 electron microscope, and electromicrographs were prepared.

3. Results

Purkinje fibers in the ostrich heart are large specialized cardiac muscle fibers. They have a much greater diameter than normal muscle fiber, about 5.5–16 μm. The arrangement of the component in P-fiber is different from cardiac muscle fibers. The great part of the fiber is composed of clear, structure less sarcoplasm, and the myofibrils tend to be confined to a thin ring around the periphery of the cells. They have one or more large nuclei centrally located within the fiber (Figure 1). P-fiber in the heart of the ostrich was distributed widely. Throughout the endocardial layers, particularly the subendocardial layer, numerous P-fibers, arranged in one or more rows, are seen (Figure 2). In the myocardium, between muscles bundles, the P-fibers are seen. Passing outwards from the network of P-fibers lying in the endothelial layer, there are large numbers of conduction fiber tracts. These tracts divide up, passing between the muscle fiber bundles, and eventually the finest divisions of the tracts, single Purkinje fibers, are found within the bundles in close association with the muscle fibers. These fiber tracts are usually found in association with the branches of the coronary arteries (Figure 3). The P-fibers are occasionally seen in epicardial layer but occurs lie in protrusion of epicardium below of the surface of myocardium around coronary arteries. In the right auricle of the ostrich heart, P-fibers are found mostextenely beneath the endocardium as a single fibre or well-developed tracts. These latter are found most frequently in the junction of auricle with right atrium, and macroscopically they are with fibers sheet in this junction (Figure 4).

The conducting system of the heart (sinoatrial node, arterioventricular node his bundle and its branches) consists of these fibers. The structure of P-fiber varies according to the part of conducting system in which they occur. The sinoatrial node is composed of P-fibers and intermediate fibers (the fibers which, in appearance, are intermediate between P-fibers and cardiac muscle fibers). P-fibers are located in the peripheral region of this node and passing away from it. In the initial stages of their passage away from the node, they have a thick connection tissue sheet (Figure 5).

The atrioventricular node is composed of fibers which resemble P-fiber more closely than the myocardial muscle fibers. Posteriorly, the AV-node narrows down somewhat and

Distribution and Structure of Purkinje Fibers in the Heart of Ostrich (Struthio camelus) with the Special References
on the Ultrastructure

33

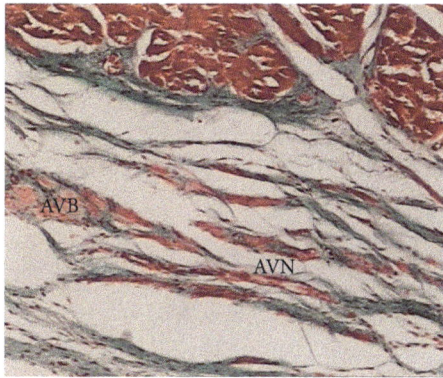

FIGURE 6: Indicative junction of the arterioventricular node (AVN) and arterioventricular bundle (AVB) in the male ostrich heart. Green Masson's Trichrome.

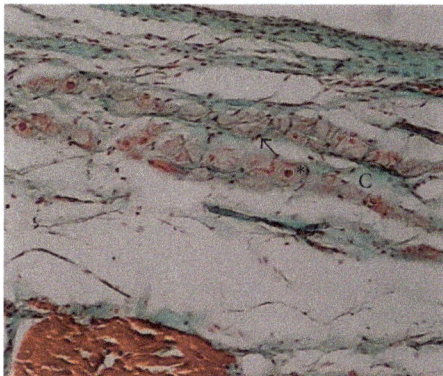

FIGURE 8: Photomicrograph from the microscopic location connecting the left stalk in the connecting area to atrioventricular valve. CT—connective tissue mass, P—Purkinje cells, E—elastic fibers, F—fibroblast cell, C—collagen fibers, P—parietal wall of the right ventricle; H&E.

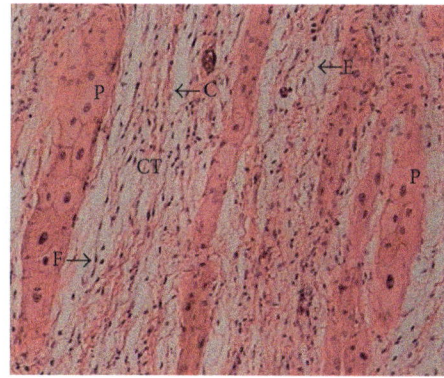

FIGURE 7: Photomicrograph from the microscopic location of the Purkinje cells(*) in arterioventricular of bundle (AVB) in the male ostrich heart. C—collagen fibers, arrow—fibroblast; Green Masson's Trichrome.

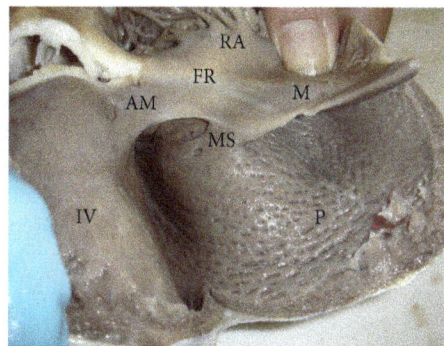

FIGURE 9: The heart of male ostrich, showing the anatomical location of the atrioventricular valve. FR—fibrous ring, RA—right atrium, IV—interventricular septum, M—right muscular valve, MS—muscular stalk, AM—connect the left stalk.

continues as the arterioventricular (His) bundle (Figure 6). This bundle itself is composed of a large number of P-fibers. Three bundle branches (right/left and recurrent) are consisting of P-fibers which are arranged as a cord (Figure 7).

The moderator bands, which are found in the right and left ventricle of the ostrich's heart, consist of only dense irregular connective tissue, and the Purkinje cells fill the core of moderator bands. There was cell-to-cell communication between Purkinje fibers within the bundle. The Purkinje cells were surrounded by connective tissue sheath (Figure 8).

The thick muscular stalk attached the peripheral edge of the muscular right AV value to the arterio-ventricular septum and rough parietal wall of the ventricle. Histologically it is composed of Purkinje fibers (Figure 9).

With the electron microscopy, the P-cell is easily distinguished. The main distinction feature is the lack of electron density and having a light appearance, due to the absence of organized myofibrils. P-cells usually have two nuclei with a mass of short, delicate microfilaments scattered randomly in the cytoplasm; they contain short sarcomeres and myofibrillar insertion plaque. They do not have T-tubules (Figure 10).

4. Discussion

The distribution of Purkinje fibers has been studied in various mammals. In mammals, the Purkinje network is distributed in the subendocardial connective tissue in ventricle. These networks conduct the cardiac excitation from right and left his bundle to the myocardium. In human and dogs, the Purkinje cell in the network is cylindrical or fusiform in shape and arranged in a parallel pattern [9]. In ungulates (sheep and goat), the chain of Purkinje cells was larger than myocytes, and 2–8 oval cells formed these network. The cells connect to each other by desmosomes and gap junction. These cells are covered by a thick sheath of reticular fibers [10]. In rat and mice, Purkinje cells were very similar to ventricular myocytes even a little smaller. They are cylindrical in shape and organized in parallel pattern, continuing with ventricular myocytes in the endocardium [11].

Various aspects of the histochemistry of Purkinje fibers have been studied by Getty [12] and Gossrau [13]. The observations of DiDio [14], Hirako [15], and Sommer and Johnson [16] on the general morphology of the ordinary and Purkinje cells of the fowl heart have been confirmed. The Purkinje cells

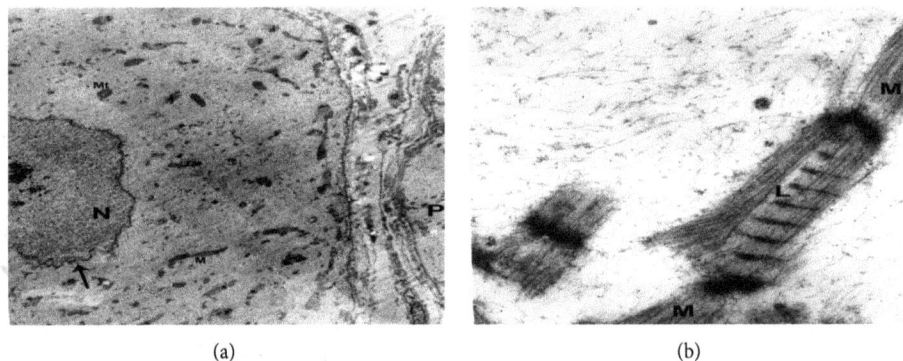

(a)

(b)

FIGURE 10: (a) Electromicrograph of two P cells (P) represents the folded membrane (arrow), nucleus (N), myofibrils (M), and mitochondrial (Mt). (b) Electromicrograph part of the P cell, showing leptomer (L), along myofibril (M).

of the fowl are similar in almost all aspects to those of mammals. They do not have step-like intercalated discs but have irregularly arranged areas of membrane apposition, which possess desmosomes, myofibrillar insertion plaques, and large pentalaminar nexuses. No transverse tubular system is present, and the sarcoplasmic reticulum is poorly developed. The few myofibrils are irregularly arranged. The Purkinje cells have been observed in this study to contain rounded aggregations of leptomeres. Leptomeres have previously been described by Hirako [15] who suggested that they represented an aberrant form of muscle fibril arising during development. This is unlikely since they are present in almost all Purkinje cells.

It seems reasonable to assume that they are associated with myofibrillar formation or destruction since they are seen in association with large masses of disorganized filaments and also in continuity with apparently normal myofibrils. The conducting system consists of the sinoatrial node, the atrioventricular node and bundles, and highly complex network of fine bundles and tracts [16]. Although SA node has been demonstrated in other birds, there is still some doubt as to its position and even its presence as a discrete node in birds [17]. Concluded from physiological evidence that an SA node was present near the termination of the right precaval vein. The node in an ostrich heart is composed of fibers, which in appearance are intermediate between muscle fibers and true Purkinje fibers. Peripherally within the node and passing away from it are several true P-fibers. The distribution of P-fibers within the auricles in ostrich heart is similar to that described by Davis [18] for the pigeon. Prakash [19] has stated that the atria are devoid of P-fibers. In our study, they are found within the myocardium in association with blood vessels, and occasionally in the epicardium [20]. Davis [18] stated that the auricular P-fibers stop short at the base of the auricles and that there are no interconnections between them and the reminder of the cardiac conducting system.

In the ostrich, the AV-node lies in the base of the auricular septum. It is very close to the junction of the auricular and ventricular septa. It is composed of fibers which resemble P-fibers more closely than the myocardial muscle fibres and that they are not identical to those of the SA-node. As Prakash [19] reported, the atrioventricular bundle in our study extend

deep into the interventricular septum and then bifurcate into right and left limb. The later branch divides up from the network of P-fibers. Prakash's [19] description does not exactly coincide with that of DiDio [14] for the pigeon and swan but resemble more closely to that of Drennan [21] for the Ostrich. Similarly, the description of Yousuf [22] for the heart of passer differs in the basic structure of this part of the conducting system.

One of the features of specific interest in the ostrich heart is the presence of moderator bands in both the right and left ventricles and in different locations. The right ventricle has one tendinous moderator band about the base of the ventricle that extends from the septum to the muscular valve. Also the moderator bands are usually about apex of the right ventricle that extend from septum to parietal wall. In the left ventricle, there were some tendinous moderator bands close to the apex that extend from septum to parietal wall and between trabeculae carneae of parietal wall. The moderator bands tend to prevent over distension and serve as the pathway for the passing of Purkinje fibers across the lumen of the cavity forming a part of the conducting system. Anatomically, the location of the moderator band in the right and left ventricle of the ostrich heart is different from the other animals. There were no papillary muscle in the right ventricle of the ostrich and the moderator band attaches directly to the ventral surface of muscular valve from interventricular septum.

Since in human [23, 24] the moderator band extends between interventricular septum and ventricular free wall and in domestic animals [25] and ungulates [26] these bands extend from interventricular septum to the papillary muscle and there is no connection to the vulvar cusps. In the Ostrich left ventricle, despite the presence of papillary muscle, these bands have no connection to them. Histologic structure of moderator band showed that they have muscular tissue in various proportions with connective and conductive tissue [27]. It is similar to that of the human [28], ungulate [26], sheep [24, 29] and goat [24] hearts but in the carnivores, a real moderator band was never found [26]. Whatever is the size and shape, the moderator band must be regarded as the shortest pathway from interventricular septum to the free wall of left and right ventricle in the ostrich and other animals. The Purkinje fibers are large in size and similar in

Distribution and Structure of Purkinje Fibers in the Heart of Ostrich (Struthio camelus) with the Special References on the Ultrastructure

35

most cellular characteristics in ostrich and dog [30], but in the ostrich, there is no perinuclear clear area. There is a little glycogen in these cells, but in human and mammals these cells are rich in glycogen [4]. The Purkinje fibers are organized into the bundle with cell-to-cell communication and little lateral communication. In the ostrich, there is a sheath of connective tissue around the Purkinje cells, but in human and mammals there is no fibrous sheath around these cells [25]. This organization of the bundle fibers increases the spread of propagated impulses and inhibits the transverse spread. In the right ventricle, there is one musculotendinous moderator band about the base of the ventricle, which extends from the interventricular septum to the muscular stalk of the muscular valve. It was single and sometimes branched [4].

Acknowledgment

The authors are grateful to the research council of the Shiraz University for providing financial assistance.

References

[1] O. Eliška, "Purkynje fibers of the heart conduction system—history and the present time," Časopis Lekaru Ceskych, vol. 145, no. 4, pp. 329–335, 2006.

[2] S. Tawara, Das Reizleitungssystem des Saugetierhezens, Verlag von Gustav Fischer, Jena, Germany, 1906.

[3] B. F. Hoffman, P. F. Cranefield, J. H. Stuckey, and A. A. Bagdanas, "Electrical activity during the P-R interval," Circulation Research, vol. 8, pp. 1200–1211, 1960.

[4] L. C. Armiger, F. Urthaler, and T. N. James, "Morphological changes in the right ventricular septomarginal trabecula (false tendon) during maturation and ageing in the dog heart," Journal of Anatomy, vol. 129, no. 4, pp. 805–817, 1979.

[5] S. Forsgren, L. E. Thornell, and A. Eriksson, "The development of the Purkinje fibre system in the bovine fetal heart," Anatomy and Embryology, vol. 159, no. 2, pp. 125–135, 1980.

[6] S. Forsgren and L. E. Thornell, "The development of Purkinje fibres and ordinary myocytes in the bovine fetal heart. An ultrastructural study," Anatomy and Embryology, vol. 162, no. 2, pp. 127–136, 1981.

[7] M. Tadjalli, S. R. Ghazi, and P. Parto, "Gross anatomy of the heart in Ostrich (Struthio camelus)," Iranian Journal of Veterinary Research, vol. 10, no. 1, pp. 21–27, 2009.

[8] L. G. Luna, Manual of Histologic Staining Methods of the Armed Forces Institute of Pathology, American Registry of Pathology, New York, NY, USA, 3rd edition, 1968.

[9] P. F. Cranefield, A. L. Wit, and B. F. Hoffman, "Conduction of the cardiac impulse. 3. Characteristics of very slow conduction," Journal of General Physiology, vol. 59, no. 2, pp. 227–246, 1972.

[10] T. Shimada, T. Ushiki, and T. Fujita, "Purkinje fibers of the heart," Shinyaku to Chiryou, vol. 42, pp. 11–13, 1992.

[11] Sawazaki, Hikaku Sinzougaku, Asakura Shoten, 1985, (Japanese).

[12] R. Getty, Sisson and Grossman's, the Anatomy of the Domestic Animals, WB Saunders, 5th edition, 1975.

[13] R. Gossrau, "The impulse conducting system of the birds—histochemical and electron microscopical investigations," Histochemie, vol. 13, no. 2, pp. 111–159, 1968.

[14] L. J. DiDio, "Myocardial ultrastructure and electrocardiograms of the hummingbird under normal and experimental conditions," Anatomical Record, vol. 159, no. 4, pp. 335–352, 1967.

[15] R. Hirako, "Fine structure of Purkinje fibers in the chick heart," Archivum Histologicum Japonicum, vol. 27, no. 1, pp. 485–499, 1966.

[16] J. R. Sommer and E. A. Johnson, "Cardiac muscle—a comparative ultrastructural study with special reference to frog and chicken hearts," Zeitschrift für Zellforschung und Mikroskopische Anatomie, vol. 98, no. 3, pp. 437–468, 1969.

[17] E. Mangold and T. Kato, "Zur vergleichenden Physiologie des His'schen Bündels," Pflügers Archiv, vol. 160, pp. 91–131, 1914.

[18] F. Davis, "The conducting system of the bird's heart," Journal of Anatomy, vol. 64, pp. 129–146, 1930.

[19] R. Prakash, "The heart and its conducting system in the common Indian fowl," Proceedings of the National Institute of Sciences of India, vol. 22, pp. 22–27, 1956.

[20] R. C. Truex, Comparative Anatomy and Functional Consideration of the Cardiac Conducting System, Elsevier, Amsterdam, The Netherlands, 1961.

[21] M. R. Drennan, "The auriculo-ventricular bundle in the bird's heart," British Medical Journal, vol. 1, pp. 321–322, 1927.

[22] N. Yousuf, "The conducting system of the heart of the house sparrow, Passer domesticus indicus," Anatomical Record, vol. 152, no. 3, pp. 235–249, 1965.

[23] A. K. Abdulla, A. Frustaci, J. E. Martinez, R. A. Florio, J. Somerville, and E. G. J. Olsen, "Echocardiography and pathology of left ventricular "false tendons,"" Chest, vol. 98, no. 1, pp. 129–132, 1990.

[24] M. Deniz, M. Kilinç, and E. S. Hatipoglu, "Morphologic study of left ventricular bands," Surgical and Radiologic Anatomy, vol. 26, no. 3, pp. 230–234, 2004.

[25] R. Depreux, H. Mestdagh, and M. Houcke, "Comparative morphology of the trabecula septomarginalis in terrestrial mammals," Anatomischer Anzeiger, vol. 139, no. 1-2, pp. 24–35, 1976.

[26] D. Lotkowski, M. Grzybiak, D. Kozłowski, K. Budzyn, and W. Kuta, "A microscopic view of false tendons in the left ventricle of the human heart," Folia morphologica, vol. 56, no. 1, pp. 31–39, 1997.

[27] P. Parto, M. Tadjalli, and S. R. Ghazi, "Macroscopic and microscopic studies on maderator bands in the heart of ostrich (struthio camelus)," Global Veterinaria, vol. 4, no. 4, pp. 374–379, 2010.

[28] R. I. Clelland, "Note on a moderator band in the left ventricle and perforate septum ovale in the heart of a sheep," Journal of Anatomy and Physiology, vol. 32, no. 4, p. 779, 1898.

[29] G. E. Sandusky Jr. and S. L. White, "Scanning electron microscopy of the canine atrioventricular bundle and moderator band," American Journal of Veterinary Research, vol. 46, no. 1, pp. 249–252, 1985.

[30] R. C. Truex and W. M. Copenhover, "Histology of the moderator band in man and other mammals with special reference to the conduction system," American Journal of Anatomy, vol. 80, no. 2, pp. 173–201, 1947.

6

Determinants and Congruence of Species Richness Patterns across Multiple Taxonomic Groups on a Regional Scale

Jörn Buse[1,2] and Eva Maria Griebeler[1]

[1] Department of Ecology, Institute of Zoology, Johannes Gutenberg-University Mainz, Becherweg 13, 55099 Mainz, Germany
[2] Ecosystem Analysis, Institute for Environmental Sciences, University of Koblenz-Landau, Fortstrasse 7, 76829 Landau, Germany

Correspondence should be addressed to Jörn Buse, joernbuse@gmx.de

Academic Editor: Beth Okamura

Applying multiple generalized regression models, we studied spatial patterns in species richness for different taxonomic groups (amphibians, reptiles, grasshoppers, plants, mosses) within the German federal state Rhineland-Palatinate (RP). We aimed (1) to detect their centres of richness, (2) to rate the influence of climatic and land-use parameters on spatial patterns, and (3) to test whether patterns are congruent between taxonomic groups in RP. Centres of species richness differed between taxonomic groups and overall richness was the highest in the valleys of large rivers and in different areas of southern RP. Climatic parameters strongly correlated with richness in all taxa whereas land use was less significant. Spatial richness patterns of all groups were to a certain extent congruent but differed between group pairs. The number of grasshoppers strongly correlated with the number of plants and with overall species richness. An external validation corroborated the generality of our species richness models.

1. Introduction

Europe has undergone a period of environmental change and loss of biodiversity over the last decades [1, 2]. A high level of biodiversity may help to preserve a range of options to adapt under changing environmental conditions such as climate and land-use change. Hence, studies of spatial patterns of species richness and its environmental determinants are required. Broad-scale patterns (i.e., global or continental extent and large grain size) in species richness are relatively well studied, and the determining mechanisms of patterns cover a wide range from energy and water availability [3–5] to historic climate and climate stability as predictors of present patterns [6, 7]. Distribution models are frequently applied to understand the relationship between spatial patterns in species occurrence and environmental variables, (e.g., [8–11]). While broad-scale species richness patterns are mainly determined by energy and water availability, (e.g., [3–5, 12–14]), regional patterns (except of few taxonomic groups, e.g., [15, 16]) were less frequently studied and underlying mechanisms are widely unknown. The strength of the impact of environmental variables on species distributions

may differ with spatial extent and grain size [17–20] and a simple downscaling of the results found at broad spatial scales is not wise. The poor knowledge is particularly true for less studied taxonomic groups, for example, mosses, which encompass hundreds of species, but data availability on this group has only recently been improved at a national level (e.g., [21] for mosses in Germany).

Studies at the meso- or microscale (i.e., intermediate or small spatial extent and grain size) are needed to analyse the impact of water and energy availability on species richness patterns and to evaluate potential effects of climate change. Hortal et al. [19] showed for mammal assemblages that climatic gradients are stronger predictors of geographic ecological richness, that is, at broad spatial extent (areas between 1,000 and 10,000 km^2), whereas other features such as habitat type become more important for the ecological richness of mammals at smaller spatial extent (areas between 100 and 1,000 km^2, cf. also [22]). However, for other taxonomic groups, it remains unclear how much variation of regional species richness patterns is determined by climate and/or land use. Sensitivity of taxonomic groups to climate variables at a regional scale might help to assess

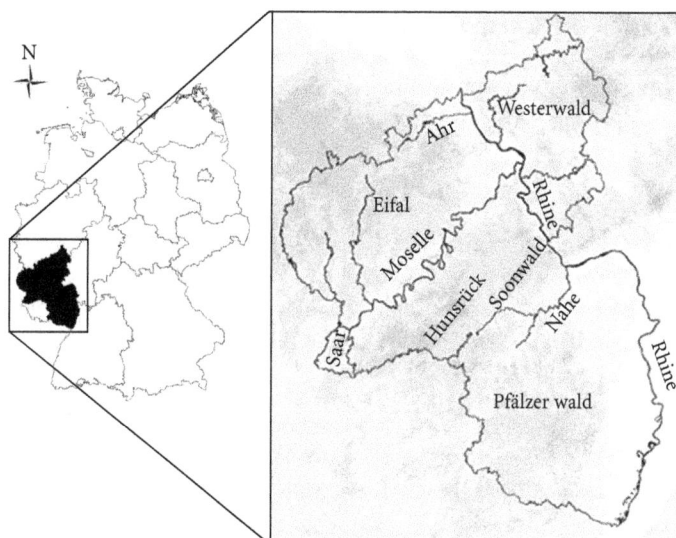

FIGURE 1: Overview map of the study region Rhineland-Palatinate (Germany) showing important rivers and mountain ranges. Palatinate is the southeastern part of Rhineland-Palatinate located southeast of the River Nahe.

effects of climatic changes on species distributions. Furthermore, conservation strategies are usually developed and implemented at smaller regional scales. Knowing whether particular taxonomic groups show congruence in spatial patterns, as this was reviewed by Heino [23] for aquatic ecosystems, is an important information for conservation.

Here, we investigate regional patterns in species richness and their environmental determinants for the poorly studied grasshoppers and mosses, and for amphibians, reptiles, and herbaceous plants in the federal state Rhineland-Palatinate in Germany. These five taxonomic groups are the only groups for which suitable data on the spatial distribution and on environmental parameters is available for the whole study region and for an adjacent region that we used to validate our results. The main questions answered in this study were as follows.

(1) Where are the centres of species richness located within the Rhineland-Palatinate?

(2) Which climatic and land-use parameters determine spatial pattern of species richness in this federal state?

(3) How congruent are regional patterns in species richness between different taxonomic groups?

We expected to find contrasting distribution patterns and different parameters being important for individual taxonomic groups. Land-use variables were expected to influence species richness distributions of all selected groups more strongly than climatic variables.

We used multiple regression models (generalized linear model, GLM) to analyse the relationship between species richness patterns and environmental variables and to identify the ecological determinants of the observed patterns. In order to determine whether the different taxonomic groups show congruence in spatial patterns of species richness at the regional scale, we analysed species richness data on the regional distribution of all five taxonomic groups.

2. Material and Methods

2.1. Research Area. We studied species richness patterns in the German federal state Rhineland-Palatinate (RP, Figure 1). RP is located in the southwest of Germany and covers a total area of 19,853 km². RP is the federal state with the largest woodland cover (ca. 42% of its total area) in Germany. Due to intensive land use in some parts of RP, the woodland cover is not homogenously distributed within the federal state; particularly in the traditional vineyard regions, the proportion of woodlands is very small. RP is characterised by several low mountain ranges up to 800 m a.s.l. Important xerothermic sites are located in the larger valleys, for example, of the rivers Rhine, Moselle, Ahr, and Nahe, where relict and "island" populations of several endangered thermophilic reptile and insect species are found.

2.2. Modelling Species Richness. Multiple generalised linear regression models (GLMs) were established to describe the spatial distribution of species richness for each taxonomic group and to describe overall species richness (all taxonomic groups pooled). All models are based on species occurrence (presence-only) data and different environmental variables. Data on species occurrence and environmental variables were available on a grid base at the resolution of ordinance survey maps (OSM, 1 : 25,000). Species richness values were calculated for each grid cell based on presence-only data of species. Land cover, landscape heterogeneity, and climate variables were used as predictors in the GLMs. All models also accounted for spatial autocorrelation in the residuals [24].

Each grid cell comprised an area of approximately 130 km². The total number of grid cells covering RP was 194. As grid cells that are located at the borders with France or Belgium are less intensively studied than others in the centre of the federal state and environmental and species data from the French and Belgian parts of border areas were not available to us, we included only those grid cells where more than 50% of the covered area is located in RP. In total, we omitted ten grid cells, resulting in 184 grid cells being used in our analyses.

2.3. Species Occurrence Data. Data on the distribution of different taxonomic groups in RP were derived from literature, databases, and from the State Agency for Environment of Rhineland-Palatinate. We collected literature data on reptiles [46, 47, unpublished database of the State Agency for Environment in RP, (1960–2008)], amphibians [46, 47, unpublished database of the State Agency for Environment in RP, (1960–2008)], grasshoppers (Orthoptera) [25, (1980–2000)], vascular plants including ferns [26], and mosses [21, (1980–2007)]. For each of the grid cells, we counted the number of species per taxonomic group. Species records were simple presence data, and absence of a species in grid cells shows only that no records were available. However, our presence-only data come close to true presence-absence data, because of the comprehensive and intensive monitoring of the studied groups by experts and volunteers in RP. Although monitoring intensity of mosses is still relatively low, we decided to roughly test the first dataset on this group that is for the first time available for Germany [21]. To test our models on an independent dataset, we additionally collected analogous data on the distribution of the five taxonomic groups in the adjacent federal state Baden-Württemberg (BW, n = 291 grid cells). For this model validation, we used the following literature sources: Günther [47] for amphibians and reptiles, Maas et al. [25] for grasshoppers, Haeupler and Schönfelder [26] for vascular plants and ferns, and Meinunger and Schröder [21] for mosses. These five taxonomic groups were selected, because species distribution data were available for both our study region RP and the validation region BW. We initially aimed to study further taxonomic groups such as dragonflies and butterflies, but data on the distribution of these insects was only available for BW and only existed for small geographic areas of RP.

2.4. Environmental Data. We selected 13 land-cover variables from the CORINE Land Cover 2000 dataset which is based on satellite images (see Table 1). CORINE provides additional land-cover classes for Germany, but these do not occur in our study area (e.g., peat bogs, salt meadows, etc.).

Landscape heterogeneity based on land-cover data was calculated using Simpson's diversity index D:

$$1 - D = 1 - \sum (p_i)^2, \tag{1}$$

where p_i is the proportion of the ith land-cover type in a grid cell.

We used Simpson's index as a measure for landscape heterogeneity because it calculates the smaller proportions of land-cover types (e.g., small waters) in our dataset more reliably than the Shannon diversity index commonly used in other studies.

An additional topographic measure of spatial landscape heterogeneity was calculated using digital elevation data (SRTM-3). We extracted from this dataset the maximum (HTMAX, Table 1) and minimum elevation (HTMIN) for each grid cell and calculated the difference between minimum and maximum as a measure of variability in elevation (HTDIFF). The topographic variables HTMAX and HTMIN were also included in the analyses.

We also used four climate variables (Table 1) from a dataset that was explicitly established at OSM resolution for Germany [27]. This dataset is based on climate data from 2342 weather stations (German Weather Service, DWD) distributed throughout Germany. In the underlying climate model, basic climate data are homogenised and corrected for the mean elevation of the grid cells (see [27] for further explanations). We used data from the period 1961 to 1990— as for the species records—from this climate dataset to estimate mean monthly and annual values for the climatic situation in each grid cell.

Additionally, plant species richness was used as a predictor in the grasshopper model because several grasshopper species, for example, *Calliptamus italicus* (L., 1758) and *Tetrix bipunctata* (L., 1758), feed exclusively on plants that occur in habitats with high plant diversity, such as dry grasslands. Different types of grassland were not distinguished in the CORINE dataset.

2.5. Development of Species Richness Models. To calculate species richness of the taxonomic groups and overall species richness, we standardised species richness for each group, because species numbers differed strongly between groups (Table 1). We set the maximum observed species number derived from all grid cells to 100 percent for each species group and divided species numbers in each of the grid cells by this maximum. For the overall species richness, percentages in grid cells obtained for taxonomic groups were averaged. Congruence was studied between each of the taxonomic groups as well as between the taxonomic groups and combined species richness (i.e., overall species richness without the group tested). In all cases, percentages were subsequently arc-sin transformed before they were used in the multiple regression analyses [28].

Hierarchical partitioning was carried out to derive the independent contribution of the predictor variables (environmental variables) to the response variable (species richness). This statistical technique is even applicable when data are highly correlated and was used here to eliminate unimportant predictor variables [29]). In this preselection procedure, we first analysed the land cover parameters (here without water related parameters, MAR, FLW, STW; Table 1). The water-related and climate variables were analysed together in a second step, because of a limited number of variables that may be tested in the partitioning procedure. We then fitted a multiple regression model (GLM with poisson error distribution, or quasipoisson in the case of overdispersion) using all the variables that had individually contributed more than 10% to the variance in the response variable in the hierarchical partitioning

TABLE 1: Variables used for the multiple regression modelling and multiple regression models established for species richness of different taxa. Listed are the beta coefficients and significance levels for the parameters in the final models established for five taxonomic groups. The explained variance in species richness for each taxonomic group is also shown. For spatial validation of models, the Spearman rank correlation coefficients r of predicted and observed species richness values and the respective significance level are shown. $*P < 0.05$, $**P < 0.01$, $***P < 0.001$, n.s.: nonsignificant parameters in the final model that improve the explanatory power of the model (AIC).

Variables (units)	Abbreviation	Amphibians	Reptiles	Grasshoppers	Plants	Mosses
Maximum observed number of species/grid cell	—	16	8	48	1180	400
CORINE 2000 land-cover classes						
Deciduous forests (%)	DEW	—	—	—	—	—
Coniferous forests (%)	COW	—	—	—	−1.406*	—
Mixed forests (%)	MIX	—	—	—	—	—
Nonirrigated arable land (%)	NIA	—	—	—	−0.909*	−0.237**
Vineyards (%)	VIY	—	—	—	—	—
Orchards (%)	ORC	—	—	—	—	—
Meadows and pastures (%)	MEP	—	—	—	—	—
Natural grasslands (%)	NAG	—	—	—	—	—
Heathlands (%)	HEA	—	—	—	—	—
Shrublands (%)	SHR	—	—	—	—	—
Marshes (%)	MAR	—	—	4.061*	—	—
Flowing water bodies (%)	FLW	—	—	—	3.052*	—
Standing water bodies (%)	STW	—	—	—	—	—
Topography						
Minimum elevation within grid cell (m)	HTMIN	—	—	—	—	−0.001**
Maximum elevation within grid cell (m)	HTMAX	—	—	0.075**	—	—
Landscape heterogeneity						
Habitat heterogeneity[b]	SIMPSON	0.374[n.s.]	—	—	—	—
Range of elevation within grid cell (m)	HTDIFF	—	—	—	—	—
Climatic parameters 1961–1990[27]						
Mean annual temperature (°C)	TMPYEAR	0.065*	—	—	—	—
Mean sum of annual precipitation (mm)	PRECYEARSUM	—	—	—	—	—
Index of aridity[a]	IOA	—	—	—	—	—
Mean temperature of the coldest month (January) (°C)	TMPJAN	—	—	1.269**	—	−0.064[n.s.]
Spatial autocorrelation	SAC	0.053**	0.159***	0.409***	0.078***	0.005***
Number of plant species		not tested	not tested	0.071***	—	not tested
Model accuracy and validation						
Goodness of fit (deviance change)		38.36%	33.76%	72.73%	62.34%	72.6%
Deviance change explained by environment		44.34%	—	45.42%	17.45%	7.05%
Deviance change explained by SAC		55.66%	100%	54.48%	82.55%	92.95%
ANOVA (versus Null model)		$F = 11.826***$	$F = 14.145***$	$F = 69.621***$	$F = 79.886***$	$F = 124.61***$
Spatial validation of the SAC model		0.758***	0.646***	0.752***	0.648***	0.721***
Spatial validation of the model without the SAC term)		0.606***	—	0.494***	0.326***	0.600***

[a] mean temperature in July/mean annual sum of precipitation.
[b] Simpson's index of diversity using the relative proportion of the land-cover classes in grid cells.

procedure. The significance of quadratic functions of the parameters in the GLMs was also tested. We used Pearson correlations to account for collinearity among preselected predictor variables. If there were highly correlated predictor variables ($r > 0.7$) in the model, one of the variables was removed. Finally, we run an automatic stepwise procedure to delete backwards nonsignificant predictor variables from the model. Deletion of variables from the model was based on the AIC value of the respective model until a minimal AIC value was reached.

2.6. Residual Spatial Autocorrelation of the Models. If model residuals show spatial autocorrelation, this may bias parameter estimates and can increase type I error rates [24]. Therefore, for each species group, we first used the R-package "ncf" to carry out spatial autocorrelation analyses of residuals of each of the five species richness models. Spline correlograms were plotted to visualise the estimated spatial dependence of the data as a continuous function of distance [30]. We calculated 95% confidence intervals of the estimated function using a bootstrap algorithm with 1000 resamples. Next, we added a spatial autocovariate (SAC) to our species richness models to test for significant effects of spatial autocorrelation on estimated beta values and their significance [24, 31, 32]. The autocovariate was calculated as the average species richness value of the direct neighbours for each grid cell using the R-package "spdep." Hierarchical partitioning was used to quantify the contribution of the spatial autocovariate and the environmental variables to the model.

The beta value of the added spatial autocovariate (SAC) was significant for all taxonomic groups. We thus included SAC as a further predictor variable in our species richness models. As the SAC variable itself is not informative with respect to the ecological determinants of species richness, we additionally tested for significant correlations between environmental variables and the SAC variable.

2.7. Model Accuracy and Validation. Model accuracy was evaluated with an ANOVA (*F*-test) which tests for significant differences from the null model based on the change in the deviance. The deviance change itself (measured in %) was used as a direct measure of the overall fit of the regression model.

Our final models for species richness were validated on an independent analogous dataset that we established for BW, a federal state adjacent to RP. Independent analogous datasets are considered as the best means of validating any predictive species distribution model [33–35]. Regression models were rated by correlating predicted and observed values of species richness in BW with the nonparametric Spearman's rank correlation test.

The development of all species richness models was carried out with the free software R version 2.9.2. All GIS work was conducted in ArcGIS 9.3 [36].

3. Results

3.1. Patterns of Species Richness within Taxonomic Groups. The number of amphibian species ranged from 5 to 16 (mean = 10.4) per grid cell, the number of reptiles from 2 to 8 per grid cell (mean = 5.4 species), and the number of grasshoppers from 8 to 48 per grid cell (mean = 26.5) in RP. Plant species numbers varied between 181 and 1180 (mean = 728.2), and the number of moss species between 63 and 400 (mean = 226.0).

There were three main centres of species richness for amphibians: Westerwald, the Lower Nahe valley, and the Upper Rhine valley (Figure 2(a)). The Middle Rhine valley represented a major centre of reptile species richness, with a minor centre in the low mountain areas of the Soonwald and along the Nahe river (Figure 2(b)). Grasshopper richness decreased from the north of RP to the south, with the exception of the sun-exposed valleys of the large rivers (Rhine, Moselle, and Nahe), where species richness was higher. The Palatinate is the most important centre of grasshopper diversity (Figure 2(c)). Plant species richness is mainly restricted to the Middle Rhine valley and to a lesser extent to the Nahe river and Lower Saar valley (Figure 2(d)). Mosses show a distribution pattern that is different from all other studied groups with two clearly distinct centres of species richness, both of which are located in the mountainous areas of RP. Large numbers of moss species exist in the northern Eifel mountains, in the southern parts of the Hunsrück, in the Saar-Nahe mountains, and to some extent in the Pfälzer Wald (Figure 2(e)).

Overall species richness increased from the north to the south but was generally highest in the valleys of large rivers (Figure 2(f)). Centres of overall species richness were found in the Middle-Rhine region, the Saar-Moselle region, the Nahe region, and in different parts of Palatinate.

Patterns in species richness differed between taxonomic groups but were to a certain extent congruent (Figure 2, Table 2). All tested groups were significantly pair-wise correlated in terms of species richness (Table 2). While amphibians and reptiles were only weakly correlated, grasshoppers correlated strongly with plant species richness (Spearman r = 0.684, P < 0.001, Spearman rank correlation, Table 2). Grasshoppers seem to be a good indicator group for combined species richness of all other studied taxonomic groups (Spearman r = 0.716, P < 0.001, Spearman rank correlation). This observation is also confirmed in BW (model validation region), where grasshopper richness also correlated strongly with overall species richness (Spearman r = 0.514, P < 0.001, Spearman rank correlation).

3.2. Species Richness Models. For all taxonomic groups, spline correlograms of residuals of standardized species richness as well as raw species richness itself showed positive spatial autocorrelation within the first two OSM grid cells and a negative autocorrelation at greater distances. No negative spatial autocorrelation of the residuals was found at large distances for the reptile model. However, a weak negative trend in residuals was observed for all other models for larger distances (>10 grid cells).

The multiple regression models including the SAC variable that were derived for each of the taxonomic groups explained between 34 (reptiles) and 73% (grasshoppers) of the observed variability in species richness (Table 1). Climate variables were significant for amphibians, grasshoppers, and mosses, but not for reptiles and plants. Landscape heterogeneity variables (SIMPSON, HTDIFF) were only significant for amphibians. Topographical variables (minimum or maximum elevation, HTMIN, HTMAX) significantly correlated with species richness of grasshoppers and of mosses.

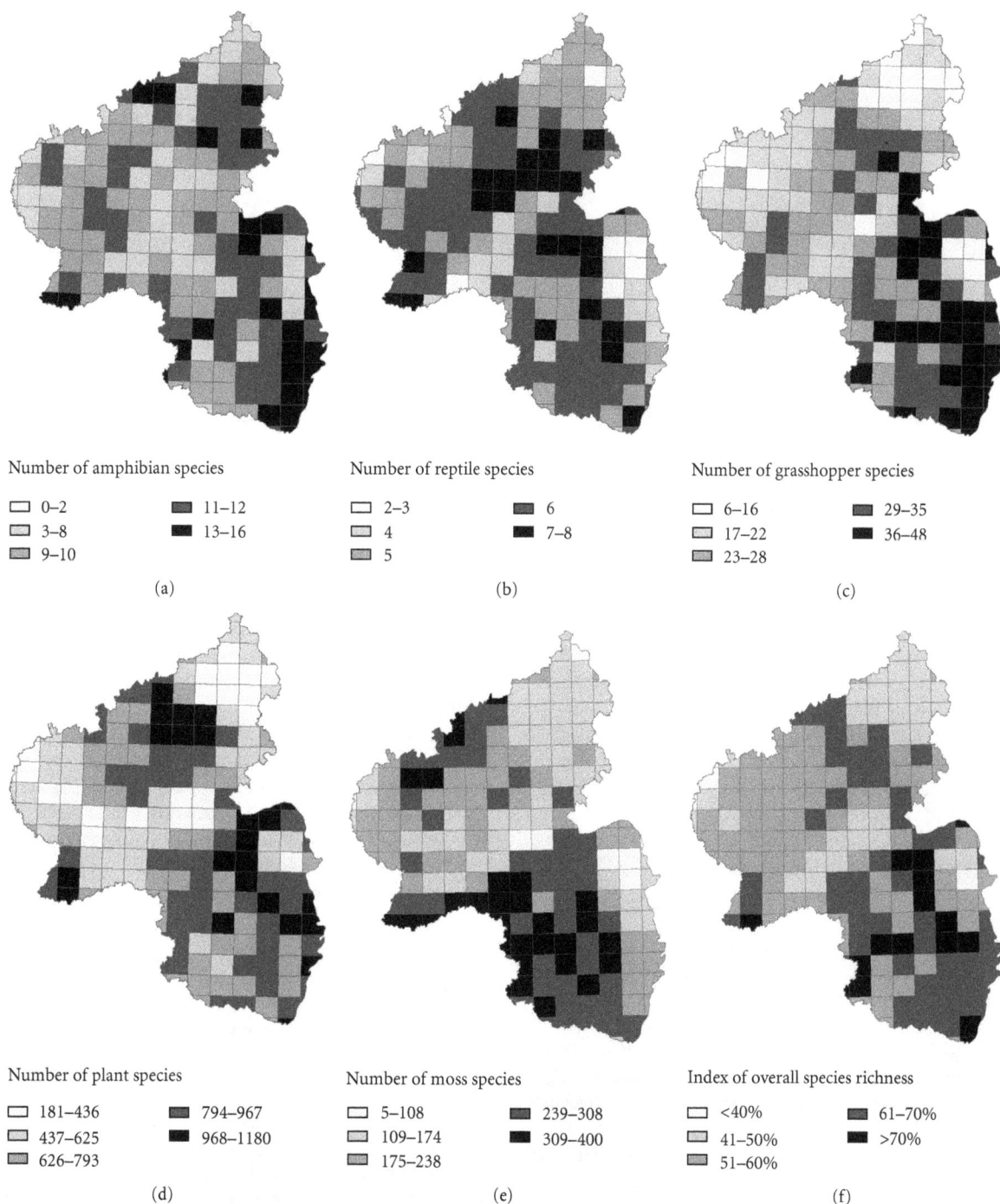

Number of amphibian species

☐ 0–2   ◼ 11–12
☐ 3–8   ◼ 13–16
▨ 9–10

(a)

Number of reptile species

☐ 2–3   ◼ 6
☐ 4   ◼ 7–8
▨ 5

(b)

Number of grasshopper species

☐ 6–16   ◼ 29–35
☐ 17–22   ◼ 36–48
▨ 23–28

(c)

Number of plant species

☐ 181–436   ◼ 794–967
☐ 437–625   ◼ 968–1180
▨ 626–793

(d)

Number of moss species

☐ 5–108   ◼ 239–308
☐ 109–174   ◼ 309–400
▨ 175–238

(e)

Index of overall species richness

☐ <40%   ◼ 61–70%
☐ 41–50%   ◼ >70%
▨ 51–60%

(f)

FIGURE 2: Species richness patterns of amphibians (a), reptiles (b), grasshoppers (c), plants (d), mosses (e), and overall species richness (f) among the tested taxonomic groups in Rhineland-Palatinate at the OSM resolution.

Landscape heterogeneity (SIMPSON) significantly affected only amphibian species richness (Table 1). TMPYEAR (mean annual temperature) was the most important predictor of amphibian species richness and explained 23% of the deviance change compared to the null model. As the SAC itself does not contain any useful ecological information, we carried out additional tests to rate correlations between SAC and environmental variables. Important variables found

indirectly determining amphibian species richness via SAC were the cover percentage of standing water in the grid cells (STW, $r_{Pearson} = 0.406$, $P < 0.001$), the mean temperature in January (TMPJAN, $r_{Pearson} = 0.482$, $P < 0.001$), and the maximum elevation within the grid cells (HTMAX, $r_{Pearson} = -0.511$, $P < 0.001$).

None of the environmental variables analysed did directly influence reptile species richness, but the effect of the

TABLE 2: Spearman rank correlations for the species richness of the different taxonomic groups in Rhineland-Palatinate (modelled region) and Baden-Württemberg (the adjacent region used for model validation). We used arcsin-transformed data to calculate correlations. $***P <$ 0.001, $**P < 0.01$, $*P < 0.05$, n.s.: not significant.

	All other species[a]	Amphibians	Reptiles	Grasshoppers	Plants
Rhineland-Palatinate					
Amphibians	0.455***	—			
Reptiles	0.420***	0.187*	—		
Grasshoppers	0.716***	0.489***	0.452***	—	
Plants	0.678***	0.450***	0.365***	0.684***	—
Mosses	0.452***	0.219 **	0.348***	0.393***	0.421***
Baden-Württemberg					
Amphibians	0.431***	—			
Reptiles	0.467***	0.333***	—		
Grasshoppers	0.514***	0.345***	0.387***	—	
Plants	0.275***	0.417***	0.030 n.s.	0.323***	—
Mosses	0.282***	0.005 n.s.	0.440***	0.186**	0.007 n.s.

[a] Combined species richness of all other studied taxonomic groups.

SAC variable was significant. Positive correlations existed between SAC and the proportion of woodland (DEW, $r_{Pearson} = 0.308$, $P < 0.001$; MIX, $r_{Pearson} = 0.255$, $P < 0.001$) and between SAC and the proportion of orchards (ORC, $r_{Pearson} = 0.168$, $P = 0.022$). SAC was negatively correlated with the proportion of meadows and pastures (MEP, $r_{Pearson} = -0.357$, $P < 0.001$), the mean sum of annual precipitation (PRECYEARSUM, $r_{Pearson} = -0.234$, $P = 0.001$), and the proportion of marshes (MAR, $r_{Pearson} = -0.226$, $P = 0.002$). Topographical landscape structure impacted reptile richness indirectly via SAC, which decreased slightly with increasing minimum elevation levels in the grid cells (HTMIN, $r_{Pearson} = -0.194$, $P = 0.008$).

The number of grasshopper species correlated positively with the mean temperature in January (TMPJAN) and with maximum elevation (HTMAX, Table 1). SAC was, for example, significantly negatively correlated with the proportion of meadows and pastures (MEP, $r_{Pearson} = -0.664$, $P < 0.001$). This negative correlation was not considered as a single predictor in the final model but should be discussed because of the strength of the relationship (Figures 3(a) and 3(b)).

Plant species richness was strongly positively correlated with grasshopper species richness (Table 2). Plant species richness itself correlated negatively with the proportion of coniferous forests (COW), but positively with increasing proportions of flowing water bodies (FLW). The SAC of plant species richness was, for example, significantly positively correlated with mean annual temperature (TMPYEAR) and aridity (see the relationship between plant species richness and aridity (IOA) in Figure 3(c)) but negatively correlated with mean annual precipitation.

The main parameters influencing moss species richness patterns in the final model were mean temperature in January (TMPJAN), the proportion of nonirrigated agricultural areas, and the minimum elevation in the grid cell. However, these three parameters explained only 7% of the variability in species richness, whereas the remaining variability (93%)

was explained by the SAC. The SAC for moss species richness was significantly correlated to a couple of different environmental and climatic variables. The most important of these were mean annual temperature (TMPYEAR) and aridity (IOA), which had a negative relationship with the SAC. The final model explained 73% of the variation in moss species richness. A quadratic relationship between the SAC and mean annual precipitation (PRECYEARSUM) was obtained from our dataset (Figure 3(d)). The highest species numbers (>350 species) were found in regions with an annual precipitation ranging between 650 and 1100 mm. The richness of moss species was lower in regions showing more than 1100 mm and less than 600 mm annual precipitation.

A change from high levels of plant species richness at lower elevations to higher levels of moss species richness at higher elevations was found in RP, as the moss species to plant species ratio increased with increasing mean elevations (Pearson correlation, $r = 0.537$, $P < 0.001$).

3.3. Evaluation of the Fitted Species Richness Models on an Independent Dataset. We validated our final species richness models established for RP (both with and without the spatial autocovariate) by applying them to analogous external datasets existing for BW. Predicted and observed values of species richness in all five taxonomic groups significantly correlated with each other in BW (Spearman rank correlation coefficient), but accuracy of predictions varied between taxonomic groups (Table 1). When SAC was included in the models, the correlation was the highest for both amphibians (Spearman correlation, $r = 0.758$, $P < 0.001$) and grasshoppers ($r = 0.752$, $P < 0.001$), followed by mosses ($r = 0.721$, $P < 0.001$), and it was the lowest for both reptiles ($r = 0.646$, $P < 0.001$) and plants ($r = 0.648$, $P < 0.001$). All models established for RP showed only a moderate transferability to BW when the SAC term was removed from the models (Table 1). In this case, an

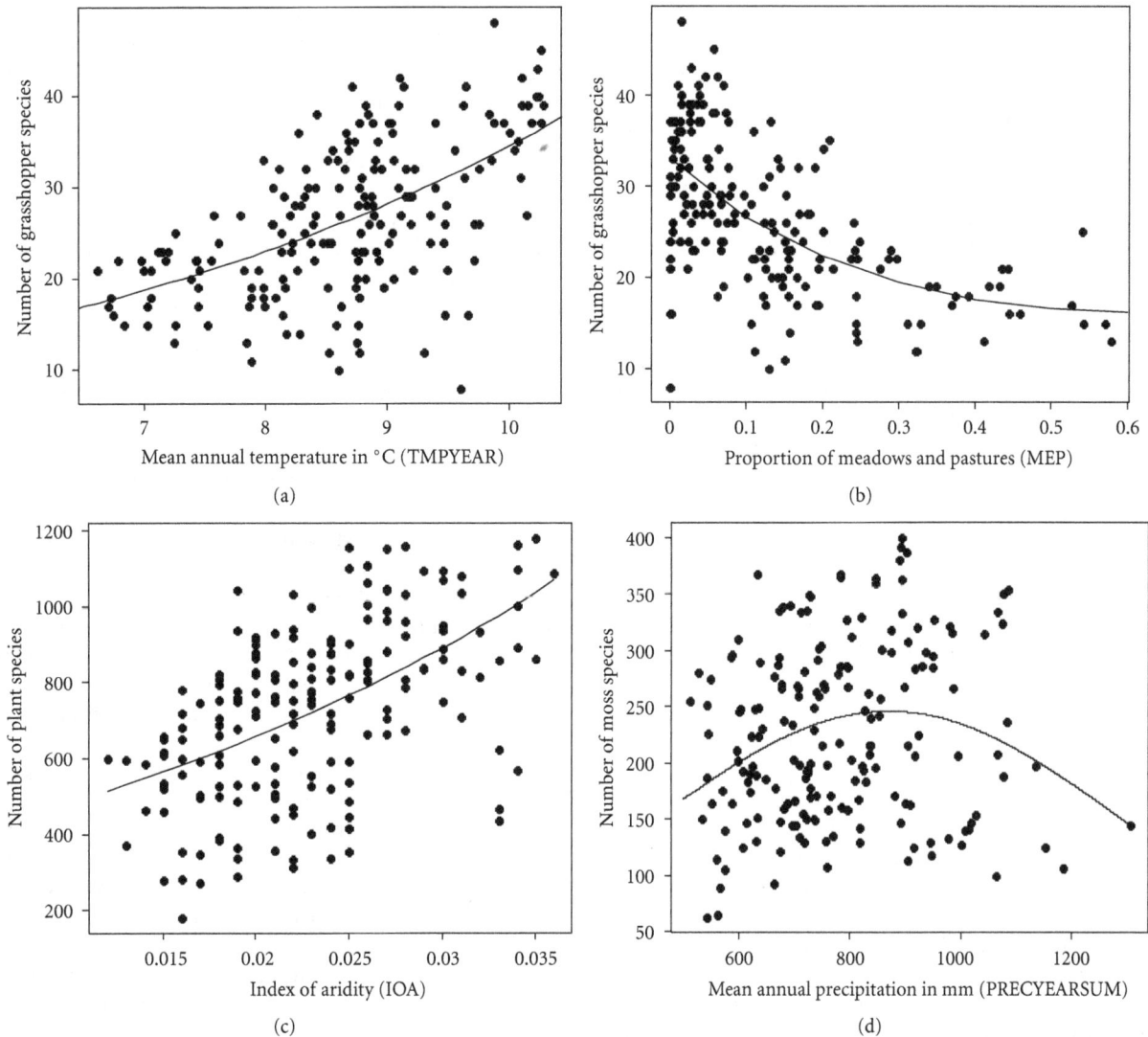

FIGURE 3: Correlations of species richness for different taxa with environmental parameters in 184 grid cells: (a) number of grasshopper species and mean annual temperature, GLM (with poisson errors), 32% deviance change, $P < 0.001$, (b) number of grasshopper species and proportion of meadows and pastures, GLM (with poisson errors), 36% deviance change, $P < 0.001$, (c) number of plant species and index of aridity, GLM (with quasipoisson errors), 23% deviance change, $P < 0.001$, (d) number of moss species and mean sum of annual precipitation, GLM (with quasipoisson errors), 6% deviance change, $P < 0.01$.

overestimation of low species numbers and underestimation of large species numbers was observed for all tested models.

4. Discussion

4.1. Model Accuracy. Our models which took land-cover, topographical, and climatic variables into account explained between 34 and 73% of the variation in species richness of the five taxonomic groups. This level of explanative power coincides with other studies on species richness at a micro- or mesoscale (max. 65% explained variance, e.g., [15, 16] with grid cell sizes 5 × 5 km), but studies of species richness patterns on a macroscale (= large spatial extent and large grain size) showed much higher levels (e.g., [37, 38] which used provinces or countries). The same pattern in which

models at a macroscale prove to be more accurate is apparent when correlations of species numbers between different taxonomic groups are studied. For China, Qian [37] showed that amphibian (reptile) and plant species richness is higly correlated ($r_{Pearson} = 0.9$ for amphibians; $r_{Pearson} = 0.7$ for reptiles), whereas Maes et al. [15] found a lower correlation ($r_{Spearman} = 0.3$) for the same groups in a study that was conducted at a smaller grain size in the small province of Flanders, Belgium. The strength of the correlations found in our study RP between plant species richness and amphibians ($r_{Spearman} = 0.45$) or reptiles ($r_{Spearman} = 0.36$) coincides with the amount found for the Flanders region, Belgium [15]. The causal mechanisms for varying strength in species richness correlations over spatial scales are poorly understood [39]. We believe that the differences in the strength of correlations at different spatial scales probably result from an increasing influence of dispersal constraints and population processes on species distributions at smaller grain size. Extinction and recolonisation dynamics in metapopulations might be a good example, where not all suitable patches are colonised by a species. Forces such as intensive management and habitat simplification might also bias richness patterns found in natural environments at smaller grain size [40]. There are several further reasons explaining the lower performance of the models developed for smaller grain sizes or for smaller spatial extent [18]. First, soil parameters varying at different spatial scales were not tested in our approach but have been shown to significantly influence model accuracy [16]. Second, other important variables explaining richness patterns could be still missing, as most of the models, even in other studies, rarely explain more than 50% of the total variability. Finally, a geographic variation in sampling effort may exist [41] as well as different sampling efforts concerning the taxa itself. However, species distribution models express a tradeoff between practicability and data quality as most data used for modelling are averages of a time period. Species distributions may shift, and climatic trends may arise within such a period, but this is never reflected in the data.

We used a spatial validation on an independent dataset to evaluate our models. Different validation approaches for species distribution models have been applied in the past, for example, division of the dataset into a training and a test subset, internal bootstrapping, and so forth (for an overview, see [33, 34, 42]). It is, however, suggested by the previously named authors that the use of an independent dataset is the best approach to test the predictive ability of a model. Our spatial validation revealed that the same predictor variables can be used at least for Baden-Württemberg to sufficiently predict species richness. Predictions were relatively bad when the SAC term was not included in the model that shows that spatial autocorrelation has to be considered in analyses of species distributions [24]. Further analyses should be started in RP to find regions that are underinvestigated using the discrepancy between observed and predicted species richness [43].

4.2. Spatial Distribution of Species Richness and Its Determinants. In general, the climatic variables are a major determinant of variation in species richness at the regional extent of RP. The importance of climatic variables for the distribution of species richness differed in previous studies that used a smaller grain size (e.g., grid cell sizes 5×5 km, [15, 16]). This difference could be explained by the preselection of variables that had been considered as predictors for models. Our models suggest that environmental heterogeneity (elevational gradient, landscape heterogeneity) is very important for the spatial distribution of amphibians, and to a lesser extent important for grasshoppers. Maes et al. [15] found a positive correlation between species richness and biotope diversity for plants, dragonflies, herpetofauna, butterflies, and birds. The data presented by Schouten et al. [16] reveal the same trend for grasshoppers, dragonflies, and mosses that was observed by Maes et al. [15], but not for hoverflies and

the herpetofauna. Amphibians use a broad range of habitats (from woodlands to wet meadows) but generally need access to ponds, streams, or wetlands for reproduction. Proportions of wetlands of any type are positively associated with habitat heterogeneity in RP and may contribute to higher observed amphibian species richness in more heterogeneous landscapes [44].

In our study, plant species richness was not a good indicator of reptile species richness, although plant species richness had turned out to be relatively well correlated with ants, spiders, gastropods, orthopterans, and birds at a smaller grain size in another study [45]. However, in RP, reptiles comprise a low number of species. We found several indirect correlations between environmental variables and reptile species richness. Among these were the proportion of meadows and pastures (MEPs) and the maximum elevation (HTMAX), both of which had a negative effect on reptile occurrence. Areas of RP with a high number of reptile species are typically highly structured landscapes, whereas intensive forestry and agriculture lead to species-poor reptile assemblages [46]. Higher elevations and higher annual precipitation also had an indirect negative impact on reptile species richness, indicating that the most favourable conditions for reptiles exist at low elevations and under relatively dry conditions in RP (most reptile species are xerothermic; [47]).

Only a small proportion of the variability in moss species richness was directly explained by environmental variables, whereas the largest proportion was explained by the SAC. The SAC primarily accounts for the spatial information which cannot be explained by the other predictors used in the model [24]. When SAC takes over much of the explained variability in a model, as this was the case for mosses, one is interested in the ecological information behind. The SAC in the model for mosses showed a strong negative correlation relation with mean annual temperature and aridity, indicating that these parameters are important determinants for moss species richness. Mosses are likely to be sensitive to aridity or the availability of water, because their reproduction cycle depends on high levels of humidity. The impact of the range of elevation on species richness of mosses, expressed by a positive correlation between this range and the SAC, shows that mosses have adapted to a variety of different habitats and that they thus benefit from landscape heterogeneity. Mosses are more frost resistant than herbaceous plants [48] and should be at an advantage with respect to survival at higher elevations where frost events are more frequent. This might be reflected in the moss species to plant species ratio increasing with mean elevation of the studied grid cells. Environmental correlations with SAC are of the same magnitude as the direct correlation between species numbers and environmental variables, which shows that conditions in the neighboring grid cells are often similar. Whether conditions in the surrounding cells have an effect on species richness in the focussed grid cells is not clear. Recorder effort might bias species distribution patterns [49]. In general, we are aware that recorder effort and monitoring of mosses are still far from being spatially homogenous within Germany. Present patterns analysed here are certainly somehow biased by varying recorder effort in different geographic regions.

Grasshoppers consist of several stenotopic species and are found almost exclusively in open habitats. Landscape heterogeneity may be important for grasshoppers at a microscale, that is, in datasets that use finer resolutions than CORINE or in datasets that use smaller grid cells ($25\,km^2$ grid cells in Schouten et al. [16]). Habitat heterogeneity was the strongest predictor of grasshopper species richness in the Netherlands and explained 30% of the total variation [16]. In this study, plant species richness was the strongest predictor of the spatial variation in grasshopper richness, when plant species richness was included as a further predictor in the model using abiotic variables before only. The model which takes into account plant species richness in addition to other environmental predictors and SAC explains more than 72% of the variation in grasshopper species richness. Plant and grasshopper species also showed a high correlation at landscape scale in Austria ($r_{Pearson} = 0.77$, [45]). It has been shown for Europe that the number of Caelifera species (a suborder of grasshoppers) is associated with variability in plant species richness at country level [50]. In RP, large numbers of plants were found in regions with increasing values of aridity. In Switzerland, floristic species richness is mainly influenced by a temperature gradient [51]; in our study, such a temperature gradient was highly correlated with aridity (IOA) in RP (TMPYEAR and IOA, $r_{Pearson} = 0.730$, $P < 0.001$).

4.3. How Important Are Climatic Parameters for Species Richness Patterns at Regional Scales?

Applying only climatic variables, Bakkenes et al. [52] were able to explain 42% of the variation in European plant species richness. Climatic variables have been shown to be similar important for mammals, amphibians, and birds in Europe [12]. Araújo et al. [6] have recently shown that the diversity patterns of amphibians and reptiles in Europe are better explained by the stable climatic conditions during the last 20,000 years than by the contemporary climatic situation. These observations suggest that large-scale variation in species richness seems to be mainly driven by climatic parameters [19]. Scaling down to smaller grain size ($1 \times 1\,km$), Soares and Brito [44] showed that richness patterns of amphibians and reptiles are well explained by precipitation, water surfaces per grid cells, and tree diversity cover. We found a similar trend in RP where landscape heterogeneity is a predictor of amphibian species richness, although mean annual temperature was still the strongest predictor. Species richness patterns of the five taxonomic groups studied in RP suggest that climatic variables, particularly the annual minimum temperatures, were the most important determinants of richness; this observation was in contrast to other studies with regional extent [15, 16]. One explanation is that RP comprises a stronger geographic variation in climate (TMPYEAR ranges from 6.6 to 10.3°C) and topography (mean elevation in grid cells ranges from 89 to 580 m a.s.l.) than the Flanders [15] and the Netherlands [16] analysed in these studies.

We conclude that land-cover variables are likely to become more important at small grain size than at large grain size. Climatic parameters represent large proportions of the variability in species richness at smaller grain sizes (cf. [8, 44]) and at regional extent [5]. However, a certain amount (here >30%) of unexplained variability in species richness patterns remains in the species richness data, even if spatial autocorrelation in the residuals is incorporated in the models.

4.4. Congruence of Spatial Patterns.

Even at a regional extent reliable distribution data for the majority of taxonomic groups are simply not available. The lack of data for less "popular" taxonomic groups leads to conservation strategies that are only based upon a selection of charismatic species that are easy to survey. The use of indicator taxa is considered as an approach to overcome this problem [53, 54]. Correlations between different invertebrate groups at small grain size (1 ha plots) seem to be very weak and no surrogate group for invertebrate species diversity has been found [55]. At large grain size (country level), patterns in richness and endemism of highly diverse insects such as ground beetles were found to be highly correlated with those of plants, amphibians, and reptiles [56]. For nature conservation, the congruence of spatial patterns in species richness between less studied taxonomic groups, for example, mosses and most invertebrate groups, and popular groups, such as amphibians and reptiles, still has to be tested on small spatial extent and on small grain size.

We found congruence of spatial patterns in species richness among the five taxonomic groups for RP. The highest correlation was observed between grasshopper and plant species richness. These findings contradict species richness patterns described by Duelli and Obrist [57] at a microscale (5 km transect) for several invertebrate taxa, and also patterns of plants, reptiles, and amphibians analysed on a national level in China (Qian [37] which used national provinces as units) and even at a global scale (Qian and Ricklefs [38] which used countries and regions as units). However, the congruence patterns that were found for the five taxonomical groups in RP may change at different grain size (grid cell sizes $19,000\,km^2$ and $75,000\,km^2$), as has been shown for the relationship between birds and butterflies [58].

Invertebrates are more efficient predictors of species richness patterns than vertebrates [59]. In RP, grasshopper richness correlated to species richness for all other taxonomic groups ($r_{Pearson} = 0.716$). When selecting surrogates for species richness, a critical correlation coefficient higher than 0.75 has been recommended [55]. As our correlation is even somewhat lower than this threshold, we suggest a careful use of grasshoppers as an indicator group for species richness in further studies. Nevertheless, grasshoppers are easy to survey, and their monitoring is less time consuming than of other taxa. A further advantage of grasshoppers is their restricted diversity in Europe (Fauna Europaea [60]: e.g., Switzerland 109 species, Germany 83 species, Hungary 123 species, Belgium 51 species). However, grasshopper species richness must not necessarily be an indicator for other taxonomic groups than those tested here. Particularly rare and protected species might not occur in most species-rich grid cells [61].

A multitaxa approach could be more appropriate to assess overall species richness patterns in diverse habitats.

Acknowledgments

This study was funded within the project "Klima- und Landschaftswandel in Rheinland-Pfalz (KlimLandRP)" by the "Ministerium für Umwelt, Forsten und Verbraucherschutz" of Rhineland-Palatinate, Germany. The authors thank Oliver Dürhammer for kindly providing the moss data from his database. Many thanks to Franz Badeck for his support in the usage of climate data from the TKCLIM-database. The authors also thank Rüdiger Burkhardt and Michael Altmoos (Landesamt für Umwelt in Rheinland-Pfalz) for providing distribution data for numerous species. The authors are indebted to Thorsten Assmann and an anonymous referee for their comments on an earlier draft of the manuscript.

References

[1] T. Kull and M. J. Hutchings, "A comparative analysis of decline in the distribution ranges of orchid species in Estonia and the United Kingdom," *Biological Conservation*, vol. 129, no. 1, pp. 31–39, 2006.

[2] S. Prieto-Benitez and M. Mendez, "Effects of land management on the abundance and richness of spiders (Araneae): a meta-analysis," *Biological Conservation*, vol. 144, pp. 683–691, 2011.

[3] J. A. F. Diniz-Filho, T. F. L. V. B. Rangel, and B. A. Hawkins, "A test of multiple hypotheses for the species richness gradient of South American owls," *Oecologia*, vol. 140, no. 4, pp. 633–638, 2004.

[4] O. D. Finch, T. Blick, and A. Schuldt, "Macroecological patterns of spider species richness across Europe," *Biodiversity and Conservation*, vol. 17, no. 12, pp. 2849–2868, 2008.

[5] H. Qian, "Environment-richness relationships for mammals, birds, reptiles, and amphibians at global and regional scales," *Ecological Research*, vol. 25, no. 3, pp. 629–637, 2010.

[6] M. B. Araújo, D. Nogues-Bravo, J. A. F. Diniz-Filho, A. M. Haywood, P. J. Valdes, and C. Rahbek, "Quaternary climate changes explain diversity among reptiles and amphibians," *Ecography*, vol. 31, no. 1, pp. 8–15, 2008.

[7] J. J. Wiens and M. J. Donoghue, "Historical biogeography, ecology and species richness," *Trends in Ecology and Evolution*, vol. 19, no. 12, pp. 639–644, 2004.

[8] A. Guisan and U. Hofer, "Predicting reptile distributions at the mesoscale: relation to climate and topography," *Journal of Biogeography*, vol. 30, no. 8, pp. 1233–1243, 2003.

[9] J. M. Finch, M. J. Samways, T. R. Hill, S. E. Piper, and S. Taylor, "Application of predictive distribution modelling to invertebrates: odonata in South Africa," *Biodiversity and Conservation*, vol. 15, no. 13, pp. 4239–4251, 2006.

[10] L. Boitani, I. Sinibaldi, F. Corsi et al., "Distribution of medium- to large-sized African mammals based on habitat suitability models," *Biodiversity and Conservation*, vol. 17, no. 3, pp. 605–621, 2008.

[11] J. Müller, J. Pöllath, R. Moshammer, and B. Schröder, "Predicting the occurrence of Middle Spotted Woodpecker Dendrocopos medius on a regional scale, using forest inventory data," *Forest Ecology and Management*, vol. 257, no. 2, pp. 502–509, 2009.

[12] R. J. Whittaker, D. Nogués-Bravo, and M. B. Araújo, "Geographical gradients of species richness: a test of the water-energy conjecture of Hawkins et al. (2003) using European data for five taxa," *Global Ecology and Biogeography*, vol. 16, no. 1, pp. 76–89, 2007.

[13] A. Baselga, "Determinants of species richness, endemism and turnover in European longhorn beetles," *Ecography*, vol. 31, no. 2, pp. 263–271, 2008.

[14] A. Schuldt and T. Assmann, "Environmental and historical effects on richness and endemism patterns of carabid beetles in the western Palaearctic," *Ecography*, vol. 32, no. 5, pp. 705–714, 2009.

[15] D. Maes, D. Bauwens, L. De Bruyn et al., "Species richness coincidence: conservation strategies based on predictive modelling," *Biodiversity and Conservation*, vol. 14, no. 6, pp. 1345–1364, 2005.

[16] M. A. Schouten, P. A. Verweij, A. Barendregt, R. M. J. C. Kleukers, V. J. Kalkman, and P. C. De Ruiter, "Determinants of species richness patterns in the Netherlands across multiple taxonomic groups," *Biodiversity and Conservation*, vol. 18, no. 1, pp. 203–217, 2009.

[17] M. Altmoos, "Erkennen wir die richtigen habitate von tieren? Räumliche skalen der habitatpräferenz und skalierte habitatmodelle in fallstudien mit amphibien und heuschrecken," *Naturschutz und Landschaftsplanung*, vol. 35, no. 7, pp. 212–218, 2003.

[18] C. Rahbek, "The role of spatial scale and the perception of large-scale species-richness patterns," *Ecology Letters*, vol. 8, no. 2, pp. 224–239, 2005.

[19] J. Hortal, J. Rodriguez, M. Nieto-Diaz, and J. M. Lobo, "Regional and environmental effects on the species richness of mammal assemblages," *Journal of Biogeography*, vol. 35, no. 7, pp. 1202–1214, 2008.

[20] P. Janssen, D. Fortin, and C. Hébert, "Beetle diversity in a matrix of old-growth boreal forest: influence of habitat heterogeneity at multiple scales," *Ecography*, vol. 32, no. 3, pp. 423–432, 2009.

[21] L. Meinunger and W. Schröder, *Verbreitungsatlas der Moose Deutschlands*, Edited by O. Dürhammer, Regensburgische Botanische Gesellschaft von 1790, 2007.

[22] M. Luoto, R. Virkkala, and R. K. Heikkinen, "The role of land cover in bioclimatic models depends on spatial resolution," *Global Ecology and Biogeography*, vol. 16, no. 1, pp. 34–42, 2007.

[23] J. Heino, "Are indicator groups and cross-taxon congruence useful for predicting biodiversity in aquatic ecosystems?" *Ecological Indicators*, vol. 10, no. 2, pp. 112–117, 2010.

[24] C. F. Dormann, J. M. McPherson, M. B. Araujo et al., "Methods to account for spatial autocorrelation in the analysis of species distributional data: a review," *Ecography*, vol. 30, no. 5, pp. 609–628, 2007.

[25] S. Maas, P. Detzel, and A. Staudt, *Gefährdungsanalyse der Heuschrecken Deutschlands Verbreitungsatlas, Gefährdungseinstufung und Schutzkonzepte*, Landwirtschaftsverlag, Godesberg, Germany, 2002.

[26] H. Haeupler and P. Schönfelder, *Atlas der Farn- und Blütenpflanzen der Bundesrepublik Deutschland*, Ulmer, Stuttgart, Germany, 2nd edition, 1989.

[27] F. W. Badeck, S. Pompe, and I. Kühn, "Wetterextreme und artenvielfalt. Zeitlich hochauflösende Klimainformationen auf dem Messtischblattraster und für Schutzgebiete Deutschlands," *Naturschutz und Landschaftsplanung*, vol. 40, pp. 343–345, 2008.

[28] C. Krebs, *Ecological Methodology*, Benjamin Cummings, Menlo Park, Calif, USA, 2nd edition, 1999.

[29] R. Mac Nally and C. J. Walsh, "Hierarchical partitioning public-domain software," *Biodiversity and Conservation*, vol. 13, no. 3, pp. 659–660, 2004.

[30] O. N. Bjørnstad and W. Falck, "Nonparametric spatial covariance functions: estimation and testing," *Environmental and Ecological Statistics*, vol. 8, no. 1, pp. 53–70, 2001.

[31] N. H. Augustin, M. A. Mugglestone, and S. T. Buckland, "An autologistic model for the spatial distribution of wildlife," *Journal of Applied Ecology*, vol. 33, no. 2, pp. 339–347, 1996.

[32] M. Luoto, M. Kuussaari, H. Rita, J. Salminen, and T. Von Bonsdorff, "Determinants of distribution and abundance in the clouded apollo butterfly: a landscape ecological approach," *Ecography*, vol. 24, no. 5, pp. 601–617, 2001.

[33] A. Guisan and N. E. Zimmermann, "Predictive habitat distribution models in ecology," *Ecological Modelling*, vol. 135, no. 2-3, pp. 147–186, 2000.

[34] S. Manel, H. Ceri Williams, and S. J. Ormerod, "Evaluating presence-absence models in ecology: the need to account for prevalence," *Journal of Applied Ecology*, vol. 38, no. 5, pp. 921–931, 2001.

[35] I. P. Vaughan and S. J. Ormerod, "The continuing challenges of testing species distribution models," *Journal of Applied Ecology*, vol. 42, no. 4, pp. 720–730, 2005.

[36] Esri, *ArcGIS 9.3*, Environmental Systems Research Institute, Redlands, Calif, USA, 2008.

[37] H. Qian, "Relationships between plant and animal species richness at a regional scale in China," *Conservation Biology*, vol. 21, no. 4, pp. 937–944, 2007.

[38] H. Qian and R. E. Ricklefs, "Global concordance in diversity patterns of vascular plants and terrestrial vertebrates," *Ecology Letters*, vol. 11, no. 6, pp. 547–553, 2008.

[39] V. Wolters, J. Bengtsson, and A. S. Zaitsev, "Relationship among the species richness of different taxa," *Ecology*, vol. 87, no. 8, pp. 1886–1895, 2006.

[40] A. C. Weibull, O. Ostman, and A. Granqvist, "Species richness in agroecosystems: the effect of landscape, habitat and farm management," *Biodiversity and Conservation*, vol. 12, no. 7, pp. 1335–1355, 2003.

[41] R. L. H. Dennis and C. D. Thomas, "Bias in butterfly distribution maps: the influence of hot spots and recorder's home range," *Journal of Insect Conservation*, vol. 4, no. 2, pp. 73–77, 2000.

[42] A. H. Fielding and J. F. Bell, "A review of methods for the assessment of prediction errors in conservation presence/absence models," *Environmental Conservation*, vol. 24, no. 1, pp. 38–49, 1997.

[43] J. M. Lobo and F. Martin-Piera, "Searching for a predictive model for species richness of Iberian dung beetle based on spatial and environmental variables," *Conservation Biology*, vol. 16, no. 1, pp. 158–173, 2002.

[44] C. Soares and J. C. Brito, "Environmental correlates for species richness among amphibians and reptiles in a climate transition area," *Biodiversity and Conservation*, vol. 16, no. 4, pp. 1087–1102, 2007.

[45] N. Sauberer, K. P. Zulka, M. Abensperg-Traun et al., "Surrogate taxa for biodiversity in agricultural landscapes of eastern Austria," *Biological Conservation*, vol. 117, no. 2, pp. 181–190, 2004.

[46] A. Bitz, K. Fischer, L. Simon et al., *Die Amphibien und Reptilien in Rheinland-Pfalz*, GNOR-Eigenverlag, Landau, Germany, 1996.

[47] R. Günther, *Die Amphibien und Reptilien Deutschlands*, Gustav Fischer, Jena, Germany, 1996.

[48] N. Balagurova, S. Drozdov, and S. Grabovik, "Cold and heat resistance of five species of Sphagnum," *Annales Botanici Fennici*, vol. 33, no. 1, pp. 33–37, 1996.

[49] C. Hassall and D. J. Thompson, "Accounting for recorder effort in the detection of range shifts from historical data," *Methods in Ecology and Evolution*, vol. 1, pp. 343–350, 2010.

[50] C. E. Steck and M. Pautasso, "Human population, grasshopper and plant species richness in European countries," *Acta Oecologica*, vol. 34, no. 3, pp. 303–310, 2008.

[51] T. Wohlgemuth, "Modelling floristic species richness on a regional scale: a case study in Switzerland," *Biodiversity and Conservation*, vol. 7, no. 2, pp. 159–177, 1998.

[52] M. Bakkenes, J. R. M. Alkemade, F. Ihle, R. Leemans, and J. B. Latour, "Assessing effects of forecasted climate change on the diversity and distribution of European higher plants for 2050," *Global Change Biology*, vol. 8, no. 4, pp. 390–407, 2002.

[53] T. M. Caro and G. O'Doherty, "On the use of surrogate species in conservation biology," *Conservation Biology*, vol. 13, no. 4, pp. 805–814, 1999.

[54] E. Fleishman, J. R. Thomson, R. Mac Nally, D. D. Murphy, and J. P. Fay, "Using indicator species to predict species richness of multiple taxonomic groups," *Conservation Biology*, vol. 19, no. 4, pp. 1125–1137, 2005.

[55] S. Lovell, M. Hamer, R. Slotow, and D. Herbert, "Assessment of congruency across invertebrate taxa and taxonomic levels to identify potential surrogates," *Biological Conservation*, vol. 139, no. 1-2, pp. 113–125, 2007.

[56] A. Schuldt, Z. Wang, H. Zhou, and T. Assmann, "Integrating highly diverse invertebrates into broad-scale analyses of cross-taxon congruence across the Palaearctic," *Ecography*, vol. 32, no. 6, pp. 1019–1030, 2009.

[57] P. Duelli and M. K. Obrist, "In search of the best correlates for local organismal biodiversity in cultivated areas," *Biodiversity and Conservation*, vol. 7, no. 3, pp. 297–309, 1998.

[58] D. L. Pearson and S. S. Carroll, "The influence of spatial scale on cross-taxon congruence patterns and prediction accuracy of species richness," *Journal of Biogeography*, vol. 26, no. 5, pp. 1079–1090, 1999.

[59] C. Moritz, K. S. Richardson, S. Ferrier et al., "Biogeographical concordance and efficiency of taxon indicators for establishing conservation priority in a tropical rainforest biota," *Proceedings of the Royal Society B*, vol. 268, no. 1479, pp. 1875–1881, 2001.

[60] Fauna Europaea, "Fauna Europaea version 2.4," 2009, http://www.faunaeur.org/.

[61] J. R. Prendergast, R. M. Quinn, J. H. Lawton, B. C. Eversham, and D. W. Gibbons, "Rare species, the coincidence of diversity hotspots and conservation strategies," *Nature*, vol. 365, no. 6444, pp. 335–337, 1993.

7

Pathogens Associated with Sugarcane Borers, *Diatraea* spp. (Lepidoptera: Crambidae): A Review

Víctor M. Hernández-Velázquez,[1] Laura P. Lina-García,[1] Verónica Obregón-Barboza,[1] Adriana G. Trejo-Loyo,[2] and Guadalupe Peña-Chora[2]

[1] *Centro de Investigación en Biotecnología, Universidad Autónoma del Estado de Morelos, Avenida Universidad No. 1001, Colonia Chamilpa, 62210 Cuernavaca, MOR, Mexico*
[2] *Centro de Investigaciones Biológicas, Universidad Autónoma del Estado de Morelos, Avenida Universidad No. 1001, Colonia Chamilpa, 62210 Cuernavaca, MOR, Mexico*

Correspondence should be addressed to Víctor M. Hernández-Velázquez, vmanuelh@uaem.mx

Academic Editor: Thomas Iliffe

The objective of this paper was to analyze information related to entomopathogenic-associated *Diatraea* spp. Gaining a better understanding of the effects of these microorganisms will help in the development of successful microbial control strategies against stem borers that attack sugarcane plants.

1. Introduction

The *Diatraea* spp. (Lepidoptera: Crambidae) complex is only found in the American continent, and it is the most important group of stem borers that principally attack maize and sugarcane, as well as other gramineous crops, including rice, sorghum, and forage grasses [1]. The sugarcane borer (SCB) *D. saccharalis* Fab. is the most economically important pest in South America [2, 3], whereas the neotropical corn stalk borer (NCB) *D. lineolata* Walk is primarily found in Central America [4], *D. magnifactella* Dyar and *D. considerata* Heinrich are found in Mexico [1], and the southwestern corn borer *D. grandiosella* Dyar is found in the United States [5].

Unfortunately, commercially available insecticides are not efficient for the control of *Diatraea* spp. for a variety of reasons, mainly because of the continuous presence of the host plants in fields throughout the year, the concomitant presence of mature and immature forms of the insect, and the cryptic feeding habits of the insect [6]. An alternative strategy is integrated pest management with biological control as the first defense, which includes the use of parasitoids and entomopathogens.

In this paper, we present a perspective on the attempts to control *Diatraea* spp. using pathogens in the Americas. We also discuss the status of recent attempts to use pathogens in the field.

2. Fungi

One promising field for research is the use of entomopathogenic fungi as biological control agents of insect pests in sugarcane plants. Approximately 80% of the etiological agents involved in insect diseases are fungi, which encompass 90 genera and more than 700 species [7]. A number of fungi (Hypocreales: Clavicipitaceae) including *B. bassiana*, *B. brongniartii*, *M. anisopliae*, *P. fumosoroseus*, *Hirsutella* sp., *Cylindrocarpon* sp., and *Nomuraea rileyi* have been isolated from *Diatraea* spp. in the Americas from Argentina to the USA (Table 1). Under certain climatic conditions, *B. bassiana* has been reported to cause natural epizootics on *D. grandiosella* [8].

The life cycle of entomopathogenic fungi in the arthropod hosts is initiated with the germination of conidia that contacts the host integument and produces a germ tube that penetrates the host through a combination of physical pressure and enzymatic degradation of the cuticle. The fungus initially colonizes the host through a yeast phase. Host death usually results from a combination of nutrient

TABLE 1: Entomopathogens fungi from *Diatraea* spp.

Host species	Entomopathogen	Country	Reference
D. grandiosella *D. crambidoides*	*B. bassiana* (1–8%)	Mississippi, United States	Inglis et al., 2000 [5]
D. grandiosella	*B. bassiana* (epizootics)	Texas High Plains, United States	Knutson and Gilstrap, 1990 [8]
D. saccharalis	*B. bassiana* *M. anisopliae* *P. fumosoroseus*	Venezuela	Zambrano et al., 2002 [9]
D. saccharalis	*M. anisopliae*	PE, Brazil	Alves, et al. 2002 [10]
D. saccharalis	*B. bassiana*	Pinar del Río, La Habana, Matanzas, Villa Clara, Cienfuegos and Camagüey, Cuba	Estrada et al., 2004 [11]
D. saccharalis	*B. bassiana* *M. anisopliae* *Nomuraea rileyi* *Iseria* sp.	Tucumán, Argentina	Yasen de Romero et al., 2008 [12]
D. saccharalis	*M. anisopliae* (Ma2 and Ma3)	Mexico	Angel-Sahagún et al., 2005 [13]
D. saccharalis	*B. bassiana* ARSEF: 1489 1832 1834 2629 3020 3858 5500 5502	Pernambuco Brazil Brazil Brazil Pernambuco Brazil Pernambuco, Brazil Brazil Santa Fe, Argentina Santa Fe, Argentina	Humber et al., 2009 [14]
D. saccharalis	*B. brigniartii* ARSEF: 1830	Sao Paulo, Brazil	Humber et al., 2009 [14]
D. saccharalis	*Cylindrocarpon* sp. ARSEF: 8043 8044	Colima, Mexico	Humber et al., 2009 [14]
D. saccharalis	*M. anisopliae* ARSEF: 3290 3291 3292 3298	Colima, Mexico Colima, Mexico Colima, Mexico Colima, Mexico Colima, Mexico	Humber et al., 2009 [14]
D. magnifactella	*M. anisolpliae* *B. bassiana* *Hirsutella* sp.	Mexico	Hernández and Velázquez 2004 [15]
D. grandiosella	*Nosema* sp. isolate 167	Marshal County, (Mississippi, USA)	Inglis et al. 2000 [5]
D. grandiosella	*Nosema* sp. isolates 295, 504	Oktibbeha County, (Mississippi, USA)	Inglis et al. 2000 [5]

TABLE 1: Continued.

Host species	Entomopathogen	Country	Reference
D. grandiosella	Nosema sp. isolates 181, 513, 522	Washington County, (Mississippi, USA)	Inglis et al. 2000 [5]
D. saccharalis	Nosema sp.	Colombia	Lastra and Gómez 2000 [16]
D. saccharalis	Granulovirus (DsGV)	Southern United States	Pavan et al. (1983) [17]
D. saccharalis	Densovirus (DsDNV)	Guadeloupe	Meynadier et al. (1977) [18]
D. magnifactella	B. thuringiensis	Mexico	Fonseca-González et al. (2011) [19]

depletion, invasion of organs, and the action of fungal toxins. Hyphae usually emerge from the cadaver. The mummified corpse of the insect remains in the environment for several weeks and, in the case of stem borer, it keeps remains protected inside the stem. Therefore, it is more common to detect and isolate fungi compared to other pathogens that destroy the host, such as *Bacillus thuringiensis (Bt)*, viruses, or nematodes.

Entomopathogenic fungi are widely distributed in all regions of the world; these species have wide genetic variation among the different isolates. Pathogenicity and virulence to different species, as well as enzymatic and DNA characteristics, vary among different isolates [14, 20, 21]. Therefore, it is important to evaluate as many geographic isolates as possible from different areas to select the most suitable isolate based on its virulence and growth at high temperatures. Several research groups have verified the pathogenicity and virulence of Hypocreales fungi, such as *M. anisopliae* and *B. bassiana* (Table 2), which have become important biocontrol agents used for the microbial control of *Diatraea* spp. It is possible that a limited selection of available isolates of *B. bassiana* could identify highly virulent strains from each of the different *Diatraea* species.

Field evaluations have been performed using *M. anisopliae* and *B. bassiana* against several insect pests of sugarcane, including *D. saccharalis* in Brazil. Application of *M. anisopliae* at a rate of 1×10^{13} spores per hectare caused 58% mortality of *D. saccharalis*, and *B. bassiana* at 3.7×10^8 spores per milliliter reduced *D. saccharalis* damage by 45% (Alves et al., 1984 and 1985, cited in Legaspi et al. [22]). For effectiveness in the field, it is important to consider the contact between the spores and the host, formulation, and the virulence of the pathogens.

3. Microsporidia

Microsporidia (Eukaryota: Fungi) is the most ubiquitous group among insect populations [23, 24]. Microsporidia are tiny unicellular organisms (from 2 to 40 μm in diameter), that are opportunistic and obligate intracellular parasites and attack different groups of invertebrate and vertebrate animals. Microsporidia generally produce chronic diseases and reduce the physiological and reproductive ability of their hosts. Many species of microsporidia infect arthropods, especially insects such as Lepidoptera and Coleoptera [23–25].

In general, these microorganisms live as parasites in cells of the midgut epithelium, where they complete their development, and the cycle starts when the infective states of microsporidia (spores) arrive at the digestive tube and colonize this region all the way to the excretory system. The spores germinate because of the acid intestinal pH, the microorganisms penetrate the midgut cells, and the intestinal activity is paralyzed between 14 to 21 days later because the insect cannot assimilate nutrients. *Nosema locustae* infects the adipocytes of fat body, which interferes with the adequate function of the insect's intermediary metabolism and competes with the insect for energetic reserves. Microsporidia produce effects that depend on the species and concentration; however, they generally produce weakness and eventually lead to death [31, 32].

Only some species in this group have the possibility to be potentially relevant for natural or classical control. There are many studies worldwide on the pathogenic effects of the microsporidia *N. pyrausta* (Paillot) and *Vairimorpha necatrix* (Kramer) on borer organisms such as *Ostrinia nubilalis* (Hübner) (European corn borer), *Lymantria dispar* (Gypsy moth), and the grasshopper [5, 32]. In this group, only the microsporidium *N. locustae* has been registered as a microbial insecticide for the control of grasshoppers in grasslands [33].

In many areas of the USA and Europe, *Nosema* is the main agent used for the control of grasshoppers; however, in Latin America, few studies have been performed with *Nosema* [33]. For example, Inglis et al. [5] reported six isolates of *Nosema* on larvae in the winter diapause stage from *D. grandiosella* Dyer collected during 1998 in corn stems from three locations in Mississippi, USA. However, the frequency of infection in the field was very low (1, 3, and 15% in the counties of Marshal, Oktibbeha, and Washington, resp.), and no isolates were found in *D. crambidoides* (Grote) (Table 3).

When the mortality produced by the *Nosema* isolates was assessed in the laboratory using larvae that had stayed in environmental-like natural winter diapause conditions, variations in mortality between 0 and 55% were observed in larvae, and variations between 7 and 29% were observed in pupae; homogenization of the dead larvae revealed a large amount of *Nosema* spores. However, in the surviving adults, a large number of larvae were positive for the *Nosema* spores when they were analyzed under light microscopy using staining techniques and electron microscopy.

Pathogens Associated with Sugarcane Borers, Diatraea spp. (Lepidoptera: Crambidae): A Review

51

TABLE 2: Entomopathogenic-fungi bioassays on *Diatraea* spp.

Host species	Pathogen	Bioassay	Result	Reference
D. saccharalis	*B. bassiana*	Immersing third instar larvae into a suspension of 10^8 conidia/mL	TL_{50} (conidia isolates Bb1 and 5): 4.3 days. Mortality dry mycelium preparations: 21.3% (Bb1) 82.5% (Bb5) at 7 days after inoculation	Arcas et al., 1999 [26]
D. saccharalis	*B. bassiana* Mycotrol strain GHA	Sprayed first, second and third instar larvae with 1 mL	LD_{50} in spores/mm^2 1st instar: 72.1 2nd instar: 384.3 3rd instar: 777.0 Mean days of survival 1st instar: 4.6 2nd instar: 4.8 3rd instar: 6.4	Legaspi et al., 2000 [27]
D. grandiosella	*B. bassiana* Eight isolates	Dipped in conidial suspension (dosage not specified)	Mortality from 10 to 21%	Inglis et al., 2000 [5]
D. saccharalis	*B. bassiana* ATCC 20872 from *Solenopsis invicta*	Third instar larvae sprayed with suspension of 10^8 conida/mL	TL_{50} 3.02 to 4.10 days	Marques et al., 2000 [28]
D. sacccharalis	*B. bassiana* ATCC 20872 from *Solenopsis invicta*	Third instar larvae sprayed with 3 mL of yeast-like cells or conidia	LC_{50} 5.6×10^6 yeast/mL and 4.8×10^6 conidia/mL	Alves et al., 2002 [29]
D. saccharalis	*B. bassiana* IBCB from Brazil	1 cm larvae sprayed with 1 mL	CL_{50} 1.58×10^7 conidia/mL at 6 days after inoculation	Wenzel et al., 2006 [30]

TABLE 3: Entomopathogenic-microsporidia bioassays on *Diatraea* spp.

Host specie	Pathogen	Bioassay	Results	Reference
D. grandiosella	*Nosema* sp isolate 167	Surface of diet (squares), concentration not mentioned. Larvae and pupae	Mortality of 1–15%	Inglis et al., 2000 [5]
D. grandiosella	*Nosema* sp isolate 295, 504	Surface of diet (squares), concentration not mentioned. Larvae and pupae	Mortality of 1–15%	Inglis et al., 2000 [5]
D. grandiosella	*Nosema* sp isolate 181, 513, 522	Surface of diet (squares), concentration not mentioned. Larvae and pupae	Mortality of 1–15%	Inglis et al., 2000 [5]
D. saccharalis	*Vairimorpha necatrix*	1st instar larvae, Surface of diet	Microsporidiosis LC_{50} = 48.4 spores/mm^2 Gut damage LC_{50} = 8941 spores/mm^2	Fuxa, 1981 [32]
D. grandiosella	*Nosema* sp isolate 506	5-day-old larvae and pupae, leaf pieces ~5 cm^2 with 10 μl inoculum (10^1 to 10^7 spores)	Median infective dose of 2 $\times 10^3$ spores per larva, pupae small, decreased egg production	Inglis et al., 2003 [34]

In these experiments, the larvae were reared in the winter diapause stage, and the temperature for *Nosema* was not optimal for the development of infection; in the field during the winter, the prediapause larvae migrate to the base of the corn stem just below the surface of the soil [5], and

it is possible that *Nosema* can produce infection in natural conditions as the temperature inside the cane is higher than the external temperature.

Phoofolo et al. [35] observed a similar behavior, and they reported that the *O. pyrausta* infection by *Nosema* is chronic. Although no immediate mortality is produced, the longevity and fecundity of the adults are reduced. Likewise, Phoofolo mentioned the possibility of a response to other factors of mortality that regulate the population dynamics of borers, such as low temperature, the host plant, and crowding, which have chronic effects on *Nosema* infection. Therefore, although the pest did not die directly because of microsporidia infection, the population size was eventually reduced.

This effect has been previously observed by Fuxa [32], when he evaluated the susceptibility of larvae in the first and third instars of six species of Lepidoptera, including *D. saccharalis*, to the microsporidium *Vairimorpha necatrix*. He described two routes of infection that resulted in mortality; one resulted from the chronic effects produced by the exposure of larvae to low doses of spores, which led to lethal septicemia (microsporidiosis) just before pupating, and the other route was due to the intake of a large number of spores, which apparently damaged the gut because of the introduction of a large number of spore polar filaments. In this study, Fuxa [32] concluded that this pathogen could be promising for the remaining five species, although not for *D. saccharalis*, as he had observed direct mortality by damage to the gut and indirect mortality by septicemia and because a high concentration of spores was necessary.

Inglis et al. [34] reported that one strain of *Nosema* (506), which had been previously isolated [5], can infect other species of insects in the Crambidae family, including *O. nubilalis* and *D. crambidoides* (Grote), but not other species of Noctuidae. However, they also observed that the infection can be transmitted transovarially, although at a very low frequency (a difference with other *Nosema* species that can be transmitted frequently in the vertical rout, such as the genus *Ostrinia*).

Solter et al. [36] researched the vertical and horizontal transmission of seven species of microsporidia, including two strains of *Nosema* sp. (isolated from their natural hosts in the field *D. saecharalis* and *Eoreuma loftini*); although five of these strains were transmitted at a low percentage horizontally and vertically, two *Nosema* strains did not behave similarly.

In assessing the infectivity percentage under laboratory conditions when applying high concentrations of spores, they observed very low mortality in the larvae of *D. sacharalis* and *E. loftini* (2 and 4%, resp.), although the same strains produced high mortality at low concentrations in the other species studied, including *O. nubilalis*. They explained the low mortality, even at high doses, as a function of low horizontal and vertical transmission because few infective or abnormal spores were produced.

In this case, it is important to clarify that they could obtain live infected larvae without mortality; however, they only measured mortality and did not measure other parameters, such as fecundity or larvae hatching in the next generation. As mentioned by Fuxa [32] and Inglis et al. [5], infection does not always lead to mortality, and occasionally the effects are observed long term and can be inferred from a reduction in fecundity and susceptibility to other stress situations.

Lastra and Gómez [16] implemented a system to produce natural enemies of *D. saccharalis* in CENICAÑA (Clombian Sugarcane Research Center, Colombia) and began their colony with larvae collected in the field. Their most significant result was the detection of one "protozoa," possibly *Nosema*, identified as the causal agent for the diminution in larvae production. They observed refractile spores in the fluids of the malpighi tubes or hemolymph using phase contrast microscopy in healthy adults. Macroscopically, the diseased larvae were dwarfed and white.

In addition, different pathogenic microorganisms have been isolated from *Diatraea* sp. in experimentally transmitted infections in the laboratory; however, there is little information about the natural presence of entomopathogens belonging to the microsporidia group in this species of borer. It is important to perform a systematic search for infected larvae, pupae, or adults of *Diatraea*, and studies need to include microscopy techniques to complement the mortality bioassays because infections could be asymptomatic. As many microsporidia do not produce insecticide activity quickly and because many species have complex life cycles that involve more than one host, very few attempts have been made to implement microsporidia and use them as agents of biological control.

4. Nematodes

Pathogenic nematodes of the Heterorhabditidae and Steinernematidae families live in the soil where they are parasitic to certain soil-dwelling insects. The free-living third juvenile stage (infective juveniles, IJs) locates a suitable host and penetrates the host in different ways. Once inside the insect, the IJs initiate its development, the nutritional tract becomes functional, and the symbiotic bacterium *Xenorhabdus* or *Photorhabdus* is released through the anus and begins to multiply in the hemocoel, which kills the insect by septicemia and creates suitable conditions for the reproduction of the nematode. Nematodes feed on bacteria and the dead tissue of the host, and they pass through several generations until new IJs are produced, which emerge from the cadaver [37]. They have a ubiquitous distribution [37]; therefore, it is not unusual to find them in the soil of sugarcane plantations (Pizano et al., 1985 cited in Khan et al. [38, 39]).

However, there are few reports of nematodes infecting stem borers of the *Diatraea* genus in natural and experimental situations. In Costa Rica, several entomopathogenic species, including nematodes, have been isolated from *D. tabernella* [40]; however, the author does not mention the nematode species.

Khan et al. [38] reviewed the world bibliography on rice stem borers and found five reports of entomopathogenic nematodes associated with *D. saccharalis*, two reports of *Steinernema* (=*Neoaplectana*) *glaseri* in Brazil and three reports of *Steinernema* (=*Neoaplectana*) *carpocapsae* in the USA and Guadeloupe. However, in one study, they report that the aim of the research was the use of *D. saccharalis* to produce *S. glaseri* in controlled conditions because of its susceptibility to the nematode. Another study by Folegatti et al. [41] used entomopathogenic nematodes and *D. saccharalis*

TABLE 4: Entomopathogenic-nematode bioassays on *Diatraea* spp.

Host specie	Pathogen	Bioassay	Result	Reference
Diatraea saccharalis	*Steinernema feltiae* [*Neoaplectana carpocapsae*]	Several concentrations for all species	100% mortality with 5000 nematodes/larvae	Sosa et al., 1993 [42]
	Heterorhabditis heliothidis		100% mortality with 5000 nematodes/larvae	
	S. glaseri		30% mortality with 5000 nematodes/larvae	
Diatraea saccharalis	*S. feltiae* *S. rarum* *Heterorhabditis bacteriophora*	500 IJs/5 larvae in Petri dishes with two moistened filter papers disks	Adults and IJs produced. Larval mortality >90%	De Doucet et al., 1999 [43]
Diatraea saccharalis	*H. bacteriophora* JPM4 isolated from soil in Larvae MG, Brazil	Larvae were exposed to 25 $\pm$ 5 IJs/100 μL for 48 h in sand-filled Petri dishes	100% larval mortality within 5 days. LT_{50} 2.1 d and LT_{95} 4.4 d	Molina et al., 2007 [44]
Diatraea saccharalis	*Heterorhabditis bacteriophora*	500 mL/L applied to larvae	93% of larvae mortality	Aguila et al., 2008 [45]

in laboratory conditions to produce *S. carpocapsae in vivo*, using larvae of the sugarcane borer as the host.

We believe that pathogenic nematodes have good potential for the biological control of *Diatraea* species because of their presence in sugarcane-producing regions (as mentioned above) and because the species *S. feltiae*, *S. glaseri*, *S. rarum*, *Heterorhabditis heliothis*, and *H. bacteriophora* have demonstrated high infectivity in *D. saccharalis* [42–45] (Table 4) as well as in *E. loftini* (some of them) [46, 47] in experimental studies. We consider that performing a systematic search for infected larvae or pupae of *Diatraea* could be helpful in finding new nematode strains and species adapted to local environmental conditions and pest species that could be used in the future.

Although we could not find any report on field trials, we are aware that there have been mass rearings of *H. bacteriophora* in Cuba since 1987 to control soil pests, such as *D. saccharalis* [45].

5. Bacteria

Bacillus thuringiensis (Bt) is a Gram-positive bacterium that has been isolated from several sources, including soil, water, phylloplane, and insect cadavers. *Bt* produces various insecticidal crystal proteins during the onset of sporulation that are toxic to insects, acari, and nematodes [48]. The steps involved in the mode of action of the proteins after their ingestion are as follows: (a) solubilization of the crystals by the highly alkaline pH of the midgut, (b) activation of the proteins by proteases, (c) binding of the toxins to specific receptors located on the microvilli membrane of the midgut columnar epithelium cells, and (d) the insertion of the toxin into the membrane, which forms a pore and induces cell lysis. There are no reports from the American continent on the isolation of Bt strains from *Diatraea* spp. larvae cadavers; however, some Bt strains and pure proteins have been evaluated against these pests.

Bohorova et al. [49] evaluated Cry1Aa, Cry1Ab, Cry1Ac, Cry1B, Cry1C, Cry1D, Cry1E, and Cry1F Bt pure proteins against four species of Lepidoptera that are pests of maize, including *D. saccharalis*. The proteins were diluted in water and added to the diet at doses of 10 and 100 mg/mL of diet, and mortality was recorded after seven days. *D. saccharalis* was susceptible to Cry1B protein at 10 mg/g, and the LC_{50} was 113.6 mg/g of meridic diet (Table 5).

Twelve Bt strains were evaluated by Rosas García et al. [50] at a dosages of 50 and 500 μg of total protein and spores per milliliter against 2-day-old *D. saccharalis* larvae. The strains used were HD1, HD2, HD9, HD29, HD37, HD59, HD133, HD137, and HD559, as well as the GM7, GM10, and GM34 native strains. The strains that killed more than 50% of larvae were selected to obtain the LC_{50}. Strains HD133, HD559, GM7, GM10, and GM34 were toxic; however, GM34 was the most toxic with an LC_{50} of 33.21 μg/mL (Table 5). PCR analysis was performed to determine the *cry1* genes of the toxic strains: HD133 *cryAa*, *cry1Ab*, *cry1C*; HD559 and GM7 *cry1Aa*, *cry1Ab*, and *cry1B*; GM10 *cry1Aa*, *cry1Ab*, *cry1Ac*, and *cry1C*; GM34 *cry1Aa*, *cry1Ab*, and *cry1Ac*.

Gitahy et al. [51] evaluated a spore-crystal complex *in vitro* from five native Bt strains (S48, S76, S90, S105, and S135) and HD1, which served as a positive control, on second instar larvae of *D. saccharalis* and mortality was recorded after 5 days. The strain S76 caused 100% mortality at 72 h, HD1 caused 69% mortality, and 3% of the other native strains caused mortality at 500 μg L^{-1} of the spore-crystal complex. The LC_{50} values of the S76 and HD1 strains were determined, and S76 was 11-fold more toxic (13.06 μg/L) than HD1 (143.88 μ/L) (Table 5). The S76 strain carries *cryAa*, *cry1Ab*, *cryAc*, *cry2Aa*, and *cry2Ab* genes, which is similar to HD1.

There are other species of bacteria with potential insecticide activity; Carneiro et al. [52] evaluated *Photorhabdus temperata*, which is a bacterium associated with *Heterorhabditis* entomopathogenic nematodes. Cells were injected with

TABLE 5: Entomopathogenic-bacteria bioassays on *Diatraea* spp.

Host species	Pathogen	Bioassay	Result	Reference
Diatraea saccharalis	*Bacillus thuringiensis*	Diet incorporated spores-crystals	LC_{50} 113.6 mg/g diet	Bohorova et al., 1997 [49]
D. saccharalis	*B. thuringiensis*		LC_{50} 33.21 μg/mL	Rosas-Garcia et al., 2004 [50]
D. saccharalis	*B. thuringiensis*	Diet incorporated spores-crystals complex	LC_{50} 13.06 μg/L	Gitahy et al., 2007 [51]
D. saccharalis	*Photorhabdustemperata*	Cells injected	LD_{50} 16.2 bacterial cells LT_{50} 33.8 h	Carneiro et al., 2008 [52]
Diatraea grandiosella	*Bacillus thuringiensis*	Diet incorporated spores-crystals	LC_{50} 5.2 mg/g diet	Bohorova et al., 1997 [49]
Diatraea grandiosella	*Bacillus thuringiensis* Biotrol BTB 183-25 Nutrilite Products Incorporated, Kansas City, Missouri, USA	Immersing stem seedling in spores suspensions: 1.25×10^8, 6.25×10^7, 2.5×10^7 and 1.25×10^7	Mean mortality on 5-day-old larvae: 32, 31, 25, 15 out of 40. Control mean mortality 0.01 Mean mortality on 10-day-old larvae: 15, 15, 12 and 6 out of 20.	Sikorowski and Davis, 1970 [53]

a volume of 10 μL of phosphate-buffered saline directly into the hemocoel of fourth instar *D. saccharalis* larvae, and the LD_{50} was 16.2 bacterial cells with an LT_{50} of 33.8 h.

Sikorowski and Davis [53] determined the susceptibility of 5- and 10-day-old *D. grandiosella* larvae with a Bt commercial product (Biotrol BTB 183-25) using 1.25×10^8, 6.25×10^7, 2.5×10^7, and 1.25×10^7 spores per milliliter (Table 5) where two-inch stem seedlings were dipped for 30 min in spore suspensions. One 10-day-old or two 5-day-old larvae were allowed to feed on each stem. In addition, a known number of spores were placed or injected into 5-day-old stem seedlings at 1/4 inch sections. Stems dipped in water were used as controls. Mortality was recorded after 48 h. With 10-day-old larvae, five replicates with 20 larvae for each treatment were used; for the 5-day-old larvae, two replicates with 40 larvae per treatment were used. With 5- and 10-day-old larvae at 1.25×10^8 spores per milliliter, a mean of 32 dead larvae out of 40 and 15 dead larvae out of 20 were recorded (Table 5), respectively. Thus, it was concluded that this species is highly susceptible to *B. thuringiensis*.

6. Viruses

Viruses that infect insects have received great attention as biological control agents because of their specificity on insect populations; they have little or no impact on the environment and are an ecological friendly alternative to chemical pesticides [58]. Entomopathogenic viruses are grouped in 33 genera within fifteen families [59]. However, only a few of these families, such as *Baculoviridae*, *Poxviridae*, and *Reoviridae*, have potential as biological control agents and have been successfully used in microbial control programs. A common characteristic among these entomopathogenic viruses is that the virions (infective unit) are occluded within a crystalline protein matrix to form an occlusion body (OB) [60], which is a unique characteristic of viruses that infect insects. In

these entomopathogenic viruses, OBs have independently evolved as a protective mechanism to environmental factors, which gives the viruses a great advantage as biological control agents [61]. Baculoviruses (BVs) and entomopoxviruses (EPVs: subfamily *Entomopoxvirinae*) have a large double-stranded DNA genome, and cypoviruses (CPVs: family *Reoviridae*, genera *Cypovirus*) contain segmented double-stranded RNA viruses.

EPVs have been reported to infecting insects of a variety of orders, such as Coleoptera, Lepidoptera, Orthoptera, and Diptera. Some EPVs have two distinct OBs spheroids and spindles. The spheroids occlude virions, whereas the spindles do not [62]. The most abundant proteinaceous component of spheroids and spindles is proteins called spheroidin and fusolin, respectively [63–66]. EPV fusolin is an enhancing factor (EF) that increases BVs infection and has been characterized as a chitin-binding protein [67]. The fusolin mechanism of action is similar to that of Calcofluor, which facilitates BV infection by disrupting or preventing the formation of the peritrophic membrane [68]. EPVs are pathogenic but are scarcely virulent; infected larvae exhibit extreme longevity and take up to 70 days to die. However, EPVs have potential as biological control agents for pest insects where BVs have not been isolated [61]. Because of the activity of spindles, the EF of EPVs can be used as a synergistic agent to increase BV infectivity or generate genetically modified organisms, such as BVs and transgenic plants.

CPVs have been mainly isolated from Lepidoptera insects. OBs are dissolved in the midgut, and virions only infect epithelial cells; therefore, CPVs are very pathogenic but act slowly and frequently to produce chronic infections [69]. At this time, no commercial bioinsecticides based on CPVs have been developed. However, Caballero and Williams [61] suggest that their greatest potential as biological control agents is through inoculative or augmentative releases.

TABLE 6: Entomopathogenic-virus bioassays on *Diatraea* spp.

Host species	Pathogen	Bioassay conditions	Result	Reference
D. grandiosella	*Autographa californica* MNPV (*Ac*MNPV)	4 and 7 day-old larvae were fed strips of corn leaf soaked in different dilutions of virus.	DL_{50}: 1.32×10^3 PIB*/larvae (for 4-day-old larvae) DL_{50}: 1.32×10^4 PIB/larvae (for 7-day-old larvae).	Davis and Sikorowsy, 1978 [54]
D. saccharalis	*Diatraea saccharalis* Granulosis Virus (DsGV)	3rd instar larvae feeding individually on small artificial diet discs treated with 2.7 μL of virus dilutions.	LD_{50}: 42 PIB/larva at 26°C LT_{50}: From 29 to 63 for 10^7 and 10^2 PIB/larvae, respectively.	Lastra and Gómez et al., 1983 [16]
D. saccharalis	*Anticarsia gemmatalis* MNPV (*Ag*MNPV)	3rd instar larvae fed with artificial diet discs treated with 5 different doses of PIB through 20 serial passages.	LD_{50}: From 7.9×10^5 to 5.3×10^2 for 1 and 20 serial passages, respectively.	Pavan and Ribeiro, 1989 [55]
D. saccharalis	*Ag*MNPV wt	3rd instar larvae individually fed small artificial diet discs containing 10^5 PIB of virus and kept individually at 10 different temperatures.	LT_{50}: From 10 to 13 for 39 and 30°C, respectively	Ribeiro and Pavan, 1994 [56]
	*Ag*MNPV-D10		From 9 to 29 for 39 and 22°C, respectively	
D. saccharalis	*Trichoplusia ni* MNPV (*Tn*MNPV) wt	3rd instar larvae fed individually with small artificial diet discs containing 10^5 PIB of virus and kept individually at 10 different temperatures.	LT_{50}: From 9 to 16 for 39 and 24°C, respectively	Ribeiro and Pavan, 1994 [56]
	*Tn*MNPV-D11		From 6 to 46 for 39 and 17°C, respectively	
D. sacccharalis	*Ag*MNPV	3rd instar larvae fed on formalin free diet disc (3 mm) inoculated with 2.7 μL of the viral solution from 10 clonal isolates of the AgMNPV-Ds20	LD_{50}: From 5.3×10^3 PIB/larvae (7A genotypic variant) to 8×10^4 PIB/larvae (33B genotypic variant)	Ribeiro et al., 1997 [57]

* PIB: Polyhedral inclusion bodies.

BVs predominantly infect insects within the Lepidoptera order, which includes important agricultural insect pests [70]. The *Baculoviridae* family includes two genera: *nucleopolyhedrovirus* (NPV), which forms large, polyhedric OBs where many enveloped virions are occluded [71], and *Granulovirus* (GV), which forms small, granular OBs that occlude only one enveloped virion each [72]. BVs are safe for humans and wildlife. Their specificity is usually very narrow and often is species specific. Because of their specificity and other characteristics, such as elevated virulence and pathogenicity, BVs are by far the most studied and extensively used as commercial biopesticides for the control of a variety of insect pests in many countries around the world [60, 61, 73].

Stem borers of the Lepidoptera order attack gramineous crops throughout the world [74, 75]. *Diatraea* stem borers (DSB) are widely distributed in the Americas and attack a wide variety of host plants, including maize and sugarcane [1]. Prediapause larvae of southwestern corn borers migrate to the base of the stalk of the corn plant below the soil surface to survive the winters [76]; cryptic habits make the chemical control of these insect pests difficult. Degaspari et al. [77] have argued that chemical control of *D. saccharalis* in Brazil is not economically feasible. To solve this problem, surveys to isolate endemic entomopathogens of stem borer populations in maize and sugarcane crops have been developed.

Inglis et al. [5] developed an exhaustive survey to isolate entomopathogens from the southern corn borer, *D. grandiosella*, and the southern corn stalk borer, *D. crambidoides*, and larvae in the diapause stage were collected from crops located in Mississippi and North Carolina. These authors did not observe OBs in any of the collected larvae and concluded that there are no naturally occurring viruses in these *Diatraea* populations [5]. According to Inglis et al. [5], Pavan and Ribeiro [55] mentioned that natural populations of the SCB in Brazil do not exhibit endemic viral pathogens.

Currently, there are only two records of endemic entomopathogenic viruses isolated from *Diatraea* spp. larvae. Pavan et al. [17] isolated a GV from the sugarcane borer (SCB), *D. saccharalis*, from sugarcane crops in the southern United States (Table 6). These authors developed bioassays with *D. saccharalis* third instar larvae (Table 6). External symptoms of GV-infected larvae were similar to those reported for other lepidopterous larvae, and the symptoms and ultrastructures were determined using electronic microscopy as well as replication of DsGV, which are typical of GVs. The LD$_{50}$ for the third instar larvae was 42.3 OBs/larva (Table 6) with 14.5 and 123.6 OBs/larva as the lower and upper limits at 95% probability, respectively [16]. These authors stated that this work was the first of a series of publications; however, there have been no additional studies produced to date, such as the molecular and biological characterization, of this GV, DsGV was introduced into Brazil [16], and Moscardi [78] mentioned that this virus is currently being applied at a small scale as an experimental product on sugarcane.

Because of their high pathogenicity combined with a limited host range, Densoviruses (DNVs) have potential as effective insecticides [79]. Meynadier et al. [80] isolated a DNV of *D. saccharalis* (DsDNV) from the Guadeloupe sugarcane borer (Table 6). Kouassi et al. [81] tested the pathogenicity of DsDNV on its host. These authors observed that the infected larvae exhibited infection symptoms from the fourth day postinfection, such as anorexia and lethargy followed by flaccidity and inhibition of molting and metamorphosis. Larvae became paralyzed and stopped feeding after 7 days. The cumulative mortality of infected larvae increased significantly and reached 60% after 12 days and 100% at 21 days postinfection [81]. Although DNVs have no potential for large scale use as a biological control agent because they have no OBs and are related to vertebrate pathogenic viruses, the genes involved with anorexia and paralysis could be used to produce transgenic BVs or plants.

Most BVs have a limited host range and may be species specific in some cases, although there are several NPVs, such as *Autographa californica* MNPV, *Anagrapha falcifera* MNPV, and *Anticarsia gemmatalis* MNPV, which have broader host ranges within the Lepidoptera order [60]. Because of the scarcity of reports on natural isolates of entomopathogenic viruses from *Diatraea* spp. populations, cross-infectivity of AgMNPV, TnMNPV, and AgMNPV has been evaluated in *D. saccharalis* and *D. grandiosella* larvae (Table 6).

Although only two reports of entomopathogenic viruses isolated from *Diatraea* stem borers exist in the Americas, entomopathogenic viruses, such as NPVs and GVs, have been reported to occur in Africa and Asia in cereal stem borers within the Crambidae family, including maize and sugarcane borers such as *Chilo* sp. [82], *Chilo sacchariphagus* [83, 84], *Ch. infuscatellus* [84, 85], *Ch. partellus* [86], and *Eldana saccharina* [87]. As the Mexican territory is the origin of some *Diatraea* stem borer species, we hypothesize that a great diversity of pathogenic viruses in *Diatraea* populations that attack sugarcane crops exists in Mexico.

Acknowledgments

This study was financed in part by the project "Selección de enemigos naturales de barrenadores de la caña de azúcar de Colima y Morelos con potencial como agentes de control biológico (SEP-PROMEP)," of Mexico. The authors also thank Mrs. Ingrid Masher for reviewing the paper and for editorial assistance.

References

[1] L. A. Rodriguez-Del-Bosque and J. W. Smith Jr., "Biological control of maize and sugarcane stemborers in Mexico: a review," *Insect Science and Its Application*, vol. 17, no. 3-4, pp. 305–314, 1997.

[2] G. Serra and E. Trumper, "Estimación de incidencia de daños provocados por larvas de *Diatraea saccharalis* (Lepidoptera: Crambidae) en tallos de maíz mediante evaluación de signos externos de infestación," *Agrosciencia*, vol. 23, no. 1, pp. 1–7, 2006.

[3] M. A. P. De Oliveira, E. J. Marques, V. Wanderley-Teixeira, and R. Barros, "Effect of *Beauveria bassiana* (Bals.) Vuill. and *Metarhizium anisopliae* (Metsch.) Sorok. on biological characteristics of *Diatraea saccharalis* F. (Lepidoptera: Crambidae)," *Acta Scientiarum*, vol. 30, no. 2, pp. 220–224, 2008.

[4] R. Reyes, "Sorghum stem borers in Central and South America," in *Proceedings of the International Workshop on Sorghum Stem Borers*, pp. 49–60, ICRISAT Center, Hyderabad, India, 1987.

[5] G. D. Inglis, A. M. Lawrence, and F. M. Davis, "Pathogens associated with southwestern corn borers and southern corn stalk borers (Lepidoptera: Crambidae)," *Journal of Economic Entomology*, vol. 93, no. 6, pp. 1619–1626, 2000.

[6] M. D. R. T. De Freitas, E. L. Da Silva, A. D. L. Mendonça et al., "The biology of *Diatraea flavipennella* (Lepidoptera: Crambidae) reared under laboratory conditions," *Florida Entomologist*, vol. 90, no. 2, pp. 309–313, 2007.

[7] R. H. R. Destéfano, S. A. L. Destéfano, and C. L. Messias, "Detection of *Metarhizium anisopliae* var. *anisopliae* within infected sugarcane borer *Diatraea saccharalis* (Lepidoptera, Pyralidae) using specific primers," *Genetics and Molecular Biology*, vol. 27, no. 2, pp. 245–252, 2004.

[8] A. E. Knutson and F. E. Gilstrap, "Seasonal occurrence of *Beauveria bassiana* in the southwestern corn borer (Lepidoptera: Pyralidae) in the Texas High Plains," *Journal of the Kansas Entomological Society*, vol. 63, no. 2, pp. 243–225, 1990.

[9] K. Zambrano, M. Dávila, and M. A. Castillo, "Detección de fragmentos de AND de hongos y su posible relación con la síntesis de proteínas de actividad entomopatógena," *Revista de la Facultad De Agronomía (LUZ)*, vol. 19, pp. 185–193, 2002.

[10] S. B. Alves, L. S. Rossi, R. B. Lopes, M. A. Tamai, and R. M. Pereira, "*Beauveria bassiana* yeast phase on agar medium and its pathogenicity against *Diatraea saccharalis* (Lepidoptera: Crambidae) and *Tetranychus urticae* (Acari: Tetranychidae)," *Journal of Invertebrate Pathology*, vol. 81, no. 2, pp. 70–77, 2002.

[11] M. E. Estrada, M. Romero, M. J. Rivero, and F. Barroso, "Natural presence of *Beauveria bassiana* (Balsamo) Vuillemin in the sugar cane (*Saccharum* sp. hybrid) in Cuba," *Revista Iberoamericana de Micologia*, vol. 21, no. 1, pp. 42–43, 2004.

[12] M. G. Yasen de Romero, A. R. Salvatore, G. López, and E. Willink, "Presencia natural de hongos hyphomycetes en larvas invernantes de *Diatraea ssaccharalis* F. en caña de azúcar en Tucumán, Argentina," *Revista Industrial y Agrícola de Tucumán*, vol. 85, no. 2, pp. 39–42, 2008.

[13] C. A. Angel-Sahagún, R. Lezama-Gutiérrez, J. Molina-Ochoa et al., "Susceptibility of biological stages of the horn fly, *Haematobia irritans*, to entomopathogenic fungi (Hyphomycetes)," *Journal of Insect Science (Online)*, vol. 5, no. 50, pp. 1–8, 2005.

[14] R. A. Humber, K. S. Hansen, and M. M. Wheeler, *Catalog of Species. ARS Collection of Entomopathogenic Fungal Cultures*, USDA-ARS Biological Integrated Pest Management Research, Ithaca, NY, USA, 2009.

[15] V. M. Hernández Velázquez and R. Lezama Gutiérrez, "Uso de entomopatógenos para el control biológico de barrenadores del tallo," in *Taller Internacional Sobre Barrenadores Del Tallo De Caña De Azúcar*, L. A. Rodríguez del Bosque, G. Vejar, and E. Cortéz, Eds., pp. 37–45, Sociedad Mexicana de Control Biológico, Los Mochis, Sinaloa, Mexico, Noviembre de 2004.

[16] B. Lastra and L. A. Gómez, "Cría y producción masiva de insectos en un programa de control biológico en caña de azúcar," in *1er Curso Taller Internacional de Control Biológico*, pp. 335–340, Memorias, Bogotá, Colombia.

[17] O. H. O. Pavan, D. G. Boucias, L. C. Almeida, J. O. Gaspar, P. S. M. Botelho, and N. Degaspari, "Granulosis Virus of *Diatraea saccharalis* (DsGV): pathogenicity, replication and ultrastructure," in *Proceedings of the International Congress of the International Society of Sugar Cane Technologists (ISSCT '83)*, vol. 2, pp. 644–659, La Havana, Cuba, 1983.

[18] G. Meynadier, P. F. Galichet, J. C. Veyrunes, and A. Amargier, "Mise en évidence d'une densonucléose chez *Diatraea saccharalis* [Lep.: Pyralidae]," *Entomophaga*, vol. 22, no. 1, pp. 115–120, 1977.

[19] A. Fonseca-González, G. Peña-Chora, A. Trejo-Loyo, L. Lina-García, L. A. Rodríguez-Del Bosque, and V. M. Hernández-Velázquez, "Virulencia de seis aislados de *Bacillus thuringiensis* nativos delestado de Morelos y evaluados sobre *Diatraea magnifactella*," in *Memoria IV Congreso Nacional de Control Biológico*, pp. 31–34, Sociedad Mexicana de Control Biológico, Monterrey, Nuevo León, México, Noviembre 2011.

[20] S. C. M. Leal, D. J. Bertioli, T. M. Butt, and J. F. Peberdy, "Characterization of isolates of the entomopathogenic fungus *Metarhizium anisopliae* by RAPD-PCR," *Mycological Research*, vol. 98, no. 9, pp. 1077–1081, 1994.

[21] S. Ali, Z. Huang, and S. Ren, "Media composition influences on growth, enzyme activity, and virulence of the entomopathogen hyphomycete *Isaria fumosoroseus*," *Entomologia Experimentalis et Applicata*, vol. 131, no. 1, pp. 30–38, 2009.

[22] J. C. Legaspi, T. J. Poprawski, and B. C. Legaspi Jr., "Laboratory and field evaluation of *Beauveria bassiana* against sugarcane stalkborers (lepidoptera: pyralidae) in the Lower Rio Grande Valley of Texas," *Journal of Economic Entomology*, vol. 93, no. 1, pp. 54–59, 2000.

[23] L. F. Solter and J. V. Maddox, "Microsporidia as classical biological control agents: research and regulatory issues," *Phytoprotection*, vol. 79, no. 4, pp. 75–80, 1998.

[24] N. Corradi and P. J. Keeling, "Microsporidia: a journey through radical taxonomical revisions," *Fungal Biology Reviews*, vol. 23, no. 1-2, pp. 1–8, 2009.

[25] J. Weiser, "Microsporidia and the society for invertebrate pathology: a personal point of view," *Journal of Invertebrate Pathology*, vol. 89, no. 1, pp. 12–18, 2005.

[26] J. A. Arcas, B. M. Díaz, and R. E. Lecuona, "Bioinsecticidal activity of conidia and dry mycelium preparations of two isolates of *Beauveria bassiana* against the sugarcane borer *Diatraea saccharalis*," *Journal of Biotechnology*, vol. 67, no. 2-3, pp. 151–158, 1999.

[27] J. C. Legaspi Jr., B. C. Legaspi, and R. R. Saldaña, "Evaluation of *Steinernema riobravis* (Nematoda: Steinernematidae) against the Mexican rice borer (Lepidoptera: Pyralidae)," *Journal of Entomological Science*, vol. 35, no. 2, pp. 141–149, 2000.

[28] E. J. Marques, S. B. Alves, and I. M. R. Marques, "Virulencia de *Beauveria bassiana* (Bals.) Vuill. a *Diatraea saccharalis* (F.) (Lepidoptera: Crambidae) após armazenamiento de conídios em baixa temperatura," *Anais da Sociedade Entomológica do Brasil*, vol. 29, no. 2, pp. 303–307, 2000.

[29] S. B. Alves, L. S. Rossi, R. B. Lopes, M. A. Tamai, and R. M. Pereira, "*Beauveria bassiana* yeast phase on agar medium and its pathogenicity against *Diatraea saccharalis* (Lepidoptera: Crambidae) and *Tetranychus urticae* (Acari: Tetranychidae)," *Journal of Invertebrate Pathology*, vol. 81, no. 2, pp. 70–77, 2002.

[30] I. M. Wenzel, F. H. C. Gionetti, and J. E. M. Almeida, "Patogenicidade do isolade IBCB66 de *Beauveria bassiana* s broca da caña-de-acúcar *Diatraea saccharalis* em condicoes de laboratório," *Arquivos Do Instituto Biologico*, vol. 72, no. 2, pp. 259–261, 2006.

[31] C. E. Lange, "Niveles de esporulación experimentales y naturales de *Nosema locustae* (Microsporidia) en especies de tucuras y langostas (Orthoptera: Acridoidea) de la Argentina," *Revista de la Sociedad Entomológica Argentina*, vol. 62, no. 1-2, pp. 15–22, 2003.

[32] J. R. Fuxa, "Susceptibility of Lepidopterous pests to two types of mortality caused by the microsporidium *Vairimorpha necatrix*," *Journal of Economic Entomology*, vol. 74, no. 1, pp. 99–102, 1981.

[33] L. F. Solter and J. V. Maddox, "Microsporidia as classical biological control agents: research and regulatory issues," *Phytoprotection*, vol. 79, no. 4, pp. 75–80, 1998.

[34] G. D. Inglis, A. M. Lawrence, and F. M. Davis, "Impact of a novel species of Nosema on the Southwestern corn borers (Lepidoptera: Cambridae)," *Journal of Economic Entomology*, vol. 96, no. 1, pp. 12–20, 2003.

[35] M. W. Phoofolo, J. J. Obrycki, and L. C. Lewis, "Quantitative assessment of biotic mortality factors of the european corn borer (lepidoptera: crambidae) in field corn," *Journal of Economic Entomology*, vol. 94, no. 3, pp. 617–622, 2001.

[36] L. F. Solter, J. V. Maddox, and C. R. Vossbrinck, "Physiological host specificity: a model using the European corn borer, Ostrinia nubilalis (Hübner) (Lepidoptera: Crambidae) and microsporidia of row crop and other stalk-boring hosts," *Journal of Invertebrate Pathology*, vol. 90, no. 2, pp. 127–130, 2005.

[37] G. O. Poinar Jr., "Taxonomy and biology of Steinernematidae and Heterorhabditidae," in *Entomopatogenic Nematodes in Biological Control*, R. Gaugler and H. K. Kaya, Eds., pp. 23–61, CRC Press, 1990.

[38] Z. R. Khan, J. A. Litsinger, A. T. Barrion, F. F. Villanueva, N. J. Fernandez, and L. D. Taylo, *World Bibliography of Rice Stem Borers 1794–1990*, International Rice Research Institute, 1991.

[39] L. Dasrat, "Discovery of an indigenous entomopathogenic nematode and its pathogenicity to two sugar cane stem borers in Guyan," http://wistonline.org/papers/proceedings/Paper16.PDF.

[40] F. Badilla, "Un programa exitoso de control biológico de insectos plaga de la caña de azúcar en Costa Rica," *Manejo Integrado de Plagas y Agroecología (Costa Rica)*, vol. 64, pp. 77–87, 2002.

[41] M. E. Folegatti, S. Batista, P. R. Kawai, and P. S. Botelho, "Nova metodologia para produção in vivo de *Neoaplectana carpocapsae* Weiser," *Nematologia Brasileira*, vol. 12, pp. 76–83, 1988.

[42] O. Sosa Jr., D. G. Hall, and W. J. Schroeder, "Mortality of sugarcane borer (Lepidoptera: Pyralidae) treated with entomopathogenic nematodes in field and laboratory trials,"

Journal of the American Society of Sugar Cane Technology, vol. 13, pp. 18–21, 1993.

[43] M. M. A. De Doucet, M. A. Bertolotti, A. L. Giayetto, and M. B. Miranda, "Host Range, Specificity, and Virulence of *Steinernema feltiae, Steinernema rarum,* and *Heterorhabditis bacteriophora* (Steinernematidae and Heterorhabditidae) from Argentina," *Journal of Invertebrate Pathology*, vol. 73, no. 3, pp. 237–242, 1999.

[44] J. P. M. Acevedo, R. I. Samuels, I. R. Machado, and C. Dolinski, "Interactions between isolates of the entomopathogenic fungus *Metarhizium anisopliae* and the entomopathogenic nematode *Heterorhabditis bacteriophora* JPM4 during infection of the sugar cane borer *Diatraea saccharalis* (Lepidoptera: Pyralidae)," *Journal of Invertebrate Pathology*, vol. 96, no. 2, pp. 187–192, 2007.

[45] P. J. Aguila, M. Vidal, M. González, L. A. Rodríguez, and E. Mesa, "Criterios ecológicos en el manejo de plagas de la caña de azúcar y cultivos varios una opción para lograr alimentos sanos," *Agroecología*, vol. 3, pp. 51–53, 2008.

[46] D. R. Ring and H. W. Browning, "Evaluation of entomopathogenic nematodes against the Mexican rice borer (Lepidoptera: Pyralidae)," *Journal of Nematology*, vol. 22, no. 3, pp. 420–422, 1990.

[47] J. C. Legaspi Jr., B. C. Legaspi, and R. R. Saldaña, "Evaluation of *Steinernema riobravis* (Nematoda: Steinernematidae) against the Mexican rice borer (Lepidoptera: Pyralidae)," *Journal of Entomological Science*, vol. 35, no. 2, pp. 141–149, 2000.

[48] E. Schnepf, N. Crickmore, J. Van Rie et al., "*Bacillus thuringiensis* and its pesticidal crystal proteins," *Microbiology and Molecular Biology Reviews*, vol. 62, no. 3, pp. 775–806, 1998.

[49] N. Bohorova, M. Cabrera, C. Abarca et al., "Susceptibility of four tropical lepidopteran Maize pests to *Bacillus thuringiensis* CryI-type insecticidal toxins," *Journal of Economic Entomology*, vol. 90, no. 2, pp. 412–415, 1997.

[50] N. M. Rosas-García, B. Pereyra-Alférez, K. A. Niño, L. J. Gaín-Wong, and L. H. Morales-Ramos, "Novel toxicity of native and HD *Bacillus thuringiensis* strains against to the sugarcane borer *Diatraea saccharalis*," *BioControl*, vol. 49, no. 4, pp. 455–465, 2004.

[51] P. D. M. Gitahy, M. T. de Souza, R. G. Monnerat, E. D. B. Arrigoni, and J. I. Baldani, "A Brazilian *Bacillus thuringiensis* strain highly active to sugarcane borer *Diatraea saccharalis* (Lepidoptera: Crambidae)," *Brazilian Journal of Microbiology*, vol. 38, no. 3, pp. 531–537, 2007.

[52] C. N. B. Carneiro, R. A. DaMatta, R. I. Samuels, and C. P. Silva, "Effects of entomopathogenic bacterium *Photorhabdus temperata* infection on the intestinal microbiota of the sugarcane stalk borer *Diatraea saccharalis* (Lepidoptera: Crambidae)," *Journal of Invertebrate Pathology*, vol. 99, no. 1, pp. 87–91, 2008.

[53] P. Sikorowski and F. M. Davis, "Susceptibility of larvae of the southwestern corn borer, *Diatraea grandiosella,* to *Bacillus thuringiensis,*" *Journal of Invertebrate Pathology*, vol. 15, no. 1, pp. 131–132, 1970.

[54] F. M. Davis and P. P. Sikorowsy, "Susceptibility of the Southwestern corn borer, *Diatraea grandiosella* Dyar (Lepidoptera: Pyralidae) to the baculovirus of *Autographa californica,*" *Journal of the Kansas Entomological Society*, vol. 51, no. 1, pp. 11–13, 1978.

[55] O. H. O. Pavan and H. C. T. Ribeiro, "Selection of a baculovirus strain with a bivalent insecticidal activity," *Memórias do Insttituto Oswaldo Cruz*, vol. 84, supplement 3, pp. 63–65, 1989.

[56] H. C. T. Ribeiro and O. H. O. Pavan, "Effect of temperature on the development of baculoviruses," *Journal of Applied Entomology*, vol. 118, pp. 316–320, 1994.

[57] H. C. T. Ribeiro, O. H. O. Pavan, and A. R. Muotri, "Comparative susceptibility of two different hosts to genotypic variants of the *Anticarsia gemmatalis* nuclear polyhedrosis virus," *Entomologia Experimentalis et Applicata*, vol. 83, no. 2, pp. 233–237, 1997.

[58] S. R. Palli, T. R. Ladd, W. L. Tomkins et al., "*Choristoneura fumiferana* entomopoxvirus prevents metamorphosis and modulates juvenile hormone and ecdysteroid titers," *Insect Biochemistry and Molecular Biology*, vol. 30, no. 8-9, pp. 869–876, 2000.

[59] C. M. Fauquet, M. A. Mayo, J. Maniloff, U. Desselberger, and L. A. Ball, Eds., *Virus Taxonomy, Eighth Report of the International Committee on Taxonomy of Viruses*, Elsevier, San Diego, Calif, USA, 2005.

[60] J. S. Cory and H. F. Evans, "Viruses," in *Field Manual of Techniques in Invertebrate Pathology*, L. A. Lacey and H. K. Kaya, Eds., pp. 149–174, Springer, Dordrecht, The Netherlands, 2nd edition, 2007.

[61] P. Caballero and T. Williams, "Virus Entomopatógenos," in *Control Biológico de Plagas Agrícolas*, J. A. Jacas and A. Urbaneja, Eds., pp. 121–135, PHYTOMA-España, Navarra, España, 2008.

[62] W. Mitsuhashi, H. Kawakita, R. Murakami et al., "Spindles of an entomopoxvirus facilitate its infection of the host insect by disrupting the peritrophic membrane," *Journal of Virology*, vol. 81, no. 8, pp. 4235–4243, 2007.

[63] M. Bergoin, J. C. Veyrunes, and R. Scalla, "Isolation and amino acid composition of the inclusions of *Melolontha melolontha* poxvirus," *Virology*, vol. 40, no. 3, pp. 760–763, 1970.

[64] R. L. Hall and R. W. Moyer, "Identification, cloning, and sequencing of a fragment of *Amsacta moorei* entomopoxvirus DNA containing the spheroidin gene and three vaccinia virus-related open reading frames," *Journal of Virology*, vol. 65, no. 12, pp. 6516–6527, 1991.

[65] D. Dall, A. Sriskantha, A. Vera, J. Lai-Fook, and T. Symonds, "A gene encoding a highly expressed spindle body protein of *Heliothis armigera* entomopoxvirus," *Journal of General Virology*, vol. 74, no. 9, pp. 1811–1818, 1993.

[66] T. Hayakawa, J. Xu, and T. Hukuhara, "Cloning and sequencing of the gene for an enhancing factor from *Pseudaletia separata* entomopoxvirus," *Gene*, vol. 177, no. 1-2, pp. 269–270, 1996.

[67] Z. Li, C. Li, K. Yang et al., "Characterization of a chitin-binding protein GP37 of *Spodoptera litura* multicapsid nucleopolyhedrovirus," *Virus Research*, vol. 96, no. 1-2, pp. 113–122, 2003.

[68] P. Wang and R. R. Granados, "Calcofluor disrupts the midgut defense system in insects," *Insect Biochemistry and Molecular Biology*, vol. 30, no. 2, pp. 135–143, 2000.

[69] T. Hukuhara and J. R. Bonami, "Reoviridae," in *Atlas of Invertebrate Viruses*, J. R. Adams and J. R. Bonami, Eds., pp. 393–434, CRC Press, Boca Raton, Fla, USA, 1991.

[70] G. W. Blissard, B. Black, N. Crook, B. A. Keddie, R. Possee et al., "Family Baculoviridae," in *Virus Taxonomy: Seventh Report of the International Committee on Taxonomy of Virus*, M. H. V. van Regenmoortel, C. M. Fauquet, D. H. L. Bishop et al. et al., Eds., pp. 195–202, Academic Press, San Diego, Calif, USA, 2000.

[71] G. F. Rhormann, "Nuclear polyhedrosis viruses," in *Encyclopedia of Virology*, R. G. Webster and A. Granoff, Eds., pp. 146–152, Academic Press, London, UK, 2nd edition, 1999.

Pathogens Associated with Sugarcane Borers, Diatraea spp. (Lepidoptera: Crambidae): A Review

59

[72] D. Winstanley and D. O. 'Reilly, "Granuloviruses," in *Encyclo-pedia of Virology*, R. G. Webster and A. Granoff, Eds., pp. 140–146, Academic Press, London, UK, 2nd edition, 1999.

[73] B. Szewczyk, L. Hoyos-Carvajal, M. Paluszek, I. Skrzecz, and M. Lobo De Souza, "Baculoviruses—re-emerging biopesticides," *Biotechnology Advances*, vol. 24, no. 2, pp. 143–160, 2006.

[74] W. F. Jepson, *A Critical Review of the World Literature of the Lepidopterous Stalk Borers of Graminaceous Crops*, Commonwealth Institute of Entomology, London, UK, 1954.

[75] J. W. Smith Jr., R. N. Wiedenmann, and W. A. Overholt, *Parasites of Lepidopteran Stemborers of Tropical Gramineous Plants*, ICIPE Science Press, Nairobi, Kenya, 1993.

[76] G. M. Chippendale, "The southwestern corn borer, *Diatraea grandiosella*, case history of an invading insect," Research Bulletin 1031, Missouri Agricultural Experiment Station, Columbia, Mo, USA, 1979.

[77] N. Degaspari, P. S. M. Botelho, and N. Macedo, "Controle quimico da *Diatraea saccharalis* em cana-de-acucar na região Centro Sul do Brasil," *Boletim Técnico. IAA/PLANALSUCAR*, vol. 3, pp. 1–16, 1981.

[78] F. Moscardi, "Use of viruses for pest control in Brazil: the case of the nuclear polyhedrosis virus of the soybean caterpillar, *Anticarsia gemmatalis*," *Memórias do Insttituto Oswaldo Cruz*, vol. 84, pp. 51–56, 1989.

[79] G. Fédière, "Epidemiology and pathology of Densovirinae," in *Parvovir Uses. From Molecular Biology to Pathology and Therapeutic Uses*, S. Faisst and J. Rommelaere, Eds., pp. 1–11, Karger, Basel, Switzerland, 2000.

[80] G. Meynadier, P. F. Galichet, J. C. Veyrunes, and A. Amargier, "Mise en évidence d'une densonucléose chez *Diatraea saccharalis* [Lep.: Pyralidae]," *Entomophaga*, vol. 22, no. 1, pp. 115–120, 1977.

[81] N. Kouassi, J. X. Peng, Y. Li, C. Cavallaro, J. C. Veyrunes, and M. Bergoin, "Pathogenicity of *Diatraea saccharalis* densovirus to host insets and characterization of its viral genome," *Virologica Sinica*, vol. 22, no. 1, pp. 53–60, 2007.

[82] E. P. Steinhaus and G. A. Marsh, "Report of diagnosis of disease insects, 1951–1961," *Hilgardia*, vol. 33, pp. 349–490, 1962.

[83] U. K. Metha and H. David, "A granulosis virus disease of sugarcane internode borer," *Madras Agricultural Journal*, vol. 67, pp. 616–619, 1970.

[84] S. Easwaramoorthy and S. Jayaraj, "Survey of granulosis virus infection in sugarcane borers, *Chilo infuscatellus* Snellen and *C. sacchariphagus* indicus (Kapur) in India," *Tropical Pest Management*, vol. 3, pp. 200–201, 1987.

[85] S. Easwaramoorthy and H. David, "A granulosis virus of sugarcane shoot borer, *Chilo infuscatellus* Snell. (Lepidoptera: Crambidae)," *Current Science*, vol. 48, pp. 685–686, 1979.

[86] S. Sethuraman and K. Narayanan, "Biological activity of Nucleopolyhedrovirus isolated from *Chilo partellus* (Swinhoe) (Lepidoptera: Pyralidae) in India," *Asian Journal of Experimental Biological Sciences*, vol. 1, pp. 325–330, 2010.

[87] A. J. Cherry, C. J. Lomer, D. Djegui, and F. Schuethess, "Pathogen incidence and their potential as microbial control agents in IPM of maize stem borers in West Africa," *BioControl*, vol. 44, no. 3, pp. 301–327, 1999.

8

Methods of Developing User-Friendly Keys to Identify Green Sea Turtles (*Chelonia mydas* L.) from Photographs

Jane R. Lloyd,[1] Miguel Á. Maldonado,[2] and Richard Stafford[3]

[1] *Department of Natural and Social Sciences, University of Gloucestershire, Cheltenham GL50 4AZ, UK*
[2] *Centro Ecológico Akumal, Akumal, 77730 Quintana Roo, Mexico*
[3] *Luton Institute of Research in the Applied Natural Sciences, Division of Science, University of Bedfordshire, Luton, LU1 3JU, UK*

Correspondence should be addressed to Richard Stafford, rick.stafford7@gmail.com

Academic Editor: Michael Thompson

Identifying individual animals is important in understanding their ecology and behaviour, as well as providing estimates of population sizes for conservation efforts. We produce identification keys from photographs of green sea turtles to identify them while foraging in Akumal Bay, Mexico. We create three keys, which (a) minimise the length of the key, (b) present the most obvious differential characteristics first, and (c) remove the strict dichotomy from key b. Keys were capable of identifying >99% of turtles in >2500 photographs during the six-month study period. The keys differed significantly in success rate for students to identify individual turtles, with key (c) being the best with >70% success and correctly being followed further than other keys before making a mistake. User-friendly keys are, therefore, a suitable method for the photographic identification of turtles and could be used for other large marine vertebrates in conservation or behavioural studies.

1. Introduction

Photographic identification of individual organisms can allow many aspects of their biology to be studied, for example, population sizes can be estimated using capture, mark, recapture techniques [1, 2], distributions and foraging ranges can be calculated and even social structure of group living individuals can be elucidated [3].

Photographic techniques can be beneficial for many reasons. They can reduce stress and handling of capturing an individual (e.g., [4], based on behaviours of turtle hatchlings) and eliminate problems such as tag loss [5, 6]. Furthermore, given the rise in popularity of digital cameras, and cameras (including those on smartphones) along with the use of Web 2.0 technologies (such as social networks), it is possible to obtain large quantities of useful scientific information from photographs taken by members of the public [7–9].

The main problem of photographic identification is the amount of time taken by highly trained personnel to correctly identify individuals [10]. Much research has shown that the use of poorly trained personnel results in misidentification of individuals [1] and for taxonomic surveys, misidentification of species [11, 12]. Naturally, an increase in number of photographs leads to increases in the total time needed to process images, and currently with the increase in camera trap studies worldwide (e.g., a SCOPUS search for scientific papers containing the word "camera trap" in the title, key words, or abstract produces 12 articles in 2002 steadily increasing to 91 in 2010), and the large number of images being uploaded to research pages of social networks, such as projects on Flickr and the iSpot identification website (reviewed in [9]), the number of images in need of identification to either species or individual level is rapidly increasing.

While patterns of markings such as spots or stripes have been shown to be unique for individuals of many species (e.g., cheetahs, [13]; zebras, [14]; whale sharks, [15]; manta rays, [16]), automatic image recognition systems are still in their infancy. Computer-aided recognition systems have been developed (e.g., numerous cetacean species, [17]; whale sharks, [15]; and even leatherback turtles, [18]), but these can still involve significant amounts of time to process an image, as well as significant time, skills, and expertise to use, modify, or develop the programs. For example, in terms of using a computer-aided identification system, a trained user would take around 10 minutes to identify an individual whale shark, using the technique developed by Arzoumanian et al. [15]. While more rapid techniques exist to identify smaller animals, such as amphibians, these techniques tend to require careful photography of the animal in artificial conditions, such as a lightbox, or while being restrained by an assistant, after the animal has been dried of excess water [19]. This problem of recognition in natural environments exists not just for wildlife, but also for human recognition systems, with far greater monetary resources for their development at their disposal [20]. Where facial or iris recognition systems have been developed successfully, they involve an individual staying still and/or facing the camera at an angle of no more than 20° [20]. Although many facial recognition systems have been deployed—for example, in airports, to detect possible terrorists—the effectiveness of these techniques has been questioned in the literature, mainly due to concerns over poor lighting, the angle at which the photograph is taken and the resolution of the face in the image [21]. All of these issues also relate to the capture of photographs of wildlife, especially from automated traps or in difficult conditions such as the photographer being in a wildlife hide or underwater. Because of this, the automatic identification of any individual (human or animal), and automatic identification of many species, from photographic records is currently best described as an area of ongoing research [19, 20].

An alternative approach for rapidly identifying individuals is the development of simple keys, akin to taxonomic keys used to identify species. For "charismatic" marine organisms, it is possible to cheaply and effectively utilise volunteers, and if keys were user-friendly and involved only limited training, then identification could be conducted from photographs accessed remotely over the internet (e.g., as done for species identification of bees, [9]).

In this paper, we examine effective mechanisms to develop simple keys to identify individual green sea turtles (*Chelonia mydas* L.) from a population in Akumal Bay, Mexico. Photographs are taken from foraging turtles, rather than nesting turtles, as occur in most studies (but see [22–24], e.g., of photo identification whilst swimming or foraging). We demonstrate that user accuracy increases with inclusion of obvious discriminatory features early on in the key. Furthermore, incorporation of statistical techniques and the removal of strict dichotomy of the keys can help minimise the number of steps involved to make an identification. Accurate identification of photographs from rapidly trained volunteers occurs if both of these processes are combined.

2. Methods

Daily snorkelling surveys between February 2nd and July 2nd, 2009 were used to capture >2500 photographs of 54 turtles in Akumal Bay, Mexico (20°394.896'N, 87°313.542'W). From these photographs, key characteristics of the head and neck, which differed between individuals, were determined

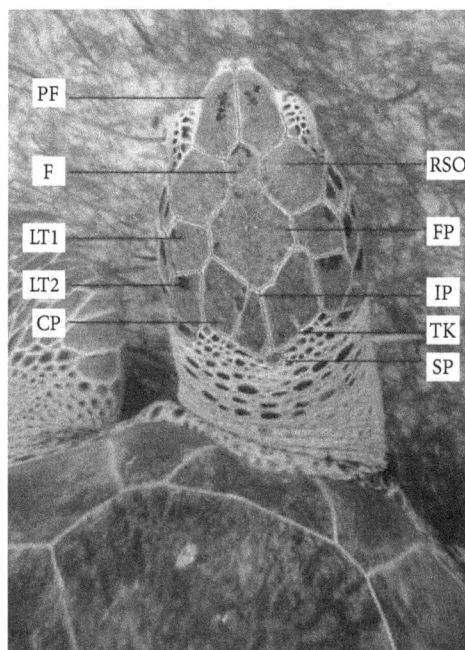

FIGURE 1: Key diagnostic features used to distinguish individual turtles from photographs of their heads. Refer to Table 1 for further details of abbreviations.

(Table 1; Figure 1) and were used as the basis for developing keys.

2.1. Minimising the Number of Steps in a Key. To develop a key with the minimum number of dichotomous steps, a matrix was developed where each turtle was listed against the classification criteria in Table 2. For each criterion either a 1 or 0 was placed for each turtle. To develop the key, the variance was calculated for each criterion (over all turtles). The highest variance corresponded to the criterion that would separate the turtles into the most equal-sized groups (i.e., had the most equal numbers of turtles with or without the characteristic). This criterion's presence or absence then became the first step in the key. By repeating the process independently for the two new subgroups of turtles divided by the first step, the next steps in the key were generated, and this process was continued until all turtles separable by the list of criteria were accounted for. Since each step splits the remaining turtles into the most equal-sized two groups, this process resulted in a minimal number of steps occurring.

2.2. Allowing the Most Obvious Factors to Be Accounted for First. To allow the most obvious distinguishing characteristics to be accounted for first in a key, but still try to minimise the number of steps taken, each characteristic was given a "priority" based on its ability to be accurately identified from a photograph (Table 2). The matrix developed above was then multiplied by this priority value for the appropriate characteristic. For example, differentiating the number of parietal scales was considered the most obvious step and was given the highest priority of 13. As such, instead of values

TABLE 1: Definitions of scale characteristics, anatomical terminology, and short hand used.

	Name	Shorthand	Definition
Characteristics	Tick	TK	Where the white boundary surrounding scales extends lineally into a scale
	Spot	SP	Small dark patch of skin, located at PB or on neck
	Dot	DOT	Lighter coloured small "dot" inside a scale
	Fronto parietal extra scale	FPxtra	Subscale located inside FP
	Interparietal meets the frontoparietal scale	IP joins FP	The IP scale meets FP
Anatomical terminology	Anterior	ANT	Towards the head
	Posterior	POS	Towards the tail
	Medial	MED	Towards the centre
	Left	L	Left
	Right	R	Right
	All	ALL	"Tick" passes through entire scale
Scales	Parietal	P	
	Interparietal	IP	
	Frontoparietal	FP	
	Frontal	F	
	Prefrontal	PF	
	Supraocular	SO	
	Temporal	T	
	Parietal base	PB	Point where left and right parietal scales join and meet the neck

TABLE 2: List of factors used to distinguish turtles in the developed keys. Priority indicates a ranking system from 1 to 13, where 13 indicates the most obvious characteristic to separate individuals, this ranking was used as a weighting factor in the development of the priority key. *For the combined key all these factors were weighted as 8.

Factor	Priority ranking (1–13)
Number of parietal scales	13
Parietal scales with tick	12
Frontoparietal scale with tick	11
Frontal scale with tick	10
Extra frontoparietal scale	9
2 temporal scales	8*
4 temporal scales	7*
5 temporal scales	6*
Prefrontal scale with tick	5
Supraocular scale with tick	4
Interparietal scale meets frontoparietal scale	3
3 temporal scales	2*
Number of spots at parietal scale base	1

of 1 or 0, this characteristic had values of either 13 or 0 for presence or absence. Using the refined matrix, the same process as above was used to develop the key. Multiplying by the "priority" value would therefore increase the variability of the characteristic across all turtles, meaning it was more likely to be selected early in the above process. From herein this is referred to as the "priority key", with the key in the section above referred to as the "nonpriority key."

2.3. Combining Steps: Nondichotomous Keys. The priority key initially involved many choices related to the number of parietal scales. Here, we combined these steps from strictly dichotomous to a choice of: for example, "how many parietal scales are present?," with possible answers ranging from 1 to 5. To avoid skewing the balance of variability (and hence order) developed in the priority key, the matrix developed for the priority key was unchanged, but the question of "how many parietal scales are present?" was asked at the point where the first occurrence of any questions relating to numbers of parietal scales occurred. These ideas were also maintained for the number of spots at the parietal base, the number of temporal scales, and the positions of frontoparietal scale ticks. Other than this, the process for developing this key—herein referred to as the "combined Key"—was identical to that of the priority key.

2.4. Testing the Efficiency and Success Rate of the Keys. The usability and time taken to identify individuals were determined for all the three keys using a group of undergraduate biosciences students. The students were given a short presentation (~30 mins) explaining the anatomy of a sea turtle's head scales. Unique characteristics of the scales and the terminology of the scale names and other scientific terms used in the keys were explained. Each test user was provided with a photographic identification guide, a table of definitions of key terms, and a reference photograph of many characteristics (as per Figure 1). Each student was required to identify turtles from photographs, using one of the three keys. In total, 27 students took part in the survey, with nine students using each key. Each key was tested with nine different groups of photographs (with each group of photographs having five images of different turtles to identify, therefore there were 45 images and in total 135 identifications made by the 27 students); hence students using the same key had different groups of photographs, and these groups of photographs were identical for each of the keys used, making the group of photographs a repeated measure in the analysis. Photographs were selected randomly, but clarity of the image and ability to see the key distinguishing features of the head and neck were ensured, since the aim of the study was to identify which keys were most user-friendly—not to show that they could distinguish individuals from a range of qualities of photograph. While each group of photographs did not contain more than one image of the same turtle, different photosets did contain different images of the same turtle. Students were asked to record their steps through the key, to determine if and when they went wrong and also to record the time taken to identify each individual. Repeated measures ANOVAs were calculated for each of the dependent variables (1) number of turtles correctly identified, (2) proportion of the way through the key before a mistake was made (arcsine transformed), and (3) time taken to identify the individuals in the five images.

3. Results

All three keys are available for visualisation online at https://public.me.com/richardstafford1.

All of the three keys produced could identify most individuals uniquely. Of the individuals the keys failed to distinguish, some individuals were consistently inseparable by all keys. The computer-generated keys could not make individual distinctions for 15 individuals, and the combination key had 17 individuals not totally separable. However, despite not being able to identify every individual, the combination key could identify all but 20 of the 2,593 photographic records of turtles obtained in the study period (a > 99% success rate for photographs obtained), hence only individuals rarely seen (i.e., photographed no more than twice) were not identified by the key. Furthermore, the turtles not identified by differences in head markings could be further separated by use of morphological differences on their carapace, although, for simplicity, these features are not included in the current key (see discussion).

The combination key had a shorter total number of steps ($n = 214$), than the priority ($n = 364$) or the nonpriority key ($n = 353$). While the combination key was shorter than the other keys, it was not fully dichotomous, thus the number of steps was reduced by combining several steps together.

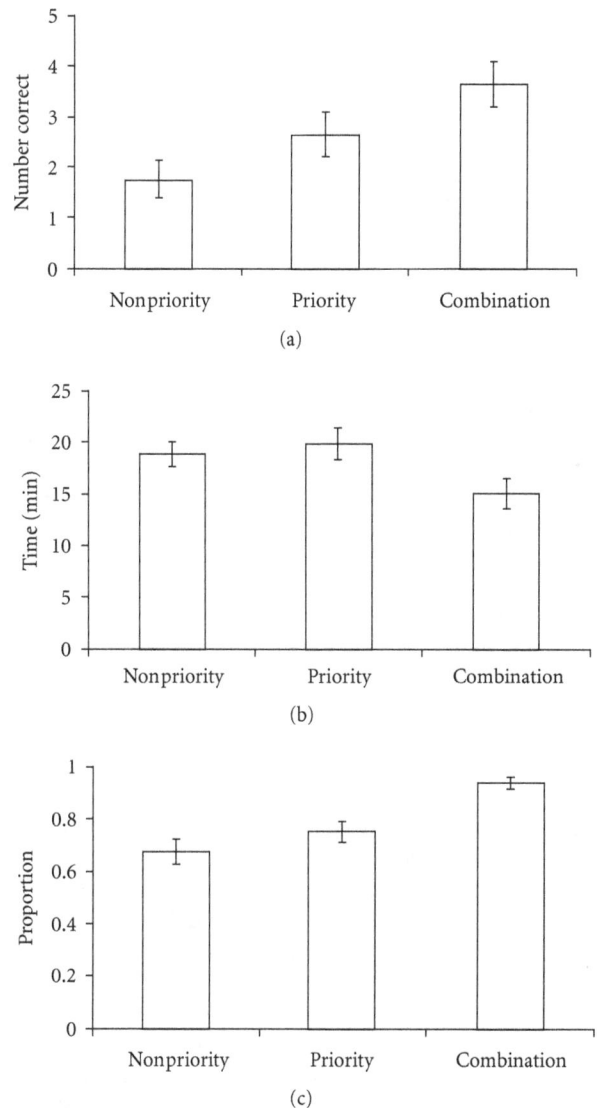

FIGURE 2: Mean ($\pm$S.E. $n = 27$) (a) success rate of correctly identifying five individuals, (b) time taken to complete five identifications, (c) proportion of key successfully navigated during identification.

The ability of students to correctly identify individuals and the proportion of the way through the key that students navigated before making mistakes both differed significantly between the different keys (Table 3); however, the time taken to identify individuals using the different keys was not significantly lower (Table 3). In terms of correct identification and proportion of the key navigated successfully, the combination key outperformed the other two keys (Figure 2). Post-hoc Tukey tests revealed that significant differences occurred between the combination key and the other two keys for both the number of correct identifications and for the proportion of the way through the key ($P < 0.05$). No significant differences were found between the priority and nonpriority keys ($P > 0.05$).

TABLE 3: Summary of repeated measures ANOVAs to identify differences between the three keys in terms of correct identifications, time taken, and the proportion of the key navigated before a mistake was made.

	d.f.	Number of correct identifications			Time taken to identify five individuals			Proportion of the key successfully navigated		
		M.S.	F	P	M.S.	F	P	M.S.	F	P
Key	2	8.04	5.13	**0.0139**	57.1	3.29	0.0545	0.748	17.5	**<0.001**
Error	24	37.56	1.56		416.7	17.36		1.028	0.042	

4. Discussion

Firstly, this study demonstrates that a large percentage of foraging turtles can be identified from photographs. While this has been previously reported (e.g., [24, 25]), these results indicate the applicability of the technique to the resident Akumal Bay population. The use of photographs creates much less disturbance to foraging turtles and eliminates problems associated with tagging [4–6, 26], such as tag loss, increased risk of infections, impaired swimming performance, or the need to capture a turtle while nesting in order to tag it [25–28]. Therefore the use of this technique may allow a greater understanding of the natural foraging behaviour of this species.

The results also demonstrate that effective keys can be made to identify turtles from photographs. Keys show an improvement in their performance (as judged by successful numbers of identifications made by end users) if (1) obvious distinguishing factors are included early on in the key and (2) the length of key is reduced, by combining steps. There was a significant difference between the performance of the combination key and the nonpriority key for all metrics studied, indicating the importance of these two improvements. For example, while the nonpriority key was itself designed to minimise the number of dichotomous steps, the mean correct identification rate by students using the keys was <50%, compared to the >80% from the combination key—with fewer steps on average and important characteristics listed first.

While the current keys cannot identify all individuals uniquely, and have a lower successful identification rate by volunteers than some previous studies (e.g., [24]), these shortcomings in fact highlight the importance of the processes undertaken in this study. The current study has focussed solely on the head and neck of turtles, rather than on more obvious characteristics (in some cases) such as shell markings. Largely, this has been done to ensure simplicity in the design of the keys by the computer-based, or semi-computer-based methods. By including features such as shell characteristics, not only can all turtles be uniquely identified, but also the success rate may be able to be improved, if some of the obvious shell characteristics were included early in the design of the key. The choice of characteristics to include in a key, and the priorities assigned to these characteristics, is clearly subjective, and will ultimately influence the structure of the keys. Given the relatively high proportion of the population which could not be separated uniquely using the keys (31%), but the low proportion of total photographs

(<1%) which could not be identified by any of the keys, it is highly probable that the choice of characteristics used to differentiate between individuals is determined by studying the most commonly seen individuals. Such an approach is suitable for small and relatively closed populations, but may not be so robust if population sizes are larger (i.e., >100 individuals), or populations are open.

It is important that the proportion of correct identifications is established, as this has implications for any subsequent work, for example, on capture mark recapture analysis. Based on genetic marker studies, where exact identification of individuals can be problematic, a successful identification rate of >94% is needed to ensure that population estimates will be reliable [29]. While this is higher than the identification rate obtained here, it is likely that, with keys using features from the entire turtle, these values could be obtained and have been in other turtle photo-identification studies [24]. Furthermore, as per the use of genetic markers, statistical processes applied to the identification may be able improve identification rates, especially for individuals that show high territoriality [30].

The methods presented here for development of a key are easy to transfer to other species, as long as a suitable list of distinguishing characteristics can be drawn up and ranked in order of the most obvious features. As such, the process may be more transferable than computer identification systems, which can require, for example, positions of spots to be inputted [13, 15] and are not directly transferable to stripes [14] or to numbers of scales [18, this study].

While it is easily possible to manually insert a few (<10% of the total population size) new individuals into appropriate positions in a key, (semi)automating the construction of the key through extraction of maximum variance, meaning that if many new individuals migrate into a population (e.g., if more than 10% individuals in a population change, either through immigration or emigration), it would be easy and not especially time consuming to reconstruct the key from scratch (i.e., by repeating the entire key creation process with all current individuals). Such a reconstruction of the key from scratch when large changes to a population occur would ensure that the key remains simple to use and continues to use the most obvious features first in its design. Moving from a paper-based key to a computer based system, as presented here, would also allow for such changes to be managed simply and cost effectively.

Since many studied populations of marine vertebrates consist of relatively small numbers of individuals, keys such as these can be very suitable techniques for behavioural studies or estimating numbers of individuals (through capture-mark-recapture methods—given the assumptions on accurate identification mentioned above can be ensured). However, if the population size of interest is very large, then keys would become very large and would need to resort to minor discriminating features to identify individuals. As such, the procedures detailed here are best for examining smaller populations (<100 individuals).

While fully automated computer identification methods for species or individuals should be developed, for example, similar to the fully automated astronomy recognition system available on Flickr (http://www.flickr.com/groups/astromery), in the mean time, user-friendly keys could be an important step in allowing effective processing of "citizen science" collected photographic data. While individuals of some species (e.g., species with characteristic spots such as leopards, cheetahs or even manta rays) can be identified solely by these characteristics [13, 16], other species, such as sea turtles, require a number of characteristics to be used for individual identification. As such, development of identification keys—where key characteristics are identified first—would be a useful step in creating automatic identification systems for these animals, akin to the feature-matching applications of face recognition technology [20].

References

[1] P. T. Stevick, P. J. Palsbøll, T. D. Smith, M. V. Bravington, and P. S. Hammond, "Errors in identification using natural markings: rates, sources, and effects on capture-recapture estimates of abundance," *Canadian Journal of Fisheries and Aquatic Sciences*, vol. 58, no. 9, pp. 1861–1870, 2001.

[2] C. W. Speed, M. G. Meekan, and C. J. A. Bradshaw, "Spot the match: wildlife photo-identification using information theory," *Frontiers in Zoology*, vol. 4, article 2, 2007.

[3] S. Gowans, B. Würsig, and L. Karczmarski, "The social structure and strategies of delphinids: predictions based on an ecological framework," *Advances in Marine Biology*, vol. 53, pp. 195–294, 2007.

[4] M. R. Rice, G. H. Balazs, and L. Hallacker, "Diving, basking and foraging patterns of a sub-adult green turtle at Punalu'u Hawaii," in *Proceedings of the 18th International Symposium on Sea Turtle Biology and Conservation Mazatlán*, F. A. Abreu-Grobois, R. Briseño, R. Márquez, and L. Sarti, Eds., pp. 229–231, Mexico, Mexico City, 2004.

[5] G. H. Balazs, "Factors affecting the retention of metal tags on sea turtles," *Marine Turtle Newsletter*, vol. 20, pp. 11–14, 1982.

[6] T. A. Henwood, "Losses of Monel flipper tags from Loggerhead sea turtles, *Caretta caretta*," *Journal of Herpetology*, vol. 20, pp. 276–279, 1986.

[7] D. M. Aanensen, D. M. Huntley, E. J. Feil, F. Al-Own, and B. G. Spratt, "EpiCollect: linking smartphones to web applications for epidemiology, ecology and community data collection," *PLoS ONE*, vol. 4, no. 9, Article ID e6968, 2009.

[8] C. L. Kirkhope, R. L. Williams, C. L. Catlin-Groves et al., "Socialnetworking for biodiversity: the BeeID project," in *Proceedings of the iSociety Conference*, C. A. Shoniregun, Ed., pp. 637–638, London, UK, 2010.

[9] R. Stafford, A. G. Hart, L. Collins et al., "Eu-social science: the role of internet social networks in the collection of bee bio-diversity data," *PLoS ONE*, vol. 5, no. 12, Article ID e14381, pp. 1–7, 2010.

[10] T. L. Cutler and D. E. Swann, "Using remote photography in wildlife ecology," *Wildlife Society Bulletin*, vol. 27, no. 3, pp. 571–581, 1999.

[11] K. S. Genet and L. G. Sargent, "Evaluation of methods and data quality from a volunteer-based amphibian call survey," *Wildlife Society Bulletin*, vol. 31, no. 3, pp. 703–714, 2003.

[12] J. L. Dickinson, B. Zuckerberg, and D. N. Bonter, "Citizen science as an ecological research tool: challenges and benefits," *Annual Review of Ecology, Evolution, and Systematics*, vol. 41, pp. 149–172, 2010.

[13] M. J. Kelly, "Computer-aided photograph matching in studies using individual identification: an example from Serengeti cheetahs," *Journal of Mammalogy*, vol. 82, no. 2, pp. 440–449, 2001.

[14] G. Foster, H. Krijger, and S. Bangay, "Zebra fingerprints: towards a computer-aided identification system for individual zebra," *African Journal of Ecology*, vol. 45, no. 2, pp. 225–227, 2007.

[15] Z. Arzoumanian, J. Holmberg, and B. Norman, "An astronomical pattern-matching algorithm for computer-aided identification of whale sharks *Rhincodon typus*," *Journal of Applied Ecology*, vol. 42, no. 6, pp. 999–1011, 2005.

[16] A. M. Kitchen-Wheeler, "Visual identification of individual manta ray (*Manta alfredi*) in the Maldives Islands, Western Indian Ocean," *Marine Biology Research*, vol. 6, no. 4, pp. 351–363, 2010.

[17] G. R. Hillman, B. Würsig, G. A. Gailey et al., "Computer-assisted photo-identification of individual marine vertebrates: a multi-species system," *Aquatic Mammals*, vol. 29, pp. 117–123, 2003.

[18] E. J. Pauwels, P. M. de Zeeuw, and D. M. Bounantony, "Leatherbacks matching by automated image recognition," *Lecture Notes in Computer Science*, vol. 5077, pp. 417–425, 2008.

[19] O. V. Kelly, *Automated digital individual identification system with application to the northern leopard frog Lithobates pipiens*, Ph.D. thesis, Idaho State University, Pocatello, Idaho, USA, 2010.

[20] S. Z. Li and A. K. Jain, *Handbook of Face Recognition*, Springer, London, UK, 2nd edition, 2011.

[21] K. Gates, *The Past Perfect Promise of Facial Recognition Technology*, ACDIS, University of Illinios, Ill, USA, 2004.

[22] G. Schofield, K. A. Katselidis, and J. D. Pantis, "Assessment of photo-identification and GIS as a technique to collect in-water information about loggerhead sea turtles in Laganas Bay, Zakynthos Greece," in *Proceedings of the 24th Annual Symposium on Sea Turtle Biology & Conservation*, San Jose, Costa Rica, 2004.

[23] M. G. White, *Marine ecology of loggerhead sea turtles Caretta caretta (Linnaeus, 1758) in the Ionian Sea: observations from Kefalonia and Lampedusa*, Ph.D. thesis, University College Cork, Ireland, UK, 2007.

[24] G. Schofield, K. A. Katselidis, P. Dimopoulos, and J. D. Pantis, "Investigating the viability of photo-identification as an objective tool to study endangered sea turtle populations," *Journal of Experimental Marine Biology and Ecology*, vol. 360, no. 2, pp. 103–108, 2008.

[25] J. Reisser, M. Proietti, P. Kinas, and I. Sazima, "Photographic identification of sea turtles: method description and validation, with an estimation of tag loss," *Endangered Species Research*, vol. 5, no. 1, pp. 73–82, 2008.

[26] K. A. Bjorndal, D. Bolten, C. Lagueux, and A. Chaves, "Probability of tag loss in green turtles nesting at Tortuguero, Costa Rica," *Journal of Herpetology*, vol. 30, no. 4, pp. 567–571, 1996.

[27] D. Buonantony, *An analysis of utilizing the Leatherback's pineal spot for photo-identification*, M.S. thesis, Duke University, NC, USA, 2008.

[28] G. H. Balazs, "Factors to consider in the tagging of sea turtles," in *Research and Management Techniques for the Conservation of Sea Turtles*, K. L. Eckert, K. A. Bjorndal, F. A. Abreu-Grobois, and M. Donnelly, Eds., pp. 101–109, IUCN/SSC Marine Turtle Specialist Group, 1999.

[29] E. Petit and N. Valiere, "Estimating population size with non-invasive capture-mark-recapture data," *Conservation Biology*, vol. 20, no. 4, pp. 1062–1073, 2006.

[30] R. Stafford and J. R. Lloyd, "Evaluating a Bayesian approach to improve accuracy of individual photographic identification methods using ecological distribution data," *Computational Ecology and Software*, vol. 1, no. 1, pp. 49–54, 2011.

9

The Maryland Amphibian and Reptile Atlas: A Volunteer-Based Distributional Survey

Heather R. Cunningham,[1] Charles A. Davis,[1] Christopher W. Swarth,[2] and Glenn D. Therres[3]

[1] *The Natural History Society of Maryland, P.O. Box 18750, Baltimore, MD 21206, USA*
[2] *Jug Bay Wetlands Sanctuary, 1361 Wrighton Road, Lothian, MD 20711, USA*
[3] *Wildlife and Heritage Service, Maryland Department of Natural Resources, 580 Taylor Avenue, Annapolis, MD 21401, USA*

Correspondence should be addressed to Glenn D. Therres, gtherres@dnr.state.md.us

Academic Editor: Richard Stafford

Declines of amphibian and reptile populations are well documented. Yet a lack of understanding of their distribution may hinder conservation planning for these species. The Maryland Amphibian and Reptile Atlas project (MARA) was launched in 2010. This five-year, citizen science project will document the distribution of the 93 amphibian and reptile species in Maryland. During the 2010 and 2011 field seasons, 488 registered MARA volunteers collected 13,919 occurrence records that document 85 of Maryland's amphibian and reptile species, including 19 frog, 20 salamander, five lizard, 25 snake, and 16 turtle species. Thirteen of these species are of conservation concern in Maryland. The MARA will establish a baseline by which future changes in the distribution of populations of native herpetofauna can be assessed as well as provide information for immediate management actions for rare and threatened species. As a citizen science project it has the added benefit of educating citizens about native amphibian and reptile diversity and its ecological benefits—an important step in creating an informed society that actively participates in the long-term conservation of Maryland's nature heritage.

1. Introduction

Amphibian and reptile species are among the most threatened groups of vertebrate animals [1, 2]. Factors that lead to population declines are habitat alteration and loss, invasive species, disease, environmental pollution, commercial collection, and climate change [1, 3]. The lack of thorough understanding of regional distribution patterns of amphibian and reptile populations can limit our ability to predict how species will respond to these factors [4]. An additional challenge to the protection and conservation of amphibian and reptile species (also called herps) is the overall negative perception by the public towards these organisms [5]. There is a pervasive attitude that these organisms are unimportant [5]. However citizen science projects, defined as projects where citizens participate in scientific research [5], have the potential to advance the protection of amphibian and reptile species. Specifically, citizen science-based atlas projects can efficiently assemble

distribution information across large spatial scales while increasing environmental awareness in the general public about the ecological importance of herpetofauna. Through participation in atlas projects citizens play an important role in the long-term protection and conservation of amphibians and reptiles.

Currently 93 native species of amphibians and reptiles occur in Maryland (20 anurans, 21 salamanders, 27 snakes, 19 turtles, and six lizards). The diversity of native herpetofauna is, in part, an outcome of the three physiographic provinces in Maryland: Appalachian, Piedmont, and Coastal Plain. Some species are restricted to particular provinces. For example, the Red-bellied Watersnake (*Nerodia erythrogaster*) and Eastern Tiger Salamander (*Ambystoma tigrinum*) are restricted to the Coastal Plain, and the Mountain Earthsnake (*Virginia valeriae pulchra*) and Green Salamander (*Aneides aeneus*) are found exclusively in the Appalachian Province. In 1975 Harris Jr., [6] compiled the most recent set of comprehensive maps for Maryland herpetofauna.

That publication maps the historic distribution of reptile and amphibian species in Maryland from the early 1900s through the mid 1970s using sightings and locality records of specimens collected and held in private collections, universities, museums, and with the Natural History Society of Maryland [7].

Since the Harris publication [6], Maryland has become much more urbanized, and natural lands have diminished. Additionally, land management practices, animal disease distribution, water pollution abatement practices, or climate have altered habitat suitability and population fitness of Maryland's herpetofauna. For example, researchers in Maryland found that particular timber management practices, specifically cutting and burning of small patches of forest can result in decreased local diversity of herpetofauna [8]. The amphibian chytrid fungus, *Batrachochytrium dendrobatidis*, which is responsible for the disease chytridiomycosis, now occurs in the Chesapeake and Ohio Canal National Historic Park, where researchers documented this pathogen in two species of stream-associated salamanders [9]. Changes to climate will inherently alter the hydrologic cycle [10]—a direct concern to habitat quality of regional herpetofauna. Further, Maryland researchers observed that in urban environments amphibians are attracted to stormwater retention ponds where they become exposed to trace metal contamination from accumulated runoff [11]. In addition to changing environmental conditions, relative to other states in the United States, Maryland ranks in the top 25th percentile for number of native amphibian and reptile species at risk of extinction [12]. Maryland has 11 species of reptiles and eight amphibians that have state conservation status of endangered, threatened, or in-need-of-conservation [13]. All of these factors reinforce the critical need to understand the current distribution patterns of amphibian and reptile species within the state.

Ecological atlases are one of the many types of citizen science projects that have been successfully conducted throughout the world [14]. Atlases show the distributions of organisms such as birds, butterflies, or plants for a given geographic area. Specifically, they are presence-only data sets of spatially explicit species occurrence data [15]. Most atlases use a predefined grid for sampling, employ a sampling protocol, have minimum requirements for the submission of data [16], and rely on citizen scientists to collect the data. The experience level of atlas volunteers can range from amateurs, with little scientific training, to scientists who specialize in the species of interest.

Atlas projects ultimately produce maps of the distribution of focal species. Repeated atlas efforts and comparisons can detect species' distributional shifts. Atlas data can inform scientists about these shifts and provide important information for focused studies on the causes. The first ecological atlas using a systematic approach to collecting field data was conducted in the United Kingdom for plants [17, 18]. In North America, many breeding bird atlases have been conducted in accordance with guidance provided by the North American Breeding Bird Atlas Committee [19] and subsequently published. Additionally, atlases for reptiles and amphibians have been completed or are in progress in several

states in North America (see review [4]). In recent years, the methodology for citizen involvement and data collection for atlas projects has evolved with advances in new technology (e.g., the internet, social media, and digital photography). Current projects use the internet extensively as a means to gather and distribute information [4].

The Maryland Amphibian and Reptile Atlas project (MARA) is a citizen science project with the goal to document the current distribution of all amphibian and reptile species in Maryland using a systematic and repeatable approach during the five-year period, 2010–2014. Another important goal is to provide current information on the location and status of rare or threatened species. MARA data will inform management strategies for the immediate and long-term conservation and protection of Maryland's herpetofauna. Surveying large areas, such as an entire state, requires significant volunteer assistance and provides an opportunity to recruit and train novice participants, and thereby raises general ecological literacy by increasing awareness, skills, understanding, and knowledge of the natural world.

The Maryland Department of Natural Resources (MDNR) and the Natural History Society of Maryland (NHSM) cosponsor the project. The MDNR is the chief government agency responsible for conservation of the herpetofauna of the state. The NHSM, established in 1929, supports the community of amateur and professional naturalists within the state and during the intervening years has organized and published research on Maryland herpetology.

2. Approach

Conducting the atlas project entails four principal challenges: defining the survey methods, preparing data handling strategies, recruiting volunteers, and managing the volunteer network. The methodological foundation for MARA is based on herpetofauna atlases conducted in other states and two Maryland breeding bird atlases: 1983–1987 and 2002–2006 [4, 20–22]. Volunteer recruitment and management today can take advantage of online social networking to establish and maintain the research community for the project. We are implementing the atlas project by using these strategies.

2.1. The Atlas Survey Grid. Building an atlas on a grid base helps to meet the objective of using a systematic and repeatable method. The MARA uses a grid based on US Geological Survey 7.5-minute topographic quadrangle maps (called quads) divided into six equal blocks (Figure 1). The blocks are designated as northeast, northwest, central east, central west, southeast, and southwest. Each atlas block covers approximately 25 km². The state of Maryland includes 1,300 of these blocks within all or portions of 260 quads. This is the same grid system that was used successfully in the two previous breeding bird atlases conducted in Maryland [20, 21].

2.2. Search Effort Targets. In 2009, prior to initiating the MARA project, a pilot study was conducted in one Maryland

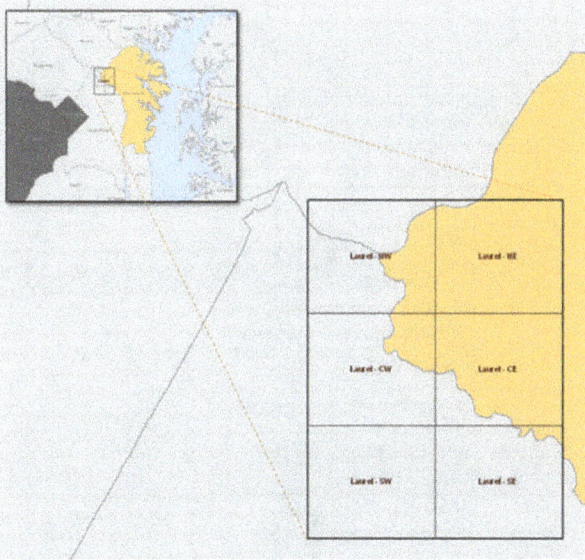

FIGURE 1: Example of a US Geological Survey 7.5-minute topographic quadrangle map divided into six atlas blocks.

county to develop the procedures and goals [23]. The pilot study helped to determine possible discovery rates of organisms per unit of survey effort. Providing an estimate of search effort is one way of standardizing results, assuming that the relationship between sampling effort and sample size is consistent [16]. In an earlier Maryland study, researchers documented that time-area searches were as or more effective than intensive trapping methods for documenting herpetofaunal presence [24]. To help assure dispersed geographic coverage across the state, the MARA Steering Committee established two goals for adequate coverage based on number of species discovered (at least ten species per atlas block and 25 species per quad) and the amount of time spent actively searching (at least 25 hours of active searching within each quad). Once these thresholds are reached in a block or quad, then surveyors should move to another less thoroughly searched area. In some blocks, which are highly urbanized, or dominated by one habitat type (i.e., Chesapeake Bay), reaching the ten species threshold may not be possible. In those instances, once the quad thresholds are met, surveyors are encouraged to move to a new quad. Cumulative time spent searching in a particular block is captured within the database as a record of effort for future comparisons. These guidelines balance the probable species numbers to be detected with the expected volunteer capacity to search the state within the project interval.

2.3. *Time Period of Survey.* Deciding the interval of the survey is strictly an estimate of the time to complete the effort based on the expected recruitment and effort of volunteers, as well as the anticipated detection rate of the herps. Ideally, this is the narrowest time interval possible so that the results will be least affected by any changes in distribution that may happen during the survey period. We chose a five-year period based on assessments of other state herpetological atlases

(e.g., for Maine [25]), experiences from the pilot study, and experience with five-year survey periods used during the two Maryland bird atlases [20, 21].

2.4. *Acquiring the Data.* Data collected during active searching or incidental observations are recorded on a standard data sheet. Among the information recorded on the data sheet is locality and observer information; additionally the data sheet contains a checklist of all native species in the state by common names based on Crother et al. [26]. Surveyors complete one data sheet per year for each atlas block surveyed. Herpetological occurrence records are obtained by fundamentally two approaches: "active searching" or "incidental observations." Active searching is the main source of atlas data and involves intentional looking for reptiles and amphibians. There are no standardized methods used during active searching. Data are collected through listening surveys for calling frogs and toads, searching various habitats, turning over logs and cover boards, scanning ponds for turtles, turning over rocks along streams, and so forth.

We measure active searching by recording the amount of time a surveyor conducts searches in a given atlas block on each date of survey work. Incidental observations are sightings that observers make when they are not engaged in a formal, active searching survey. Incidental records are added to the database just as active searches, except no survey time is noted.

2.5. *Additional Identity Verification Information.* For most common species, no additional information is required although photographs for each occurrence record are encouraged similar to procedures of the Carolina Herp Atlas [4]. However, verification is required for certain species and surveyors complete an additional data form for those observations. Verification is required for rare species to obtain exact location information or for difficult-to-identify species so that we can be certain that identifications are accurate and correct. For example, photographic or written verification is required for Common Five-lined Skink (*Plestiodon fasciatus*) and Broad-headed Skink (*P. laticeps*) which are closely-related species and require close inspection of head scales to differentiate them. Similarly, the two species of gray treefrogs (*Hyla versicolor* and *H. chrysoscelis*) require vocal recordings because these identical species can only be distinguished by their calls. Verification is required for all records of eggs or larvae of any species. Photodocumentation is the main practice to verify species identity, but written documentation can be considered. In addition to those species for which verification is required, we encourage photodocumentation of each species found within a quad. Employing verification reduces misidentification of species, a serious error that can occur in citizen scientist projects [16].

2.6. *Verification Committee.* All photos, audio recordings of anuran calls, and verification forms are reviewed by a verification committee who are experienced with Maryland's herpetofauna (Figure 2). Photos and audio recordings are

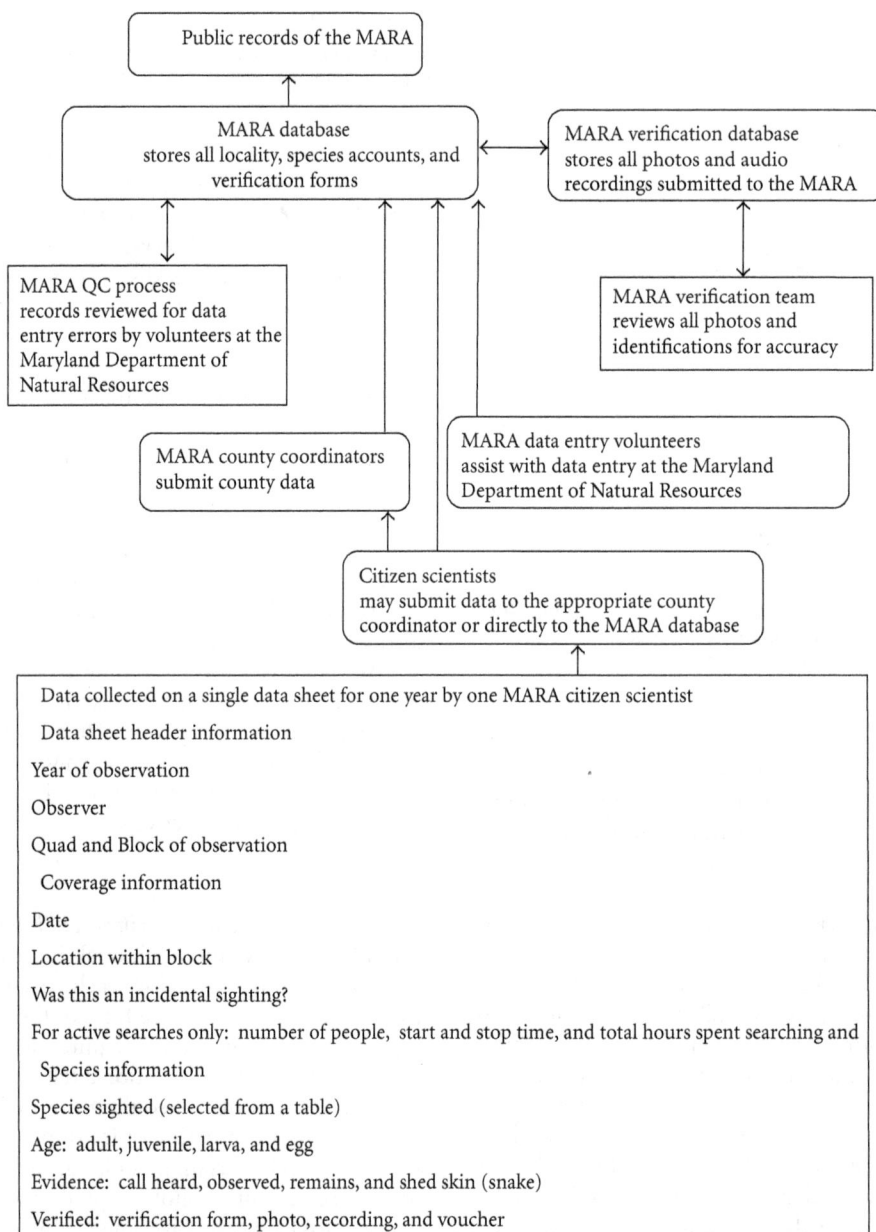

FIGURE 2: Data recorded and flow through the Maryland Amphibian and Reptile Atlas from collection to public release.

reviewed independently by each member of the committee via an online system. Only the photo or sound recording is provided to the reviewer with no additional information (no species name or observer name), except the county of occurrence. If the first three committee members to vote agree on the species identity, the record is confirmed; otherwise, the record is placed in a process to resolve the disputed record. Records are labeled "confirmed," if sufficient evidence was provided to the verification committee; "accepted" if no verification was required; or "unconfirmed" if the species required verification but none was provided or the verification committee could not positively determine the species' identity.

2.7. Database. Data are managed through a central MARA database, which was developed and maintained by the MDNR (https://webapps02.dnr.state.md.us/mara/default .aspx). Surveyors can enter data and submit verification photographs through online access to the database (Figure 2). County coordinators can access summary statistics here, view additional data summaries, and grant access for surveyors. Similarly, the verification committee accomplishes their duties through an associated portal to the submitted photographs. The central database is updated regularly with the decisions of the verification committee. Summary tables of the data are available on line to the volunteers, county coordinators, steering

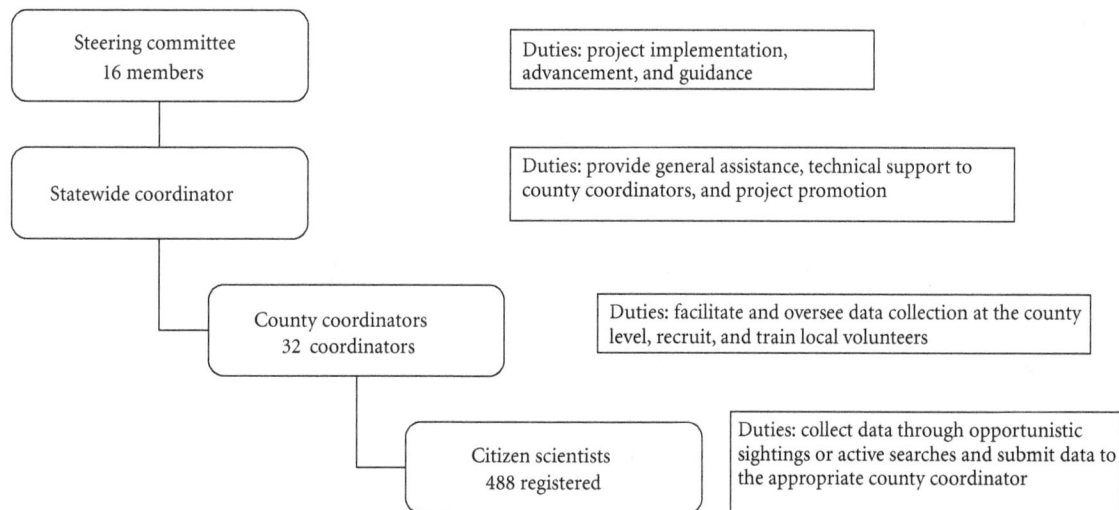

FIGURE 3: Organizational structure of the Maryland Amphibian and Reptile Atlas project.

committee members, and the public. This real-time data helps volunteers and coordinators plan for the field season and check on previous records. Allowing volunteers to examine the collected data is an important educational aspect of citizen science [27]. The MARA database will be used to produce distribution maps for all the reptile and amphibian species documented during the atlas project.

2.8. State Project Steering Committee. Recruiting, organizing, and mobilizing volunteers to achieve the project goals and objectives required oversight at statewide and local levels (Figure 3). Oversight of the MARA project is guided by a steering committee of 16 members. Successful citizen science projects require a development team comprising multiple partners and disciplines [27]. Three members cochair the steering committee, including representatives of the two primary project sponsors. The initial function of the steering committee was to develop the protocols of the distributional survey and recruit volunteer coordinators. As the atlas project got underway, the role of the steering committee shifted to project implementation and resolving technical issues. The steering committee meets monthly.

2.9. State Project Coordinator. To assist the county coordinators with recruiting, training, and motivating volunteers, and to develop strategies for collecting field data, a statewide coordinator was hired in August 2010. The statewide coordinator also conducts outreach efforts to promote the project, and to recruit and train additional volunteers. Additionally, the statewide coordinator produces a monthly project newsletter, educational materials relevant to the MARA project, and maintains the MARA website and social networking site. Federal wildlife grant money provided to the MDNR has been used to fund the statewide coordinator through a contract with the NHSM.

2.10. County Coordinators. Maryland contains 23 counties, and the MARA project chose to organize field data collection at this level. The county coordinators are integral to the success of the statewide effort. Coordinators work at the local level to recruit volunteers and to oversee the collection of field data within their counties. One to two coordinators were recruited for each county.

The county coordinators employ various strategies to achieve adequate coverage within their counties. They all recruit and coordinate volunteers to collect field data. A few counties are coordinated by employees of county park and recreation agencies that have access to networks of volunteers interested in reptiles and amphibians. These coordinators rely heavily on that network to obtain data. Other coordinators rely on a few dedicated volunteers to collect most of the field data within their county.

For the county coordinators, an annual meeting has been held each February beginning in 2010 to prepare for the upcoming field season. Topics of discussion at these meetings include how to recruit volunteers, strategies for achieving adequate coverage of blocks and quads, field techniques for finding reptiles and amphibians, and success stories presented by some of the county coordinators themselves. This meeting is an important annual check on progress of the project and a rally for the upcoming field season.

2.11. Recruiting Volunteers. Volunteers were recruited to the MARA project using various methods. Articles describing the MARA project and the need for volunteers were published in newspapers and nature club newsletters. Volunteers were also recruited at wildlife and nature festivals hosted at nature centers. MARA information was displayed before organizations such as the Maryland Association of Environmental and Outdoor Education in 2011 and 2012. Additionally, the two previous Maryland breeding bird atlases provided an existing network of citizen scientists experienced in atlas methods from which to recruit volunteers for the MARA through appeals at the Maryland Ornithological Society annual conferences in 2011 and 2012 and articles in their newsletter. Volunteers are also recruited through Facebook and Volunteer Match (http://www.volunteermatch.org/).

2.12. Training Volunteers. To aid the volunteers, several resources were developed to explain data collection and to ensure that data are assigned to the proper atlas block and quad. A handbook [28] was developed and provided to volunteers. The handbook explains the purpose of the project, grid system, data forms, techniques for finding reptiles and amphibians, and recommended references for species identification. The handbook also discusses health and safety precautions that are relevant to amphibian and reptile surveys. Measures for protecting habitat, animals, and surveyors are discussed in the handbook, covered at MARA training sessions and in the monthly project newsletter.

Paper copies of atlas blocks and a digital overlay of the grid system for use with Google Earth were made available to the volunteers. These and other resources are available on the project website (http://marylandnature.org/mara/) to guide data collection. In addition, county coordinators frequently held training sessions at the beginning of each field season to offer hands-on or technical training to volunteers. In some counties, coordinators held public hikes during which volunteers gained practical training while actively surveying for amphibians and reptiles.

2.13. Retaining Volunteers. As a means of retaining volunteers, two outreach products were developed. A monthly newsletter was initiated in November 2010. The newsletter encourages submissions from MARA participants. Photographs of interesting amphibian and reptile species encountered are highlighted in the newsletter in addition to accounts of experiences of MARA volunteers in the field. When county coordinators receive the newsletter by email, they send it to their local volunteers, and the newsletters are available to the general public through the project website. We established a Facebook page (https://www.facebook.com/MDHerpAtlas) in October 2011 to provide a forum for online exchange of project information and to encourage volunteers to communicate with each other. Volunteers are invited to post photographs and amphibian and reptile sighting information on the Facebook page.

3. Preliminary Results

3.1. Volunteer Recruitment. From January 1, 2010 to February 22, 2012, 488 citizen scientists have registered with the MARA database and contributed observations. This is a conservative estimate of the total number of actual contributors to the MARA, since this includes neither landowners that have granted access to private property nor those who assisted registered volunteers with active searching.

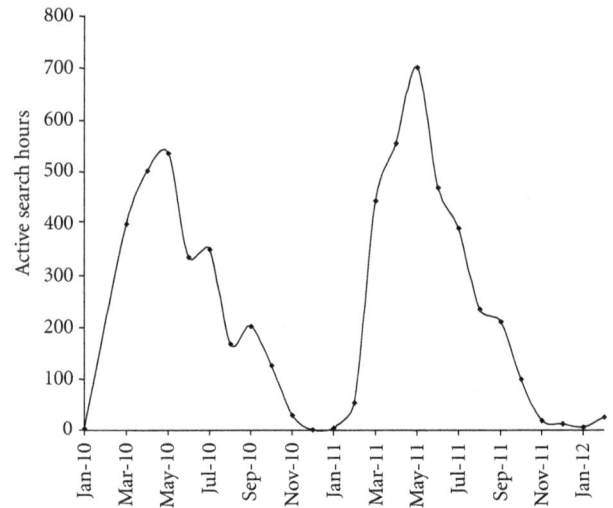

FIGURE 4: Survey hours for the Maryland Amphibian and Reptile Atlas.

3.2. Observation Counts. Surveyors have reported 13,919 occurrence records. This total includes 457 records from one county that were generated during the 2009 pilot study [23]. The mean number of occurrence records submitted per volunteer was 28.52 ± 106.13 (mean $\pm$ standard deviation, SD) with about 21% (5.89 ± 20.72) of the records being incidental observations (Table 1). Some observers contribute many more sightings than others. For instance, three observers submitted over 1,000 records each, five submitted between 500 and 1,000 each, and 39 submitted between 50 and 500 records. Sixty-nine percent of the volunteers submitted fewer than ten records each, and of those volunteers 186 submitted just one observation. Eighty percent of the records were contributed by 10% of the surveyors.

3.3. Time Effort Progress. Fifty-two percent (128/246) of quads have reached the minimum coverage goals of 25 species or 25 survey hours. Registered and unregistered volunteers have reported a total of 12,671 survey hours, but the number of survey hours reported per quad and block varied (Table 2). For instance, the total number of hours spent actively searching by all surveyors in particular quads ranged from 0 to 649 hours. The mean number of survey hours, or active searching hours, per registered participant was 12.48 ± 42.80 hours (Table 1). Survey effort occurred primarily throughout the spring and summer months, although records have been reported from every month of the year (Figure 4).

3.4. Areal Coverage Progress. Surveyors have collected data from 232 quads and 1,302 blocks (Figure 5). Thirty percent (74/246) of the quads have reached the minimum coverage goal of 25 species. Overall, surveyors observed 20.43 ± 10.44 and 8.27 ± 7.82 species per quad and block, respectively (Table 2). Forty-one percent (542/1,302) of the blocks have reached the minimum coverage goal of 10 species. On

TABLE 1: Number of occurrence records and search effort of registered MARA volunteers.

Volunteer contribution	Mean (per volunteer)	Mode	Frequency of mode	Minimum	Maximum	SD
Records submitted	28.52	1	186	0.00	1069.00	106.13
Opportunistic records	5.89	1	224	0.00	280.00	20.72
Active search hours	12.49	0	302	0.00	389.75	42.80
Quads surveyed	2.98	1	334	1.00	82.00	6.92
Blocks surveyed	5.53	1	316	1.00	284.00	18.76

TABLE 2: Species observed and person-hours of effort for blocks and quads.

Assessment	Count (quads/blocks)	Mean	Minimum	Maximum	SD
Species reported per quad	232	20.34	1	48.00	10.44
Species reported per block	1301	8.27	0	43.00	7.82
Person-hours per quad	246	51.51	0	649.00	79.55
Person-hours per block	1301	9.74	0	522.50	24.41

TABLE 3: Mean number of presence records per class/order per quad. Number of quads = 232.

Class/order	Mean	SD
Amphibia/Anura	27.43	26.03
Amphibia/Caudata	6.59	8.35
Reptilia/Squamata-Lacertilia	1.73	3.65
Reptilia/Squamata-Serpentes	12.62	13.70
Reptilia/Testudines	11.42	11.37

average, each registered participant collected data in 2.98 ± 6.92 quads and 5.53 ± 18.76 blocks (Table 1).

3.5. Species Detection Progress. Eighty-five of Maryland's 93 native amphibian and reptile species have at least one occurrence record in the MARA database. Distributional data have been recorded for 19 frog, 20 salamander, five lizard, 25 snake, and 16 turtle species. Volunteers also located 12 nonnative amphibian and reptile species. Anuran records comprise 45% (6,371/13,919) of the records in the database (Figure 6). The mean number of anuran records in surveyed quads was 27.43 ± 26.03 (Table 3). The second most commonly sighted group per quad was snakes, with salamanders and lizards sighted less frequently per quad. The most commonly reported species, per taxonomic order, were Spring Peeper (*Pseudacris crucifer*; 946 records), Red-backed Salamander (*Plethodon cinereus*; 431 records), Common Five-lined Skink (*Plestiodon fasciatus*; 212 records), Eastern Ratsnake (*Pantherophis alleghaniensis*; 617 records), and Eastern Box Turtle (*Terrapene carolina*; 665 records; Figure 3). Several records have also been collected on species listed as threatened or endangered in Maryland (Table 4). For example, volunteers submitted occurrence records for the state endangered Mountain Earthsnake (*Virginia valeriae pulchra*) from two previously undocumented locations. In addition, volunteers submitted two records for the state endangered Rainbow Snake (*Farancia erytrogramma*).

Voucher photographs or audio recordings accompanied 37% (5,203) of the submitted records. Through February 22, 2012, the verification committee has reviewed 4,406 records and determined that only 4% of the submitted records were misidentified.

4. Discussion

Public participation in the MARA resulted in a total of 13,919 occurrence records, in just 25 months. The MARA compares well with other successful herpetofauna atlases including the Georgia Herp Atlas [29] which collected 7,452 records during five years and the Carolina Herp Atlas [4] (currently underway) which collected 11,663 records during its first 36 months.

Though the number of registered participants was nearly 500 through the first two years of the MARA project, the majority of species records were submitted by less than 10% of the volunteers. This is consistent with the Carolina Herp Atlas [4]. However, the value of the few records submitted by the majority of the volunteers is also important. Their discoveries included difficult to find species, including salamanders and small snakes, and most of the nonnative species. Typically, volunteers participate in citizen scientist projects because they enjoy being able to put their skills to use in searching for and identifying species of the target group [16]. However, finding enough of these volunteers throughout the project area is the challenge in any widespread atlas effort.

The MARA project currently has a solid volunteer corps who freely contribute to the project by serving as county coordinators, steering committee members, and field workers. However, retaining volunteers for the duration of the project requires regular communication between the project management team and data collectors. This is important to ensure success of the project. Regular updates on progress are communicated via the project website, monthly electronic newsletter, and a social networking website. This communication is vital to ensure the participants remain interested and feel that the data being collected are being used [16]. The monthly newsletter and social networking site

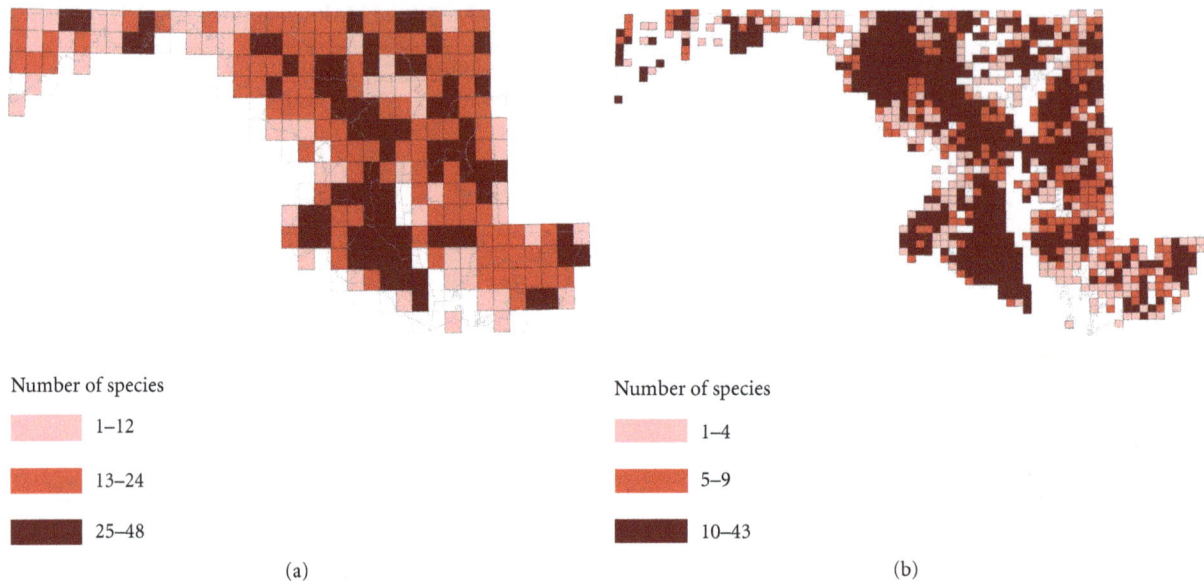

Number of species

	1–12
	13–24
	25–48

(a)

Number of species

	1–4
	5–9
	10–43

(b)

FIGURE 5: Results of the Maryland Amphibian and Reptile Atlas as of February 20, 2012. The number of species recorded in each quad (a) and block (b) is shown in a categorized series.

TABLE 4: Number of occurrence records of amphibian and reptile species submitted to the MARA that the state of Maryland lists as endangered (E) or threatened (T).

Class/order	Species	State status	Number of records
Amphibia/Anura	Barking Treefrog (*Hyla gratiosa*)	E	2
	Eastern Narrow-mouthed Toad (*Gastrophryne carolinensis*)	E	1
Amphibia/Caudata	Eastern Hellbender (*Cryptobranchus alleganiensis*)	E	1
	Eastern Tiger Salamander (*Ambystoma maculatum*)	E	5
	Green Salamander (*Aneides aeneus*)	E	5
Reptilia/Squamata-Serpentes	Mountain Earthsnake (*Virginia valeriae pulchra*)	E	3
	Rainbow Snake (*Farancia erytrogramma*)	E	2
Reptilia/Testudines	Bog Turtle (*Glyptemys muhlenbergii*)	T	27
	Green Sea Turtle (*Chelonia mydas*)	T	1
	Kemp Ridley Sea Turtle (*Lepidochelys kempii*)	E	5
	Leatherback Sea Turtle (*Dermochelys coriacea*)	E	1
	Loggerhead Sea Turtle (*Caretta caretta*)	T	14
	Northern Map Turtle (*Graptemys geographica*)	E	3

have been valuable tools to retain volunteers and increase communication among them.

To achieve consistent coverage, many atlas programs have standard benchmarks based on the number of species sampled within a given block or the hours of effort [14]. The MARA has set minimum coverage goals of 25 species or 25 survey hours per quad and 10 species per block. Currently, the MARA is on course to achieve these goals statewide by the end of 2014. We have achieved the minimum coverage goal of 25 active search hours within 52% of quads and the coverage goal of at least 10 species per block in 41% of blocks. With repeated atlases, often spanning twenty years or more, effort and the change in effort for individual blocks are crucial sources of variation that should be accounted for. When programs have no requirement for minimum effort,

studies may be biased, resulting in overreporting of rare species, underreporting of common species, and failure to report repeated sightings because they are not deemed as interesting by the observer [14]. Further, people may simply stop sampling when there are no interesting organisms to be seen.

Atlasing efforts play an important role in biodiversity conservation by providing essential data on the occurrence of species [16]. The majority of records are anuran, most likely because anurans can be detected by sight and sound. To date, salamander and lizard records are not well represented in the MARA. The cryptic nature of salamanders and strict seasonal activity patterns of particular species (e.g., ambystomid species) likely contribute to the low number of salamander records. Of the six lizard species occurring within

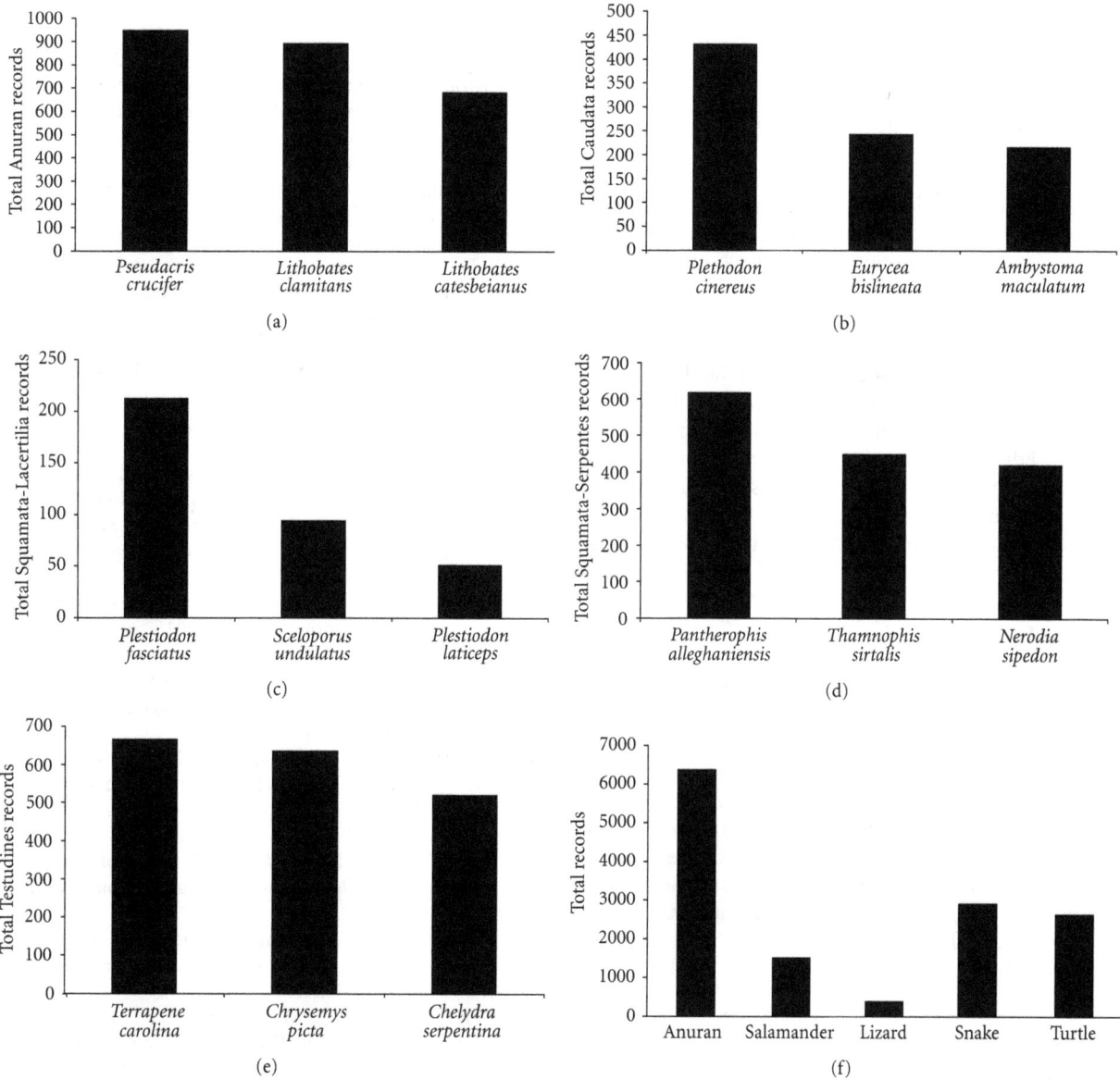

FIGURE 6: The most commonly reported species of frog (a), salamander (b), lizard (c), snake (d), turtle (e), and total number of records per group (f) through February 22, 2012.

Maryland, several have very restricted ranges which may explain the paucity of records. Several strategies have been helpful to increase records for these species. The availability of real-time data through the MARA database has allowed participants to be informed of the disparity in the record counts for the different groups. Information on how to survey for the underrepresented groups was shared with participants, resulting in an increase in record counts for those groups from 2010 to 2011.

Ultimately, the usefulness of atlas projects depends on the quality and quantity of data collected as well as the standardization of sampling methods and the appropriateness of the scale of sampling for the research question. An important function of our verification process is its capacity to quantify and correct error rates. The verification committee detected two primary error types during the early phase of the project: (1) animals were misidentified prior to data submission and (2) errors when submitting data to the online database, such as wrong photos submitted with a record or wrong species identity selected during data entry. The verification procedure enabled these errors to be identified and corrected prior to the records being finalized and the communication systems in place allowed feedback to the citizen scientists to reduce further errors. For example, in response to these findings, the MARA newsletter included, over the course of several months, articles containing information and techniques to identify the species that were found to be frequently misidentified and others discussing database usage and data entry.

Citizen science is perhaps the only practical way to achieve the geographic reach required to document ecological patterns and address ecological questions at scales relevant to species range shifts, broad-scale population trends, and impacts of environmental processes like landscape and climate change [14].

The MARA will establish a baseline by which future changes in the distribution of populations of native herpetofauna can be assessed. This project will be used to inform long-term conservation and protection strategies for Maryland's amphibian and reptile species. The MARA provides an opportunity for citizens to actively learn about native species while collecting valuable distributional data that the Natural History Society of Maryland and the Maryland Department of Natural Resources will use for the conservation and protection of Maryland's amphibians and reptiles. Educating citizens about native amphibian and reptile diversity and its ecological benefits is an important step in creating an informed society that actively participates in the long-term conservation and protection of Maryland's natural heritage.

Acknowledgments

The MARA is funded in part by State Wildlife Grant funds provided to the state wildlife agencies by US Congress and administered through the Maryland Department of Natural Resources' Wildlife and Heritage Service. Thanks go to all the MARA county coordinators, steering committee members, verification committee members, and volunteers who generously gave their time to the project. The authors extend gratitude to all the private landowners and public land managers who granted land access for surveys. Thanks are extended to Lynn Davidson and Bob Swan for developing the MARA database.

References

[1] J. W. Gibbons, D. E. Scott, T. J. Ryan et al., "The global decline of reptiles, deja vu amphibians," *BioScience*, vol. 50, no. 8, pp. 653–666, 2000.

[2] S. N. Stuart, J. S. Chanson, N. A. Cox et al., "Status and trends of amphibian declines and extinctions worldwide," *Science*, vol. 306, no. 5702, pp. 1783–1786, 2004.

[3] J. P. Collins and A. Storfer, "Global amphibian declines: sorting the hypotheses," *Diversity and Distributions*, vol. 9, no. 2, pp. 89–98, 2003.

[4] S. J. Price and M. E. Dorcas, "The Carolina Herp Atlas: an online, citizen-science approach to document amphibian and reptile occurrences," *Herpetological Conservation and Biology*, vol. 6, no. 2, pp. 287–296, 2011.

[5] J. W. Gibbons, "The management of amphibians, reptiles, and small mammals in North America: the need for an environmental attitude adjustment," in *Management of Amphibians, Reptiles, and Mammals in North America*, R. C. Szaro, Ed., USDA Forest Service General Technical Report RM-166, pp. 4–10, Rocky Mountain Forest Range Experiment Station, Fort Collins, Colo, USA, 1988.

[6] H. S. Harris Jr., "Distributional survey (Amphibia/Reptilia): Maryland and the District of Columbia," *Bulletin of the*

[7] H. S. Harris Jr., "The past history of documenting the distribution of amphibians and reptiles of Maryland and the District of Columbia," *Bulletin of the Maryland Herpetological Society*, vol. 45, no. 1, pp. 14–16, 2009.

[8] R. F. McLeod and J. E. Gates, "Response of herpetofaunal communities to forest cutting and burning at Chesapeake Farms, Maryland," *American Midland Naturalist*, vol. 139, no. 1, pp. 164–177, 1998.

[9] E. H. Campbell Grant, L. L. Bailey, J. L. Ware, and K. L. Duncan, "Prevalence of the amphibian pathogen *Batrachochytrium dendrobatidis* in stream and wetland amphibians in Maryland, USA," *Applied Herpetology*, vol. 5, no. 3, pp. 233–241, 2008.

[10] R. T. Brooks, "Potential impacts of global climate change on the hydrology and ecology of ephemeral freshwater systems of the forests of the northeastern United States," *Climatic Change*, vol. 95, no. 3-4, pp. 469–483, 2009.

[11] R. E. Casey, A. N. Shaw, L. R. Massal, and J. W. Snodgrass, "Multimedia evaluation of trace metal distribution within stormwater retention ponds in suburban Maryland, USA," *Bulletin of Environmental Contamination and Toxicology*, vol. 74, no. 2, pp. 273–280, 2005.

[12] B. A. Stein, *States of the Union: Ranking America's Biodiversity*, NatureServe, Arlington, Va, USA, 2002.

[13] Maryland Natural Heritage Program, *Rare, Threatened and Endangered Animals of Maryland*, Maryland Department of Natural Resources, Wildlife and Heritage Service, Annapolis, Md, USA, 2010.

[14] J. L. Dickinson, B. Zuckerberg, and D. N. Bonter, "Citizen science as an ecological research tool: challenges and benefits," *Annual Review of Ecology, Evolution, and Systematics*, vol. 41, pp. 149–172, 2010.

[15] A. M. Dunn and M. A. Weston, "A review of terrestrial bird atlases of the world and their application," *Emu*, vol. 108, no. 1, pp. 42–67, 2008.

[16] M. P. Robertson, G. S. Cumming, and B. F. N. Erasmus, "Getting the most out of atlas data," *Diversity and Distributions*, vol. 16, no. 3, pp. 363–375, 2010.

[17] E. H. Perring and S. M. Walters, *Atlas of the British Flora*, Botanical Society of the British Isles, T. Nelson, London, UK, 1962.

[18] P. T. Harding, "National species distribution surveys," *Monitoring for Conservation and Ecology*, pp. 133–154, 1991.

[19] North American Ornithological Atlas Committee, *Handbook for Atlasing American Breeding Birds*, Vermont Institute of Natural Science, Woodstock, Vt, USA, 1990.

[20] C. S. Robbins and E. A. T. Blom, Eds., *Atlas of the Breeding Birds of Maryland and the District of Columbia*, University of Pittsburgh Press, Pittsburgh, Pa, USA, 1996.

[21] W. G. Ellison, Ed., *Atlas of the Breeding Birds of Maryland and the District of Columbia*, Johns Hopkins University Press, Baltimore, Md, USA, 2nd edition, 2010.

[22] New Jersey Endangered and Nongame Species Program, *New Jersey's Herp Altas Project: Herp Atlas Volunteer Training Manual*, New Jersey Department of Environmental Protection, NJ, USA, 2002.

[23] G. D. Therres, C. A. Davis, and C. W. Swarth, "Grid-based Herp Atlas using active searching: a pilot project," Final Report, Maryland Department of Natural Resources, Annapolis, Md, USA, 2011.

[24] D. H. Foley III and S. A. Smith, *Comparison of two herpetofaunal inventory methods and an evaluation of their use in a volunteer-based statewide reptile and amphibian atlas project,*

Final Report, Maryland Department of Natural Resources, Wye Mills, Md, USA, 1999.

[25] M. L. Hunter Jr., J. Albright, and J. Arbuckle, Eds., *The Amphibians and Reptiles of Maine*, Maine Agricultural Experiment Station Bulletin 838, 1992.

[26] B. I. Crother, J. Boundy, F. T. Burbrink et al., *Scientific and Standard English Names of Amphibians and Reptiles of North America North of Mexico, with Comments Regarding Confidence in Our Understanding*, SSAR Herpetological Circular 37, 6th edition, 2008.

[27] R. Bonney, C. B. Cooper, J. Dickinson et al., "Citizen science: a developing tool for expanding science knowledge and scientific literacy," *BioScience*, vol. 59, no. 11, pp. 977–984, 2009.

[28] R. Gauza and D. Smith, *Maryland Amphibian and Reptile Atlas (MARA) Training Handbook*, Maryland Department of Natural Resources, Natural History Society of Maryland, 2010.

[29] J. B. Jensen, C. D. Camp, W. Gibbons, and M. J. Elliott, Eds., *Amphibians and Reptiles of Georgia*, University of Georgia Press, Athens, Ga, USA, 2008.

10

The Citizen Science Landscape: From Volunteers to Citizen Sensors and Beyond

Christina L. Catlin-Groves

Department of Natural and Social Sciences, University of Gloucestershire, Cheltenham GL50 4AZ, UK

Correspondence should be addressed to Christina L. Catlin-Groves, c.catlin.groves@googlemail.com

Academic Editor: Simon Morgan

Within conservation and ecology, volunteer participation has always been an important component of research. Within the past two decades, this use of volunteers in research has proliferated and evolved into "citizen science." Technologies are evolving rapidly. Mobile phone technologies and the emergence and uptake of high-speed Web-capable smart phones with GPS and data upload capabilities can allow instant collection and transmission of data. This is frequently used within everyday life particularly on social networking sites. Embedded sensors allow researchers to validate GPS and image data and are now affordable and regularly used by citizens. With the "perfect storm" of technology, data upload, and social networks, citizen science represents a powerful tool. This paper establishes the current state of citizen science within scientific literature, examines underlying themes, explores further possibilities for utilising citizen science within ecology, biodiversity, and biology, and identifies possible directions for further research. The paper highlights (1) lack of trust in the scientific community about the reliability of citizen science data, (2) the move from standardised data collection methods to data mining available datasets, and (3) the blurring of the line between citizen science and citizen sensors and the need to further explore online social networks for data collection.

1. Introduction

Within conservation and ecology, volunteer participation has always been an important component of research [1–5]. Within the past two decades, use of volunteers in research has begun to proliferate and evolve into the current form of "citizen science" [6, 7]. Citizen science, a term first coined by Irwin [7], is used to describe a form of research collaboration or data gathering that is performed by untrained or "nonexpert" individuals, often involving members of the public, and frequently thought of as a form of crowd-sourcing [1, 8–12].

Citizen science will usually incorporate an element of public education [2, 6, 13–15]. Silvertown [5] described the differentiation between historical and modern forms of citizen science by potential for it to be "available to all, not just a privileged few." This has been recently demonstrated by the rapid development of mobile phone technologies, in particularly the emergence and uptake of high-speed Web-capable smart phones with GPS data collection facilities and data upload capabilities [16]. This allows almost instant collection, transmission, and submission of data and provides researchers with a way to validate data (e.g., to verify the identification of an organism or the location through GPS locators) [10]. The availability of new technologies containing sensors could be argued to move citizen science into a new era whereby citizen scientists also become citizen "sensors." Collection of high-quality data can be made through the sensing capabilities of personal computing and communication technologies, making the user part of a more passive framework for data collection [17–19]. Some of the key strengths of citizen science projects lie in the ease and speed with which data can be gathered by a large number of individuals in a short time. Ordinarily constraints such as money and time would make studies unfeasible or impossible for an individual organisation [10, 15, 20]. Indeed, citizen science programmes are often more resilient to variations in financial support than other programs [19, 21, 22].

With technological connectivity peaking, the ability to select virtual "field assistants," to help gather data is within easy reach; indeed Irwin [7] said that citizen scientists can be considered as the "world's largest research team." A further step is the potential for mining data, for ecological or biological research, from the huge quantities of data which are voluntarily uploaded onto personal social media accounts for the primary reason of storage or sharing with friends. For example, there are over 26,000 images tagged with "manta ray" on Flickr (as of December 12th 2011), a species with stable patterning that can be individually identified [23]. Custom Application Programming Interfaces (API) could theoretically identify these individuals and collect GPS data (where its available). This would create vast quantities of ecological and spatial data that could be utilised in research which tracks individuals. However, despite this, citizen science projects are often limited to: (1) informal education activities or outreach to promote understanding [1, 6, 14, 24, 25]; (2) natural resource monitoring to promote stewardship [26–28]; (3) to promote social activities and action [29, 30]; (4) purely virtual whereby the entire project is ICT-mediated with no physical attribute (e.g., classifying photographs) ([31, 32], see Table 1). Table 2 provides examples of citizen science projects alongside their primary goals.

Few scientific investigative projects exist in ecology or biology using these new technologies for data collection, and where they do, they often encounter difficulties with gaining robust data [5, 6]. Even less take advantage of the rapidly increasing and evolving capabilities of Web 2.0 and social networks such as Facebook (http://www.facebook.com/), Twitter (http://www.twitter.com/) and Flickr (http://www.fli ckr.com/), through which millions of people upload and share photographs and location data, many citizen science studies also concentrate on data being collected within a very rigid framework, very similar to previous volunteer data collection whereby paper forms are replaced by online submission forms (examples include e-bird, Project Budburst, What's Invasive, and Neighbourhood Nestwatch Program). The possibility of using Web 2.0 and less rigid data collection techniques is relatively underexplored within scientific literature, even less so for biological and ecological applications.

This paper will predominantly cover the uses of citizen science for ecology, biodiversity, and biological insights. However, it may touch on various interdisciplinary citizen science programs or concepts where it is felt that it will be beneficial and may bring together other approaches which may add value. The aim is to establish the current state of citizen science within scientific literature, examine main underlying themes, and explore the possibility of utilising an untapped resource and the benefits that this can hold for the scientific community. It will also attempt to identify possible directions for further research.

2. The Citizen Science Landscape

The ability for intense monitoring by expert individuals on any subject ranging from individual or species distributions to tracking of invasive species is severely limited by both logistical and financial constraints. There are simply not enough resources, whether this is in the form of time, personnel, or money to establish large scale datasets [6, 10, 33, 34]. Citizen science circumvents many of these problems and has proven effective in a number of research areas that can have difficulty gathering large datasets. The areas in which it has, and most probably will continue to have, the greatest impact and potential are that of monitoring ecology or biodiversity at large geographic scales (see Table 3 for examples). This is particularly prevalent due to the recent proliferation of built in GPS technology and Web-capable features that many handheld devices, such as mobile phones and increasingly cameras, now have in an affordable and widely available format [10, 11].

When monitoring for rare, unusual, or declining phenomena, the scale of a large workforce over a large area will increase rates of detection in comparison to a lone researcher on a strict rotation despite having greater expert knowledge [35]. Indeed in early 2006, the rare nine-spotted ladybird (*Coccinella novemnotata*) was rediscovered during a citizen science programme designed to educate the public in biodiversity and conservation. This nine-spotted ladybird was the first discovered in eastern North America in over fourteen years, and only the sixth in the whole of North America within 10 years [36].

Traditional citizen science or volunteer programs have resulted in some of the longest ecological temporal datasets that we can access, particularly in the field of ornithology. The *Christmas Bird Count* (CBC—http://birds.audubon.org/christmas-bird-count/) was launched 1900 by the Audubon Society (in US and Canada) and provides long-term comprehensive data trends for many species for over 100 years. The British Trust for Ornithology, founded in 1932, also regularly uses data collected by amateur birdwatchers and makes up a very substantial amount of the National Biodiversity Network (http://www.nbn.org.uk/) which contains over 31 million records. The data of these programmes have helped to inform conservation actions, for example, by providing information to target conservation management at particular sites by environmental organisations [37].

Citizen science programmes conducted in the last 10 years have successfully followed the spread of invasive species or diseases, impacts of land use or climate change, and have been instrumental in understanding distributions, ranges, and migration pathways (e.g.,[38, 39]). Researchers at Cornell University, USA, have performed a large range of citizen science projects centred around avian species. Some of these projects have resulted in datasets that track the spread of conjunctivitis (*Mycoplasma gallisepticum*) in wild house finches (*Carpodacus mexicanus*)[40] and the impact of forest fragmentation on tanager populations and nesting success [41]. These efforts have led to a large database called eBird, where amateur birdwatchers can upload sightings. These citizen science data have become the basis of trends discovered through data mining and modelling techniques,

TABLE 1: Citizen science typologies as described by Wiggins and Crowston [1].

Type	Description	Example
Action	Employ volunteer-initiated participatory action research to encourage participant intervention in local concerns.	Shermans Creek Conservation Association (http://www.shermanscreek.org/)
Conservation	Address natural resource management goals, involving participants in stewardship for outreach and increased scope.	Missouri Stream Team Project (http://www.mostreamteam.org/)
Investigation	Focus of scientific research goals focussed on collecting data from the physical environment, usually underpinned by an hypothesis or research goal.	BirdTrack (http://www.bto.org/volunteer-surveys/birdtrack)
Virtual	Similar goals to the investigation project, but are entirely mediated by ICT having no physical element.	Whale FM (http://whale.fm/)
Education	Education and outreach are their primary goals, often data is not collected in a meaningful way that might be useful to other researchers. Often provides formal and informal learning resources.	Bird Sleuth (http://www.birds.cornell.edu/birdsleuth)

TABLE 2: Primary goals of citizen science projects (adapted and modified from Wiggins and Crowston [1]).

Project	URL	Primary goal	Description
Globe at Night	http://www.globeatnight.org/	Education	Learning about light pollution with use of mobile phone or Web cam and internet connection.
Fossil Finders	http://www.fossilfinders.org/	Education	Learning about Devonian Fossils through authentic inquiry-based investigation.
Bird Sleuth	http://www.birds.cornell.edu/birdsleuth/	Education	Learning about birds through inquiry-based investigation.
Missouri Stream Team Project	http://www.mostreamteam.org	Conservation	Promotes the formation of "stream teams" which monitor streams in their area.
What's Invasive	http://whatsinvasive.com/	Conservation	Locating invasive plants.
Shermans Creek Conservation Association	http://www.shermanscreek.org/	Action	Started to oppose the building of a power plant on local land, they now monitor the area and have regular talks.
ReClam the Bay	http://www.reclamthebay.org/	Action	Promotes environmental involvement by growing and maintaining baby clams and oysters to stock their local bay.
*Whale FM	http://whale.fm/	Virtual	Asks participants to listen to and classify whale song.
*Galaxy Zoo	http://www.galaxyzoo.org/	Virtual	Invites participants to classify images of galaxies.
Pathfinder	http://www.pathfinderscience.net/	Virtual	Collaborative online environment for citizen scientists.
Foldit	http://www.fold.it/	Virtual	Proving human superiority at protein folding.

*Part of Zooniverse—https://www.zooniverse.org/projects—citizen science hub for virtual citizen science projects exploiting the human ability to spot patterns and classify data where traditional statistical analysis struggles.

which have led to further more focussed studies (visit http://ebird.org/content/ebird/about/ebird-publications for more information and a full list of publications).

Datasets that have been gathered for a specific purpose will often result in unexpected phenomena or patterns emerging, that will then promote further more focussed studies. Many studies are available in scientific literature where data mining and model construction have resulted in the discovery of new patterns and processes being found in ecological systems (e.g.,[42–44]). Howard and Davis [45, 46] have published a number of peer-reviewed papers on data predominantly collected by citizen scientists, gathering useable scientific data on autumn migration flyways of monarch butterflies (Danaus plexippus). Citizen scientists record overnight roosts and report their first spring sightings to assess spring recolonisation rates.

TABLE 3: Citizen science projects and data collection/submission process(s).

Project	URL	Description	Data collection/submission process(s)
Project PigeonWatch	http://www.birds.cornell.edu/pigeon watch	A US program run by Cornell University. Participants count pigeons and record courtship behaviours observed in their neighbourhood pigeon flocks.	Virtual form submission
eBird	http://ebird.org/	Initially US-ebased but moving more into global records. eBird's goal is to maximize the utility and accessibility of the vast numbers of bird observations made each year by recreational and professional bird watchers. Has an online accessible database and visualisation facilities for the participant and other interested parties.	Virtual form submission
Ecocean	http://www.whaleshark.org/	The ECOCEAN Whale Shark Photo-identification Library is a visual database of whale shark (*Rhincodon typus*) encounters and of individually catalogued whale sharks. It asks participants to upload images and sightings of Whale Sharks.	Virtual form submission
Natures notebook	http://www.usanpn.org/participate/observe	A US program run as part of the National Phenology Network, it asks people to report the phenophases of particular species in their local areas.	Virtual form submission
BirdTrack	http://www.bto.org/volunteer-surveys/birdtrack	Partnership working between the British Trust of Ornithology, Royal Society for the protection of Birds, Birdwatch Ireland, and the Scottish Ornithologists' Club, it collects data on migration movements and distributions throughout Britain and Ireland. Has an online accessible database and visualisation facilities for the participant and other interested parties.	Virtual form submission
British Trust for Ornithology	http://www.bto.org/	Nongovernmental organisation dedicated to using volunteers who follow statistically designed sampling strategies in their research into birds.	Virtual form submission
Project Budburst	http://neoninc.org/budburst/	A US project participants observe plant phenophases. Scientists can use the data to learn more about the responsiveness of individual plant species to changes in climate locally, regionally, and nationally.	Virtual form submission and mobile application submission
What's Invasive	http://whatsinvasive.com/	Asks participants to locate invasive species by making geotagged observations and taking photos to map their spread.	Virtual form submission and mobile application submission
Neighbourhood Nestwatch	http://nationalzoo.si.edu/scbi/MigratoryBirds/Research/Neighborhood_Nestwatch/default.cfm	Participants find and monitor bird nests and record and report their observations. Researchers are especially interested in comparing how successful nests are in urban, suburban, and rural backyards.	Virtual form submission
BeeID	http://www.flickr.com/groups/beeid	A completed pilot project which asked participants to upload geo-tagged images of bees and tag them with "beeid2010" tag on the Flickr photography Website, researchers then extract these images, identify and tag with the species id.	Tagging and data mining from existing social network site with integral mobile upload facilities

One of the common features of traditional and many current projects is the formal submission process which occurs on a stand-alone Website or through one-to-one communication between researcher and citizen. The submission is often closed or inaccessible until a result is published, and even in citizen science programmes where data is shared: it is very difficult for the ordinary citizen to visualize; this is shown to have an impact on participation [6, 47]. e-Bird has gone through some lengths to overcome this. By creating an online database system which has many portals and visualisation techniques, citizen scientists and researchers alike can explore the e-Bird database [12]. In April 2006, when this newly improved Website was upgraded allowing participants to explore their own and others data, the number of individuals submitting data nearly tripled [47]. Resources such as this require the citizen scientist to make an active effort to discover the project, find the Website and input, and retrieve data. By integrating data collection into social media and fully exploiting Web 2.0, the quality, geographical range, and quantity of data collected could potentially be significantly increased, and this is something that requires further research. However, despite the lack of financial cost that social media and Web 2.0 present, it is possible that the time and effort cost might not make the process worthwhile when considering the amount of additional data gained.

3. Social Networks and Web 2.0

Web 2.0 is an ambiguous term with almost as many facets and conflicting opinions and definitions as the term citizen science, some even argue against the existence of Web 2.0 as a concept. However, for the purposes of this paper Web 2.0 can be regarded as the socially connected and interactive internet which facilitates participatory data sharing and encourages user-generated content. This medium consists of blogs, podcasts, social networking sites, wikis, crowd-sourcing tools, and "cloud-based" group working environments. Web 2.0 has been expanded to a mobile computing context with the proliferation of new technologies such as smart phones, laptops, and tablet computers [48].

The most obvious purpose for exploiting Web 2.0, which is beginning to be used by researchers, is the power of marketing and advertising, expressing branding, recruiting, retaining, and sharing, and collecting data with the citizen scientist [5, 49]. Delaney et al. [50] advocated the use of Web 2.0 capabilities for the ease of collecting and sharing data via new cloud technologies. Delaney suggests that dynamic linked databases that use online mapping technology such as Google Earth (free and familiar to citizen scientists) would prove ideal for creating a complete graphical "global" database of species. This would likely increase engagement and retention of individuals as they watch their contributions become part of the "bigger picture." In essence, social media is being adopted as part of the communication strategy for engaging individuals who collect data or participate in virtual citizen science programs; this adoption is seemingly in line with that of organisations at large to promote

products or engage audiences. This paper will not examine these factors in depth, as they are far too large to be able to cover appropriately (for more further information on this topic, (see [51–53])). Society at large is beginning to understand the increased power of "the social network effect" behind Web 2.0, which increases value to existing users in a feedback loop (e.g., more and more users begin to embrace a service, increasing its popularity, and resulting in rapidly increasing adoption) [54–56].

Figure 1 shows a brief diagram of citizen science. In addition to running programs of research that encourage users to engage in a more traditional data submission process, there is also the underexplored option of mining data from social networks and taking a more opportunistic approach. Indeed, many images, especially those taken on mobile phones, contain GPS information and can readily be searched and mapped via the integrated search facilities on Websites. The mobile interface allows the mobile phone to become a people-centric sensor which is capable of aggregating inputs from local surroundings, enabling data to be collected at a higher resolution [57]. This may be useful in plotting distributions and migration patterns or movements, both of individuals or species. Indeed, large charismatic species with stable patterning such as whales, sharks, rays, and big cats are photographed regularly by tourists and shared online, and the ability to collate and analyse these images could prove valuable to the study of their movement, social grouping, and ultimately conservation.

An emerging and particularly promising but under developed area of citizen science is that of using online social networking sites such as Facebook, Twitter, and mobile social networks such as Foursquare. Many of these have integrated image and location data upload facilities. Indeed, throughout 2011 there has been a proliferation of these facilities throughout popular social networking Websites. These features have been incorporated into basic interfaces, enabling users to simultaneously capture images; GPS tag them, add, comments, and post, to followers or friends instantly via mobile internet.

Since it's advent in 2004, Facebook (http://www.facebook.com/), the most popular social networking site, has grown to having more than 800 million active users globally, with, on average, more than 250 million photographs uploaded every day. More than 350 million people access it through a mobile phone [58]. Research by commercial online marketing and data collection agency comScore Media Metrix suggested that Facebook reached 73% of Americans in June 2011 [59]. With Flickr, the story is similar; Yahoo! announced in August 2011 that it had reached 51 million users and had, on average, 4.5 million photos uploaded every day. On the February 28th 2012 it had 176,605,443 geotagged photographs in total. With an integrated approach and the correct marketing and publicity, in addition to the increase of GPS-capable mobile devices it is likely that Flickr may become increasingly useful for gathering data, particularly for charismatic species.

The potential for scientific research is immense, particularly for image-based data collection where EXIF information can be mined using a custom API and identification can

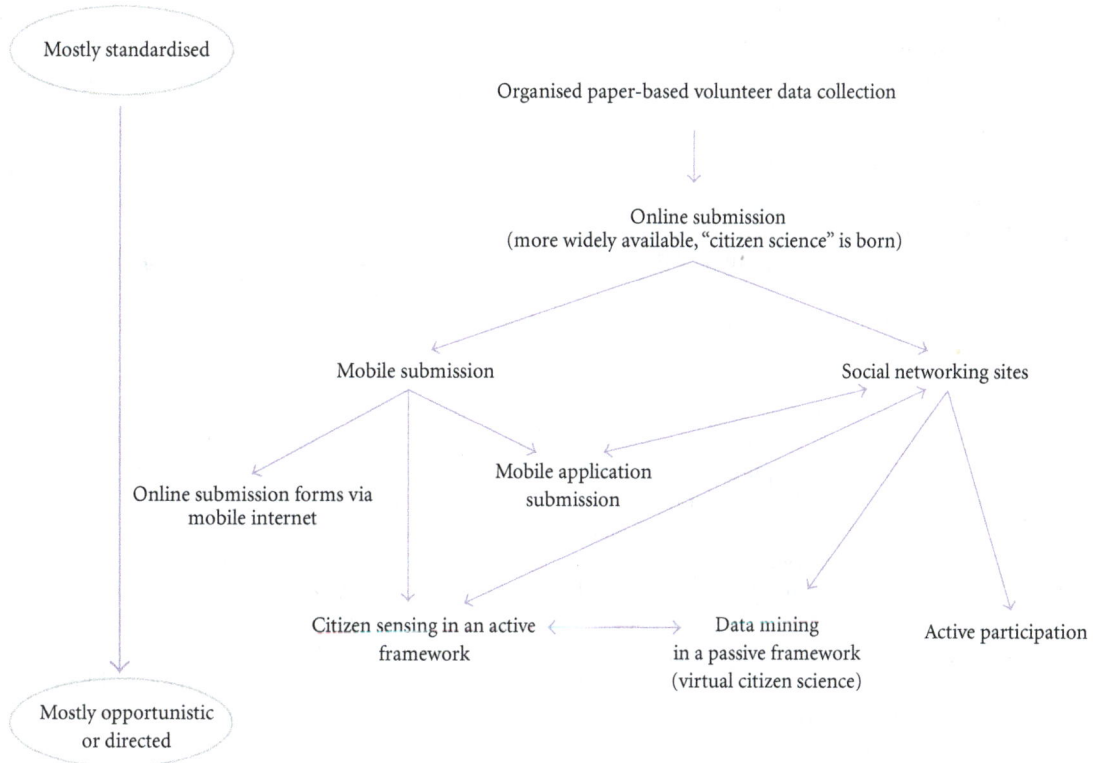

FIGURE 1: A brief diagram of citizen science, the diagram shows the proliferation in citizen science as new technologies have become available.

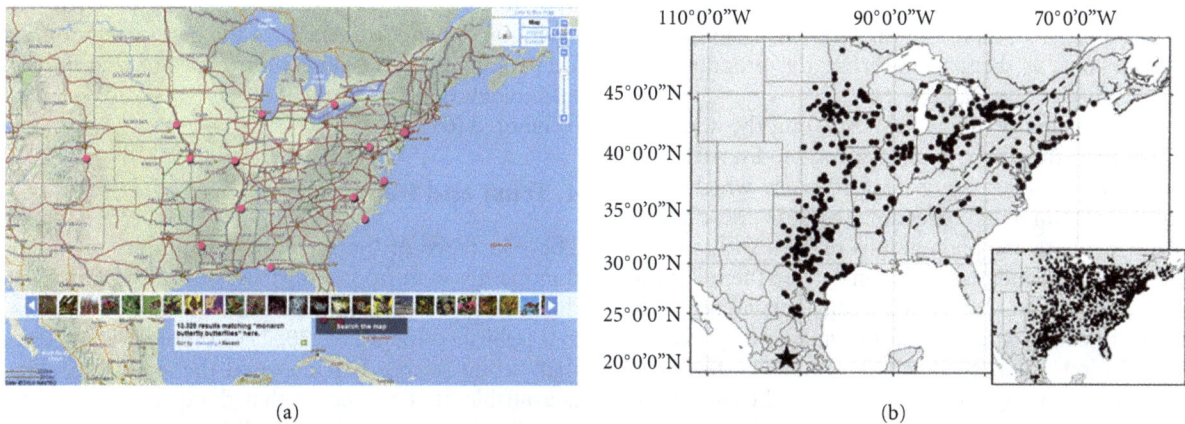

(a)   (b)

FIGURE 2: (a) Screenshot of Flickr map displaying 250 of 13,329 geo-tagged photos tagged with "monarch butterflies butterfly" on February 28th 2012. (b) Map of Journey North roost sightings from all years combined (2005–2007). Dashed line indicates division of central and eastern flyways in analysis. Roosts in Florida were not included in the analyses. Inset map shows the locations of all Journey North participants from 1997 to 2007. Star indicates location of Mexico overwintering sites [45]. Reproduced with kind permission of Springer Science and Business Media.

be verified by trained individuals or automatic recognition software [60–63]. Figure 2 shows a small example of what Flickr can do with a simple search term "monarch butterflies butterfly" (signifying a search for either butterflies or butterfly) which pulls up 13,329 geo-tagged photos within the US (February 28th 2012), 250 of which it can plot on a map on the Flickr Website. The map seemingly holds a cursory resemblance to Howard and Davis's [45] map of monarch butterfly migration roosts (created using data

from a citizen science program called Journey North which relies on a more traditional data submission process albeit via an online form—http://www.learner.org/jnorth/). Using a custom API and transposing all results onto Google Maps or other mapping software, it would be possible to limit the geo-tagged photo search by date and compare it directly with Journey North's monarch butterfly monitoring program, which has received 4078 sightings within the last year. However, without creating an API, a simple search on Flickr's advanced search facility with the search term "monarch butterfly" brought up 15,499 photographs within the same time period (using data collected on February 28th 2012). Despite being likely that a large proportion contains no useful information (i.e., not pictures of the target species) and/or is not geo-tagged, (although estimates show >40% may be geo-tagged, [64]), this suggests that if this method of data collection was further explored the number of potentially useful monarch butterfly sightings data could be greatly increased.

Currently, a general internet user's image and location uploads are predominantly limited to "events" that the user wants to share this might be "checking in" to restaurants, attractions, clubs, cinemas, or concerts, often reviewing products, or sharing visual experiences [65, 66]. Sharing these data with another user can be as simple as tagging them [66]. By exploiting social networks in this way, for ecological or biological research, many of the most common mistakes or inaccuracies that are found within volunteered data could be minimised. For example, by sharing images, and temporal and GPS data, misidentifications and location inaccuracies can be flagged and checked by trained individuals [5, 37, 64, 67, 68]. Despite this, there are very few examples of social networking sites being used actively to collect data for biological or ecological research; this may be because of confusion over copyright laws or limitations of API systems. At time of writing, there are very few examples of such usages, and those few that do exist are limited to self-contained "groups" within Flickr which search images of individual animals to export to an external catalogue for identification or use them to advertise the program and attract new submissions (Table 4).

Despite this ability to gather data quickly, they are currently underutilised for ecological or biodiversity data collection. BeeID is a program of research which used Flickr as a base for data collection [64, 67]. Researchers asked individuals to tag photographs of bees with specific searchable metatags and place location data on them if it was not already embedded. Trained individuals then confirmed species identification and marked the images as processed via the addition of a new tag. A simple custom API extracted tagged photographs from Flickr and collected the data which successfully plotted bee species distributions. Considering the project had no funding and was run by a small group of individuals with limited promotion other than on social networking sites, its success demonstrates the potential benefits of using social networking for collection of scientific data. Furthermore, the study took part before the recent integration of easily accessible location data in social networks and the continued rise of smart-phone and affordable GPS and wifi enabled camera ownership.

Another facet of Web 2.0 is the very recent addition of phone applications or "apps." These are easily integrated and simple to use; however, the release of a mobile application is not enough on its own to motivate participants and it is important to use mobile applications in an holistic approach [12]. "What's Invasive" is a very recent citizen science programme which uses a combination of a Website and custom mobile application to allow mobile devices to collect and submit information about invasive species whilst they are observing them (http://whatsinvasive.com/). Project Noah is similar in that respect but is built primarily to engage and educate individuals in addition to collecting species data through a tagging and classification system. Project Noah also incorporates "missions" to increase motivation and promote the collection of specific species sightings (http://www.projectnoah.org/).

A recently developed formatting language, Hypertext Markup Language 5 (HTML5), allows easier development across platforms and allows many of the features of mobile phone applications to be incorporated into Websites. Web pages can then be developed to contain full multimedia content that is easily accessible to popular technologies, something which some smart phones have found problematic due to limited Flash support (especially on Apple devices). In the past, this inability has limited some of the content available and increased the amount of work needed to replicate Web pages on smart-phones.

Undoubtedly, with the advent of Web 2.0 and the quickly developing technological breakthroughs, citizen science programs exploiting this technology are likely to increase exponentially in future years and should be encouraged. It is hoped that as the full potential is revealed the negative bias among the scientific community that such approaches have attracted will begin to lessen. As the population increases and we are more isolated from nature and wildlife, the use of citizen science for biodiversity studies will enable individuals to be further engaged in decision-making processes and the championing and protection of the natural environment. It is a paradigm that is evolving alongside our relationship with technology, our environment and urban ecology and cannot be ignored [69].

4. Trust and Reliability

The reluctance of the scientific community seems to predominantly stem from a mistrust of citizen science datasets due to the lack of validity assessments in academic research and published literature [70, 71]. Although many recognise that citizen science has increased the amount of data that is available, it is a concern that the quality, reliability, and overall value of these data is still preventing its adoption in many research programmes [72]. Assurance of the quality of the data is needed through rigorous scientific methods in order to allow the acceptance of citizen science data into the scientific field [20].

The literature suggests that the reliability of inherently patchy data is the most questioned aspect of citizen science. Thus, being able overcome this mistrust, a huge untapped resource of citizen scientists could be opened up, increasing the scope and insight of conducted research. Potentially, this could result in large standardised spatial and temporal datasets collected by citizen sensor networks [71]. Traditional solutions to gaining credibility are to provide reliable information or gain credentials such as qualifications; however, this works only when there are "gatekeepers" to filter information, something which is not possible with the internet on a global scale [73].

TABLE 4: Utilising Flickr for image-based citizen science programs.

Project title and URL	Description	Passive promotion	Active promotion	Active data searching	Project base
Whale Shark Identification (http://www.flickr.com/groups/ whalesharkidentification/)	To collect images to be submitted to http://www.whaleshark.org for identification from group members and other Flickr users through the search facility. It is worth noting that this Flickr Group has been formed by a volunteer and is not officially part of the project.	Y	Y Recruits members to promote	Y	N
MantaWatch (http://www.flickr.com/ groups/mantawatch/)	A place for enthusiasts to meet and a promotion tool directing people to their Website (http://mantawatch.com). Does not seem to actively recruit members or search out images of manta rays on Flickr.	Y	N	N	N
Humpback whale flukes (http://www.flickr.com/groups/ humpbackflukes/)	To collect images to be submitted to http://www.coa.edu/nahwc.htm for identification from group members and other Flickr users through the search facility. The same project also has a whale catalog (http://www.flickr.com/photos/ flukematcher/) located on Flickr so that individuals can manually match their sightings. A further more regional group (http://www.flickr.com/groups/ northatlanticflukes/) has formed due to the volume of photos uploaded.	Y	Y Recruits members to promote	Y	N
Citizen Science: Great Blue Heron (http://www.flickr.com/groups/ csgreatblueheron/)	This group aims to create a database of geo-tagged images of the Great Blue Heron, entirely run and initiated by volunteers	Y	Y Recruits members to promote	Y	Y
BeeID (http://www.flickr.com/groups/beeid/)	A completed project run by student volunteers, and overseen by a lecturer, whereby members of the public are encouraged to upload photos of UK bees (Honeybees, bumblebees, and solitary bees) to their Flickr account and "geotag" them to place them on a map, with the aim of studying distribution and phenology.	Y	Y Recruits members to promote	N	Y

The dependability of volunteer-derived data is an old problem within biology and ecology, and therefore a number of methods to help to increase the reliability of the information gathered have been developed [6, 22]. Firstly, the researchers must concisely and without jargon ask the right questions in the right way to get the quality of answer that is needed, and instructions and processes must be clear and as simple as possible [3, 9–11]. Projects are usually kept relatively simple; for example, they might include counting a few common avian species frequenting a feeding table rather than searching for rare or difficult to spot species [6, 22, 74, 75]. Projects that require higher levels of skill can be successfully developed; however, they may require additional training or longevity of participation in order to increase experience indeed, many volunteer programs document "learner" effects whereby data collectors become more accurate and correct over time [6, 10, 22, 76–80]. Some of the online citizen science programmes that Cornell University has run in the past incorporate short tests and quizzes which help in assessing a contributors' knowledge; they have also implemented an automated meso-filter which evaluates data input and evaluates it based on already known parameters, submissions which fall out of these categories are flagged for expert review, the contributor contacted, and the entry either verified or disregarded [6, 10, 81].

Although there is not enough space to review all the literature which has been published as a result of data collected through the use of citizen science participation, literature searching has resulted in the location of over 300 instances of peer-reviewed publications. This suggests that citizen science has and will continue to produce usable forms of data (See Figure 3). As with any data, datasets should be approached with caution and "cleaned" or "scrubbed" before performing analysis to remove any obvious outliers [82]. The literature suggests, however, that if the program protocols have been properly formed and tailored to the appropriate audience data does not often differ significantly from expert data collection. Delaney et al. [50] found that

TABLE 5: Comparison of avian monitoring projects focused on measuring occurrence and abundance (adapted from [10]).

Project	Method	Placement	Effort[a]	Extent	Interval	Participants
			Results from these programs have been used in over 1000 publications			
Audubon Christmas Bird Count	Count circle (24 km diameter)	Opportunistic	V (party hours)	International	Annual	59,918 (2008-09)
North American Breeding Bird Survey	Roadside survey (39.4 km; 50 stops)	Stratified random	S (3 min count)	International	Annual	2,749 (2009)
Project FeederWatch	Feeder counts	Opportunistic	V (2 days, hours, days)	International	Annual	9,750 (2009)
eBird	Online checklists	Opportunistic	V (hours, distance, and area)	International	Continuous	18,053
Bird Atlas	Systematic grid (100 km^2 blocks; 4 km^2 tetrads) and roving reports	Regular grid and opportunistic	S/V (roving, timed visits)	Britain and Ireland	Two visits (Winter: Nov./Dec. & Jan./Feb.; Breeding: April/May & June/July)	10,000–20,000
Common Birds Census (now replaced by Breeding Bird Survey)	Census plots (Farmland: 70 ha; Woodland: 20 ha)	Stratified random	S (territory mapping)	Britain	Annual (8–10) visits; late March–early July	250–300

[a]Effort is considered standardised (S) or variable (V). When standardised, the protocol specifications are presented; when variable, the effort variables that were reported during sampling are presented.

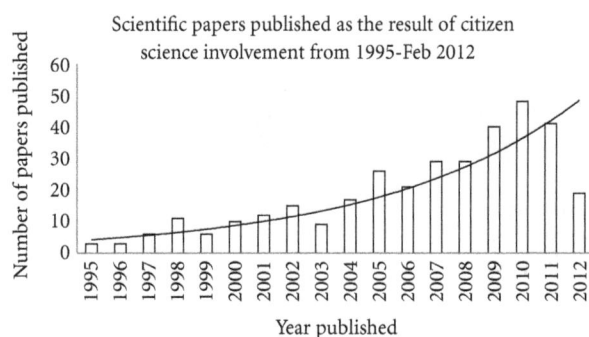

FIGURE 3: Numbers of published scientific papers using or resulting from citizen science data collection or involvement; it indicates an increasing trend.

far from overlooking data collection methods novices were "more careful" in their measurements and annotations, due to are being very aware of their novice status and shown in many studies to yield similar results to experts [22, 50, 83]. Delaney et al. [50] found experts and nonexperts did not come up with any significantly statistical differences, indeed students were found to be between 80 and 95% accurate with identification, with significant predictors of accuracy being their age and level of education. Dickinson et al. [10] reported that during Project FeederWatch between 2008 and 2009 they received 1,342,633 observations, out of those 378 records required "flagging" resulting in 158 records (54%) being confirmed, 45 identifications (16%) being corrected, and 88 reports (30%) being disregarded due to too little evidence.

Indeed, the very nature of gathering large sets of data results in decreased detrimental effects of "noise", greater statistical power and increased robustness, as statistical power is a function of sample sizes [22, 37]. Therefore, the common belief that volunteer collected data can only provide noisy and unreliable results that lack precision is generally incorrect [22, 37]. LePage and Francis [74] compared two citizen science programs with similar data collection protocols to test whether population patterns and distributions were temporally and spatially consistent. The study successfully showed that the two citizen science led studies, Christmas Bird Count and Project FeederWatch, had comparable trends and patterns across the same time periods, suggesting that the data was consistent and not significantly influenced by different methods and biases. The benefit of these larger datasets is that they allow researchers to draw broader conclusions across large spatial or temporal scales, enabling researchers to make inferences and robust cases for causation over a larger areas, and at a finer resolution, in contrast with small scale studies which cannot be "generalised" over greater areas [3, 6, 9, 11, 38].

It is, however, important to recognise that these datasets can be compromised by potential lack of precision, inherent biases, and uncertainties which are often present within these extensive studies [11, 22, 84]. For example, you may have more reports of species in areas that are highly populated by humans than in those that are sparsely populated, or more reports of species that are less cryptic than others. It is therefore a challenge to determine whether the data is correct or the reports are biased; this is the reason why many citizen science programs are so rigidly composed and use standardised protocols which are replicated across many stratified surveyed plots (see Table 5 and [11, 22, 84]). It is therefore important to ensure, in hypothesis driven studies, that sampling design does not introduce bias, and that counts are shaped by the data and not the ability of the observer to detect or record data [85]. This is partially why using such count data to establish index of abundance can be scientifically hazardous; however, by using capture-recapture algorithms, conversion to actual population estimates can be made and therefore data can be used to make a valid conclusion [86, 87].

Well known and successful UK citizen science-based programmes are those which are based in the public's back gardens. The British Trust for Ornithology's (BTO) and Royal Society for Protection of Birds' (RSPB's) garden-based citizen science programmes have been very successful in collecting biodiversity data, particularly on avian species. The "Garden BirdWatch" and "Big Garden Weigh-In" run by the BTO and the "Big Garden Birdwatch" and 'Make Your Nature Count survey' run by the RSPB are just a few of the citizen science programmes which encourage the recording of species which are visiting their gardens. For a full list of citizen science projects run by these organisations visit the BTO (http://www.bto.org/) and RSPB (http://www.rspb.org.uk/) websites.

These programmes have a number of key design similarities which help standardise the survey and mitigate against some of the perceived problems involved with nonexpert individuals collecting data. Indeed, they have proved to be reliable enough to result in published scientific papers. The Garden BirdWatch alone has resulted in 15 published scientific papers in addition to providing a strong set of baseline data (visit http://www.bto.org/volunteer-surveys/gbw/publications/papers for full list of publications).

To prevent confounding seasonal variation and to ensure continuity of recording effort citizen scientists are asked to record species within a given survey period, the Big Garden Weigh-In ran between the May 31st and June 5th in 2012 for example. To standardise effort the records are gathered over a particular time period, an hour is the most popular time, and many of the surveys require the species to be physically within the garden (not in a neighbouring garden or flying over). The Garden BirdWatch asks observers to repeat this recording at the same time and from the same place and of the same area for each recording session during the survey period.

Pseudoreplication is combated by removing the difference in the ability of the observer to identify different individuals; this is achieved by recording the maximum number of individual birds present at any one time within the garden. So if an observer sees one Blue Tit at the beginning of the survey but five in the middle and two towards the end of the survey, they would report it as five Blue Tits.

The species which are surveyed are also reduced to a range of easily identifiable species. The Garden Weigh-In reduces the number of birds under observation to 60 avian species which compose the core avian community. The Garden BirdWatch reduces the number further to the 42 most commonly recorded birds (nationally), with a further breakdown resulting in a list of the top ten which can have further detail added. The Big Garden BirdWatch reduces it further still providing a list of 20 more common species and ask observers to also record incidental records of other species that they might see on a separate sheet. The Garden BirdWatch goes one step further to collect additional data and provides a presence and absence record sheet for all species not mentioned.

The key difference between the RSPB and BTO's citizen science programmes is the method of collection. The RSPB has no paper-based submission format, but the BTO does, with a scanning machine which automates the data retrieval and decoding from the paper-based forms. The BTO suggests that the "relative proportions of participants submitting returns on paper and online are similar."

Neither of these programmes use social networks for more than publicity. In 2012, the BTO began the Cuckoo Tracking project, whereby tagged Cuckoo's were tracked during their migrations (http://www.bto.org/science/migration/tracking-studies/cuckoo-tracking). As part of the publicity, sightings were called for and the "hashtag" #heardacuckoo was created on the social network Twitter to publicise the project. Many individuals used the hashtag to report when they had indeed heard a cuckoo. If a tool such as CrowdMap (https://crowdmap.com/) was used to filter the tweet's with #heardacuckoo in them and verified by experts, could the conversion rate from publicity to actual record be higher?

5. The Shifting Paradigm: From "Knowledge-Driven" Analysis and Hypothesis Testing to "Data-Driven" Analysis

With the advent of the Web 2.0 world and the increase of the "citizen sensor network," there is a shifting paradigm from "knowledge-driven" analysis created by hypothesis-driven research to "data-driven" analysis, moving studies into more data-intensive science area [44, 83]. This is resulting in a new synthesis of disciplinary areas as new methods of analysis emerge to explore and identify interesting patterns that may not already be apparent; this is particularly prevalent when looking at data gathered over large spatial and temporal scales [44, 88, 89]. This approach offers valuable insights enabling further hypothesis for the discovery of underlying ecological processes. With such large datasets with such varying attributes; it is no wonder that all disciplines of science are seemingly beginning to merge into computer science as it enables scientists in varying fields to better understand complex systems [83, 90–93]. In order to better utilise citizen science collected datasets that provide a wide range of data over long periods, many researchers are moving into intelligent analysis. This may involve using novel probabilistic machine-learning statistical analysis in the form of computational modelling, or methods of analysis which include Bayesian or neural networking methods [90, 91, 93, 94]. Indeed, Link et al. [89] utilised a hierarchical model and Bayesian analyses to account for variations in effort on counts and to provide summaries over large geographic areas for a complex dataset provided by the Christmas Bird Count in America. They successfully revealed regional patterns of population change, which was then shown to be similar to data shown by the Midwinter Waterfowl Inventory in the US [89].

Currently, databases of species information are often disjunct, outdated, and incomplete, and data recording methods are often not standardised across organisational databases making reconciling datasets from different sources for studies often unreliable. This makes large scale data collection a necessity for research and the use of more complex methods of data collection an ever growing and underdemand area of study.

6. Conclusion

In our increasingly changing and evolving technological world, the presence of citizen scientists or citizen sensors who can contribute to science in more meaningful ways is allowing the rapid expansion of citizen science. Monitoring, anticipating, and mitigating large-scale threats to our biodiversity and natural world have also never been more prominent than they are now. In an increasingly urbanised world, successful monitoring of the environment is needed in the face of continuing climate and land-use change and the need to increase understanding of key ecological and environmental processes.

Citizen science and the exploitation of citizen science and sensor networks are probably one of the most important factors in being able to achieve this. The data is out there, just waiting to be understood, almost each and every person in the developed world and beyond has the potential to contribute to our understanding in a meaningful way. With the rapid progression of technology it is within our capabilities to begin this journey of understanding. It is, however, important to recognise the potential weaknesses that can result from poorly managing datasets and to pre-empt how the data is likely to be used and integrated beyond the original scope of the project.

It is also prudent to note something that many conservation organisations are realizing; a need to interest new generations of naturalists and enthusiasts as current recorders is an aging group with limited recruitment. By exploiting new technologies to aid recruitment of a younger generation of recorders and naturalists and educate an increasingly urbanised population, it will benefit all stake holders.

If citizen science was commonplace, how much more scientific knowledge could we discover? And in this world where people are increasingly divorced from the natural environment, how much would this influence decision making, education, and scientific thinking?

Acknowledgment

Many thanks to the University of Gloucestershire for the studentship which is supporting the author in this work. Figure 2(b) reproduced with kind permission of Springer Science and Business Media [45].

References

[1] A. Wiggins and K. Crowston, "From conservation to crowd-sourcing: a typology of citizen science," in *Proceedings of the 44th Hawaii International Conference on System Sciences (HICSS '10)*, January 2011.

[2] J. P. Cohn, "Citizen science: can volunteers do real research?" *BioScience*, vol. 58, no. 3, pp. 192–197, 2008.

[3] K. S. Oberhauser and M. D. Prysby, "Citizen science: creating a research army for conservation," *American Entomologist*, vol. 54, no. 2, p. 3, 2008.

[4] S. Droege, "Just because you paid them doesn't mean their data are better," in *Proceedings of the Citizen Science Toolkit Conference*, Ithaca, NY, USA, 2007.

[5] J. Silvertown, "A new dawn for citizen science," *Trends in Ecology & Evolution*, vol. 24, no. 9, pp. 467–471, 2009.

[6] R. Bonney, C. B. Cooper, J. Dickinson et al., "Citizen science: a developing tool for expanding science knowledge and scientific literacy," *BioScience*, vol. 59, no. 11, pp. 977–984, 2009.

[7] A. Irwin, *Citizen Science: A Study of People, Expertise and Sustainable Development*, Routledge, London, UK, 1995.

[8] J. Howe, "The rise ofcrowdsourcing," *Wired Magazine*. In press.

[9] C. B. Cooper, J. Dickinson, T. Phillips, and R. Bonney, "Citizen science as a tool for conservation in residential ecosystems," *Ecology and Society*, vol. 12, no. 2, article 11, 2007.

[10] J. L. Dickinson, B. Zuckerberg, and D. N. Bonter, "Citizen science as an ecological research tool: challenges and benefits," *Annual Review of Ecology, Evolution, and Systematics*, vol. 41, pp. 149–172, 2010.

[11] V. Devictor, R. J. Whittaker, and C. Beltrame, "Beyond scarcity: citizen science programmes as useful tools for conservation biogeography," *Diversity and Distributions*, vol. 16, no. 3, pp. 354–362, 2010.

[12] N. R. Prestopnik and K. Crowston, "Citizen science system assemblages: understanding the technologies that support crowdsourced science," in *Proceedings of the 2012 iConference*, pp. 168–176.

[13] D. Brossard, B. Lewenstein, and R. Bonney, "Scientific knowledge and attitude change: the impact of a citizen science project," *International Journal of Science Education*, vol. 27, no. 9, pp. 1099–1121, 2005.

[14] D. J. Trumbull, R. Bonney, D. Bascom, and A. Cabral, "Thinking scientifically during participation in a citizen-science project," *Science Education*, vol. 84, no. 2, pp. 265–275, 2000.

[15] A. W. Crall, G. J. Newman, C. S. Jarnevich, T. J. Stohlgren, D. M. Waller, and J. Graham, "Improving and integrating data on invasive species collected by citizen scientists," *Biological Invasions*, vol. 12, no. 10, pp. 3419–3428, 2010.

[16] D. M. Aanensen, D. M. Huntley, E. J. Feil, F. Al-Own, and B. G. Spratt, "EpiCollect: linking smartphones to web applications for epidemiology, ecology and community data collection," *PLoS ONE*, vol. 4, no. 9, Article ID e6968, 2009.

[17] N. D. Lane, E. Miluzzo, H. Lu, D. Peebles, T. Choudhury, and A. T. Campbell, "A survey of mobile phone sensing," *IEEE Communications Magazine*, vol. 48, no. 9, pp. 140–150, 2010.

[18] S. Matyas, C. Matyas, C. Schlieder, P. Kiefer, H. Mitarai, and M. Kamata, "Designing location-based mobile games with a purpose—collecting geospatial data with cityexplorer," in *Proceedings of the International Conference on Advances in Computer Entertainment Technology (ACE '08)*, pp. 244–247, ACM, Yokohama, Japan, December 2008.

[19] D. Couvet, F. Jiguet, R. Julliard, H. Levrel, and A. Teyssedre, "Enhancing citizen contributions to biodiversity science and public policy," *Interdisciplinary Science Reviews*, vol. 33, no. 1, pp. 95–103, 2008.

[20] S. A. Boudreau and N. D. Yan, "Auditing the accuracy of a volunteer-based surveillance program for an aquatic invader bythotrephes," *Environmental Monitoring and Assessment*, vol. 91, no. 1–3, pp. 17–26, 2004.

[21] A. Sheth, "Citizen sensing, social signals, and enriching human experience," *IEEE Internet Computing*, vol. 13, no. 4, pp. 87–92, 2009.

[22] D. S. Schmeller, P. Y. Henry, R. Julliard et al., "Advantages of volunteer-based biodiversity monitoring in Europe," *Conservation Biology*, vol. 23, no. 2, pp. 307–316, 2009.

[23] A. M. Kitchen-Wheeler, "Visual identification of individual manta ray (Manta alfredi) in the Maldives Islands, Western Indian Ocean," *Marine Biology Research*, vol. 6, no. 4, pp. 351–363, 2010.

[24] M. E. Fernandez-Gimenez, H. L. Ballard, and V. E. Sturtevant, "Adaptive management and social learning in collaborative and community-based monitoring: a study of five community-based forestry organizations in the western USA," *Ecology and Society*, vol. 13, no. 2, article 4, 2008.

[25] D. Penrose and S. M. Call, "Volunteer monitoring of benthic macroinvertebrates: regulatory biologists' perspectives," *Journal of the North American Benthological Society*, vol. 14, no. 1, pp. 203–209, 1995.

[26] C. Evans, E. Abrams, R. Reitsma, K. Roux, L. Salmonsen, and P. P. Marra, "The Neighborhood Nestwatch program: participant outcomes of a citizen-science ecological research project," *Conservation Biology*, vol. 19, no. 3, pp. 589–594, 2005.

[27] Y. Bhattacharjee, "Citizen scientists supplement work of Cornell researchers," *Science*, vol. 308, no. 5727, pp. 1402–1403, 2005.

[28] Missouri Stream Team Project, 2011, http://www.mostreamteam.org/.

[29] ReClam the Bay, 2005, http://www.reclamthebay.org/.

[30] Shermans Creek Conservation Association, 1998.

[31] M. J. Raddick, A. S. Szalay, J. Vandenberg et al., "Galaxy zoo: exploring the motivations of citizen science volunteers," *Astronomy Education Review*, vol. 9, no. 1, pp. 010103–010118, 2010.

[32] K. Luther, S. Counts, K. B. Stecher, A. Hoff, and P. Johns, "Pathfinder: an online collaboration environment for citizen scientists," in *Proceedings of the 27th International Conference on Human factors in Computing Systems (Chi '09)*, vol. 1–4, pp. 239–248, 2009.

[33] L. S. Fore, K. Paulsen, and K. O'Laughlin, "Assessing the performance of volunteers in monitoring streams," *Freshwater Biology*, vol. 46, no. 1, pp. 109–123, 2001.

[34] D. M. Lodge, S. Williams, H. J. MacIsaac et al., "Biological invasions: recommendations for U.S. policy and management," *Ecological Applications*, vol. 16, no. 6, pp. 2035–2054, 2006.

[35] R. Lukyanenko, J. Parsons, and Y. Wiersma, "Enhancing citizen science participation in GeoWeb projects through the instance-based data model," in *Proceedings of the Spatial Knowledge and Information*, Fernie, Canada, 2011.

[36] J. E. Losey, J. E. Perlman, and E. R. Hoebeke, "Citizen scientist rediscovers rare nine-spotted lady beetle, Coccinella novemnotata, in eastern North America," *Journal of Insect Conservation*, vol. 11, no. 4, pp. 415–417, 2007.

[37] J. J. D. Greenwood, "Citizens, science and bird conservation," *Journal of Ornithology*, vol. 148, no. 1, pp. S77–S124, 2007.

[38] V. Devictor, R. Julliard, D. Couvet, and F. Jiguet, "Birds are tracking climate warming, but not fast enough," *Proceedings of the Royal Society B*, vol. 275, no. 1652, pp. 2743–2748, 2008.

[39] A. H. Hurlbert and Z. Liang, "Spatiotemporal variation in avian migration phenology: citizen science reveals effects of climate change," *Plos One*, vol. 7, no. 2, Article ID e31662, 2012.

[40] S. Altizer, W. M. Hochachka, and A. A. Dhondt, "Seasonal dynamics of mycoplasmal conjunctivitis in eastern North American house finches," *Journal of Animal Ecology*, vol. 73, no. 2, pp. 309–322, 2004.

[41] K. V. Rosenberg, J. D. Lowe, and A. A. Dhondt, "Effects of forest fragmentation on breeding tanagers: a continental perspective," *Conservation Biology*, vol. 13, no. 3, pp. 568–583, 1999.

[42] W. M. Hochachka, R. Caruana, D. Fink et al., "Data-mining discovery of pattern and process in ecological systems," *Journal of Wildlife Management*, vol. 71, no. 7, pp. 2427–2437, 2007.

[43] D. Fink, W. M. Hochachka, B. Zuckerberg et al., "Spatiotemporal exploratory models for broad-scale survey data," *Ecological Applications*, vol. 20, no. 8, pp. 2131–2147, 2010.

[44] S. Kelling, W. M. Hochachka, D. Fink et al., "Data-intensive science: a new paradigm for biodiversity studies," *BioScience*, vol. 59, no. 7, pp. 613–620, 2009.

[45] E. Howard and A. K. Davis, "The fall migration flyways of monarch butterflies in eastern North America revealed by citizen scientists," *Journal of Insect Conservation*, vol. 13, no. 3, pp. 279–286, 2009.

[46] A. K. Davis and E. Howard, "Spring recolonization rate of Monarch butterflies in eastern North America: new estimates from citizen-science data," *Journal of the Lepidopterists' Society*, vol. 59, no. 1, pp. 1–5, 2005.

[47] B. L. Sullivan, C. L. Wood, M. J. Iliff, R. E. Bonney, D. Fink, and S. Kelling, "eBird: a citizen-based bird observation network in the biological sciences," *Biological Conservation*, vol. 142, no. 10, pp. 2282–2292, 2009.

[48] T. O'Reilly, "What is Web 2.0: design patterns and business models for the next generation of software," *International Journal of Digital Economics*, no. 65, pp. 17–37, 2007.

[49] C. Wood, B. Sullivan, M. Iliff, D. Fink, and S. Kelling, "eBird: engaging birders in science and conservation," *Plos Biology*, vol. 9, no. 12, Article ID e1001220, 2011.

[50] D. G. Delaney, C. D. Sperling, C. S. Adams, and B. Leung, "Marine invasive species: validation of citizen science and implications for national monitoring networks," *Biological Invasions*, vol. 10, no. 1, pp. 117–128, 2008.

[51] J. Bughin, "The rise of enterprise 2.0," *Journal of Direct, Data and Digital Marketing Practice*, vol. 9, no. 3, pp. 251–259, 2008.

[52] E. Constantinides, C. L. Romero, and M. A. G. Boria, "Social media: a new frontier for retailers?" in *European Retail Research*, B. Swoboda, D. Morschett, T. Rudolph, P. Schnedlitz, and H. Schramm-Klein, Eds., pp. 1–28, Gabler, 2009.

[53] W. G. Mangold and D. J. Faulds, "Social media: the new hybrid element of the promotion mix," *Business Horizons*, vol. 52, no. 4, pp. 357–365, 2009.

[54] P. Klemperer, *Network Effects and Switching Costs: Two Short Essays for the New Palgrave*, Social Science Research Network, 2006.

[55] S. J. Liebowitz and S. E. Margolis, "Network externality—an uncommon tragedy," *Journal of Economic Perspectives*, vol. 8, no. 2, pp. 133–150, 1994.

[56] P. Anderson, *What is Web 2.0? Ideas, Technologies and Implications for Education*, JISC Technology & Standards Watch, 2007.

[57] S. Gaonkar, J. Li, R. R. Choudhury, L. Cox, and A. Schmidt, "Micro-blog: sharing and querying content through mobile phones and social participation," in *Proceedings of the 6th*

International Conference on Mobile Systems, Applications, and Services (Mobisys'08), pp. 174–186, 2008.

[58] Facebook, Statistics, 2011, http://www.facebook.com/press/info.php?statistics.

[59] A. Lipsman, *The Network Effect: Facebook, Linkedin, Twitter & Tumblr Reach New Height in May*, 2011, http://blogcom-.score.com/2011/06/facebook_linkedin_twitter_tumblr.html.

[60] C. W. Speed, M. G. Meekan, and C. J. A. Bradshaw, "Spot the match—wildlife photo-identification using information theory," *Frontiers in Zoology*, vol. 4, article 2, 2007.

[61] Z. Arzoumanian, J. Holmberg, and B. Norman, "An astronomical pattern-matching algorithm for computer-aided identification of whale sharks Rhincodon typus," *Journal of Applied Ecology*, vol. 42, no. 6, pp. 999–1011, 2005.

[62] M. J. Kelly, "Computer-aided photograph matching in studies using individual identification: an example from Serengeti cheetahs," *Journal of Mammalogy*, vol. 82, no. 2, pp. 440–449, 2001.

[63] C. J. R. Anderson, N. Da Vitoria Lobo, J. D. Roth, and J. M. Waterman, "Computer-aided photo-identification system with an application to polar bears based on whisker spot patterns," *Journal of Mammalogy*, vol. 91, no. 6, pp. 1350–1359, 2010.

[64] R. Stafford, A. G. Hart, L. Collins et al., "Eu-social science: the role of internet social networks in the collection of bee biodiversity data," *PloS one*, vol. 5, no. 12, p. e14381, 2010.

[65] M. Ebner and M. Schiefner, "Microbloggging—more than fun?" in *Proceedings of the IADIS Mobile Learning Conference*, Algarve, Portugal, 2008.

[66] M. Ames and M. Naaman, "Why we tag: motivations for annotation in mobile and online media," in *Proceedings of the 25th SIGCHI Conference on Human Factors in Computing Systems (CHI '07)*, vol. 1-2, pp. 971–980, May 2007.

[67] C. L. Kirkhope, R. L. Williams, C. L. Catlin-Groves et al., "Social networking for biodiversity: the BeeID project," in *Proceedings of the International Conference on Information Society (i-Society '10)*, 2010.

[68] Y. F. Wiersma, "Birding 2.0: citizen science and effective monitoring in the web 2.0 world," *Avian Conservation and Ecology*, vol. 5, no. 2, 2010.

[69] E. Paulos, "Designing for doubt: citizen science and the challenge of change," in *Engaging Data: First International Forum on the Application and Management of Personal Electronic Information*, MIT, 2009.

[70] M. F. Goodchild, "Citizens as sensors: the world of volunteered geography," *GeoJournal*, vol. 69, no. 4, pp. 211–221, 2007.

[71] M. F. Goodchild, "Citizens as voluntary sensors: spatial data infrastructures in the world of web 2.0," *Journal of Spatial Data Infrastructures Research*, vol. 2, p. 8, 2007.

[72] A. Flanagin and M. Metzger, "The credibility of volunteered geographic information," *GeoJournal*, vol. 72, no. 3-4, pp. 137–148, 2008.

[73] T. A. Callister Jr., "Media literacy: on-ramp to the literacy of the 21st century or cul-de-sac on the information superhighway," *Advances in Reading/Language Research*, vol. 7, p. 17, 2000.

[74] D. Lepage and C. M. Francis, "Do feeder counts reliably indicate bird population changes? 21 Years of winter bird counts in Ontario, Canada," *The Condor*, vol. 104, no. 2, pp. 255–270, 2002.

[75] J. J. Lennon, P. Koleff, J. J. D. Greenwood, and K. J. Gaston, "Contribution of rarity and commonness to patterns of species richness," *Ecology Letters*, vol. 7, no. 2, pp. 81–87, 2004.

[76] Y. Bas, V. Devictor, J. P. Moussus, and F. Jiguet, "Accounting for weather and time-of-day parameters when analysing count data from monitoring programs," *Biodiversity and Conservation*, vol. 17, no. 14, pp. 3403–3416, 2008.

[77] W. L. Kendall, B. G. Peterjohn, and J. R. Sauer, "First-time observer effects in the North American Breeding Bird Survey," *Auk*, vol. 113, no. 4, pp. 823–829, 1996.

[78] F. Jiguet, "Method learning caused a first-time observer effect in a newly started breeding bird survey," *Bird Study*, vol. 56, no. 2, pp. 253–258, 2009.

[79] J. R. Sauer, B. G. Peterjohn, and W. A. Link, "Observer differences in the North-American Breeding Bird Survey," *Auk*, vol. 111, no. 1, pp. 50–62, 1994.

[80] R. E. McCaffrey, "Using citizen science in urban bird studies," *Urban Habitats*, vol. 3, no. 1, p. 16, 2005.

[81] R. Caruana, M. Elhawary, A. Munson et al., "Mining citizen science data to predict prevalence of wild bird species," in *Proceedings if the 12th ACM SIGKDD International Conference on Knowledge Discovery and Data Mining (KDD '06)*, pp. 909–915, ACM, Philadelphia, Pa, USA, August 2006.

[82] E. Rahm, "Data cleaning: problems and current approaches," *IEEE Data Engineering Bulletin*, vol. 23, p. 11, 2000.

[83] H. B. Newman, M. H. Ellisman, and J. A. Orcutt, "Data-intensive e-science frontier research," *Communications of the ACM*, vol. 46, no. 11, pp. 68–77, 2003.

[84] M. J. G. Hopkins, "Modelling the known and unknown plant biodiversity of the Amazon Basin," *Journal of Biogeography*, vol. 34, no. 8, pp. 1400–1411, 2007.

[85] C. A. Lepczyk, "Integrating published data and citizen science to describe bird diversity across a landscape," *Journal of Applied Ecology*, vol. 42, no. 4, pp. 672–677, 2005.

[86] C. Eraud, J. M. Boutin, D. Roux, and B. Faivre, "Spatial dynamics of an invasive bird species assessed using robust design occupancy analysis: the case of the Eurasian collared dove (*Streptopelia decaocto*) in France," *Journal of Biogeography*, vol. 34, no. 6, pp. 1077–1086, 2007.

[87] B. K. Williams, J. D. Nichols, and M. J. Conroy, *Analysis and Management of Animal Populations*, Academic Press, London, UK, 2002.

[88] P. H. Crowley, "Resampling methods for computation-intensive data analysis in ecology and evolution," *Annual Review of Ecology and Systematics*, vol. 23, no. 1, pp. 405–447, 1992.

[89] W. A. Link, J. R. Sauer, and D. K. Niven, "A hierarchical model for regional analysis of population change using Christmas Bird Count data, with application to the American Black Duck," *The Condor*, vol. 108, no. 1, pp. 13–24, 2006.

[90] S. Reddy, D. Estrin, and M. Srivastava, "Recruitment framework for participatory sensing data collections," in *Proceedings of the 8th International Conference on Pervasive Computing*, P. Floréen, A. Krüger, and M. Spasojevic, Eds., pp. 138–155, Springer, Berlin, Germany, 2010.

[91] Y. Jun, W. Weng-Keen, and R. A. Hutchinson, "Modeling experts and novices in citizen science data for species distribution modeling," in *Proceedings of the 10th IEEE International Conference on Data Mining (ICDM '10)*, pp. 1157–1162, December 2010.

[92] P. Cohen and N. Adams, "Intelligent Data Analysis in the 21st Century," in *Proceedings of the 8th International Symposium on Intelligent Data Analysis: Advances in Intelligent Data Analysis VIII (IDA '09)*, N. Adams, C. Robardet, A. Siebes, and J. -F. Boulicaut, Eds., pp. 1–9, Springer, Berlin, Germany, 2009.

[93] D. Z. Wang, E. Michelakis, M. N. Garofalakis, and J. M. Hellerstein, "BayesStore: managing large, uncertain data repositories

with probabilistic graphical models," in *Proceedings of the VLDB Endowment*, vol. 1, no. 1, pp. 340–351, 2008.

[94] W. A. Link and J. R. Sauer, "Seasonal components of avian population change: joint analysis of two large-scale monitoring programs," *Ecology*, vol. 88, no. 1, pp. 49–55, 2007.

11

Genetic Characterization of Six Stocks of *Litopenaeus vannamei* Used in Cuba for Aquaculture by Means of Microsatellite Loci

Anna Pérez-Beloborodova,[1] Adriana Artiles-Valor,[2] Lourdes Pérez-Jar,[2] Damir Hernández-Martínez,[1] Missael Guerra-Aznay,[2] and Georgina Espinosa-López[3]

[1] *Conservation Genetic Group, Marine Research Centre, University of Havana, Street 16 No 114, Playa Havana, CP 11300, Cuba*
[2] *Molecular Biology Laboratory, Aquaculture Division, Fisheries Research Centre, 5th Avenue and 246, Barlovento, Playa, Havana, CP 19100, Cuba*
[3] *Department of Biochemistry, Faculty of Biology, Havana University, Street 25 No. 455 between J. and I. Vedado, Havana, Cuba*

Correspondence should be addressed to Georgina Espinosa-López, georgina@fbio.uh.cu

Academic Editor: Pung-Pung Hwang

Four microsatellite loci were used to achieve genetic characterization of six stocks from *Litopenaeus vannamei* used for aquaculture in Cuba: second generation from first introduction (S2-1), first generation from the second one (S1-2), from the third one (S1-3), and the fourth one (S1-4) and the crossings from two parental population: first generation from the first with first generation from the third (S1-1 × S1-3) and first generation from the second with first generation from the third (S1-2 × S1-3). 66% (16/24) of genetic systems in total loci were in genetic disequilibrium. The four microsatellite loci were polymorphic for all six stocks. Major quantities of allelic variants correspond to locus Pvan 1758; which is at the same time that one where there are private alleles from first generation of the third. All Fst comparisons were significant. This indicates big differences between stocks. The highest values are those in which there is presence of the second introduction. This introduction and its descendants are also more consanguineous.

1. Introduction

In Latin America shrimp-producer countries, the Pacific white shrimp, *Litopenaeus vannamei*, is the most representative species, with about 90% of production. Native from East Pacific, and from the tropical American continent, this species has shown an excellent culture adaptation and has been more resistant to salinity, oxygen, and temperature fluctuations. That is why, in the last twenty years, *Litopenaeus vannamei*, has been introduced in many culture programs, and nowadays it is the second culture after *Penaeus monodon*.

In 2003 the first introduction of two stocks of White Pacific Shrimp, *Litopenaeus vannamei*, [1] was achieved in Cuba, imported from USA, Shrimp Improving System and so handling, nutrition, and health techniques are well established, as well as the assessment of genetic variation in farms that had before cultured the indigenous species *Litopenaeus*

schmitti. In total, five stocks have been introduced, and all of them have been characterized using microsatellite techniques [2–4].

Genetic studies have a capital importance in shrimp industry, in order to determine genetic variability level either in natural or cultured populations, but mainly to know when the latter could be enriched with new specimens [5]. It is moreover important to know the structure of natural population from which those specimens will be taken [6] and also to have good markers that allow population and family studies. It is clear that the priority should be given to the domestication and handling of broodstocks through the application of genetic techniques [7–9]. The aim of the present work is to characterize different cultured stocks of *Litopenaeus vannamei*, used in Cuba for aquaculture, by means of microsatellite markers.

Genetic Characterization of Six Stocks of Litopenaeus vannamei Used in Cuba for Aquaculture by Means of
Microsatellite Loci

93

2. Materials and Methods

2.1. Samples. Samples were taken from pleopods of 30 shrimps, specifically from the fourth pair, between exo-and endopodite. The same male and female quantity was taken randomly in shrimps from the Postlarvae Production Hatchery "YAGUACAM," Cienfuegos, Cuba.

Individuals from first or second generation of introduced shrimps were taken for this genetic characterization as described below. Two first parental generations from first and second introduced stocks were randomly crossed with descendants of the first generation of the third introduced stock resulting in (S1-2 × S1-3) and (S1-1 × S1-3) as it is called in this work.

Thus, characterized lots were second generation from first introduced stock (S2-1), first generation from the second one (S1-2), from the third one (S1-3), and from the fourth one (S1-4) and the crossings: first generation from the first with first generation from the third (S1-1 × S1-3) and first generation from the second with first generation from the third (S1-2 × S1-3).

2.2. Microsatellite Genotyping. DNA isolation, amplification programs, and electrophoresis procedures were carried out as described in [2]. Used loci were M-1, isolated from *L. vannamei* by [10] and Pvan 0040, Pvan 1758, and Pvan 1815 obtained from the same species by [11]. As weight allelic controls, samples previously genotyped by [2], as well as PGEM, were used.

2.3. Statistics. Allele number by locus, allele frequency, and observed and expected heterozygosity for each locus, as well as, the Hardy-Weinberg equilibrium, were determined by GeneAlEx 6.1 program [12]. A locus was considered as polymorphic if it presented at least two alleles and when the most common frequency of an allele did not exceed 95%.

The FSTAT statistical package version 2.93 [13] was used to calculate the linkage disequilibrium, and also the inbreeding coefficient within populations, F_{is} values [14], and pairwise F_{st} and P values between populations, after 1500 permutations. Significance levels were assessed through the Markov chains using 10000 Dememorisation, 100 batches, and 5000 iterations per batch using the Genepop Version 4.1.0 [15].

Assignment tests were used to clarify the belonging of individuals to the stocks. The Bayesian assignment test was performed [16] among shrimp stocks from Cuba using GeneClass 2 [17]. Simulations were run to determine the probability of assignment, using a probability of rejection set at $P < 0.05$.

Relatedness coefficient (r) according to [18] between individuals' pairs was calculated by GeneAlEx 6.1 program [12]. The equation for this coefficient taking into consideration codominant marker is

$$r = \frac{\sum_x \sum_k \sum_l (p_y - p^*)}{\sum_x \sum_k \sum_l (p_x - p^*)}, \quad (1)$$

where x represents individuals, k all loci, l allelic positions (2 for diploids, 1 for haploids), P_x the frequency of the individual x for the locus k and allelic position l, P_y the frequency of the allele in the group or individual with it is compared x, and P^* the total frequency in the population.

Relatedness coefficient must be $r \leq 0$ to unrelated individuals, $r = 0.25$ to half sibs, and $r \geq 0.5$ to full sibs [18].

Relatedness means coefficients were compared with the Kruskal Wallis test (nonparametric ANOVA) by means of graphPad InStat version 3.00 [19]

3. Results

3.1. General Parameters of Genetic Variation of Six Stocks Descendants from Original Introductions: Hardy-Weinberg Deviation of Equilibrium. As it has been previously reported, *L. vannamei* shrimps introduced from SIS are characterized [2–4].

In the present work, the genetic variation analysis was performed to six relative stocks of those introductions by means of four microsatellite loci: M1, isolated from *L. vannamei*, and Pvan 0040, Pvan 1758, and Pvan 1815 also obtained from these species. Linkage disequilibrium was analyzed, and as it is known [2, 4, 11, 20], those loci for *L. vannamei* are not linked.

In Table 1 are shown the main genetic estimation parameters, such as number of alleles (Na) and number of effective alleles (Ne), observed and expected heterozigocities (Ho and He), and so, the Hardy-Weinberg equilibrium deviations (F_{IS} values). As several stocks are compared, private alleles are also provided (Np). Note that first generation from second introduction (S1-2) is the one of less genetic variability, within first generation stocks, and even lesser than the only second generation stock analyzed. Highest variation stock belongs to crossing of first generation from the first with first generation from the third (S1-1 × S1-3). The lowest relative variation of any generations and stocks which included second introduction is remarkable and stocks which included second introduction.

Normally, the Hardy-Weinberg equilibrium conditions are not accomplished in cultures. In Table 1 it is observed that 66% (16/24) of genetic systems in total loci are significant in genetic disequilibrium by homozygotes excess.

3.2. Differentiation between Stocks. In Figure 1 frequencies of alleles of each microsatellite for all stocks are shown. Allele sizes are on range reported by other authors [2–4, 10, 20–22]. All loci were polymorphic and those with more allelic variants were Pvan 1758 (12 variants) and Pvan 1815 (9 variants), which is coincident with previous studies. Private alleles found in this paper corresponded to those loci. In the case of M1, distribution of frequencies is bimodal, with two predominant sizes, which includes the majority of stocks: 216, which contains all of them, and 202 in which it only lacks S1-4. In total there are eight allelic variants, and there are not private alleles for this microsatellite region.

TABLE 1: Genetic variability parameters and the Hardy-Weinberg deviation of equilibrium (F_{IS}) and its probability (Pfis) of four microsatellites loci isolated from *L. vannamei*, M-1 [10], Pvan 0040, Pvan 1758 and Pvan 1815 [11], in six different cultured stocks of *L. vannamei* used for aquaculture in Cuba. *n*: number of samples. Na: number of alleles, Ne: number of effective alleles, Np: number of private alleles, He: expected heterozygosity, Ho: observed heterozygosity. The associated probability was estimated using the Markov chains (10000 Dememorisation, 100 batches, and 5000 iterations per batch using Genepop program Version 4.1.0. Numbers in parenthesis are standard mean error.

Stock	N	Na	Ne	Np	Ho	He	Fis	Pfis
				Locus M1				
S2-1	19	7	4.6	0	0.89	0.78	−0.115	0.001
S1-2	19	4	2.2	0	0.62	0.55	−2.212	0.660
S1-3	16	5	3.6	0	0.69	0.72	0.078	0.000
S1-4	22	3	2.0	0	0.59	0.50	−0.147	0.142
S1-1 × S1-3	23	5	3.0	0	0.91	0.67	−0.349	0.242
S1-2 × S1-3	25	4	3.1	0	0.36	0.68	0.485	0.000
				Locus Pvan 0040				
S2-1	21	4	2.4	0	0.52	0.58	0.127	0.002
S1-2	21	2	1.3	0	0.24	0.21	−0.111	1.000
S1-3	19	4	2.5	0	0.16	0.59	0.746	0.000
S1-4	22	3	2.7	0	0.36	0.63	0.440	0.000
S1-1 × S1-3	23	4	3.9	0	0.56	0.74	0.261	0.000
S1-2 × S1-3	26	2	1.3	0	0.23	0.20	-0.111	1.000
				Locus 1758				
S2-1	21	7	3.8	0	0.67	0.54	0.121	0.00
S1-2	21	4	3.5	0	0.67	0.71	0.088	0.000
S1-3	20	9	6.2	3	0.75	0.84	0.132	0.054
S1-4	22	7	3.9	1	0.68	0.74	0.108	0.009
S1-1 × S1-3	23	6	4.5	0	0.61	0.78	0.239	0.119
S1-2 × S1-3	26	6	1.8	0	0.73	0.64	-0.116	0.563
				Locus 1815				
S2-1	21	3	2.1	0	0.14	0.52	0.737	0.00
S1-2	21	4	1.6	0	0.25	0.37	0.371	0.056
S1-3	20	6	3.8	0	0.65	0.74	0.147	0.006
S1-4	22	6	4.4	1	0.73	0.77	0.084	0.000
S1-1 × S1-3	23	8	3.8	0	0.52	0.74	0.315	0.000
S1-2 × S1-3	26	4	3.6	0	0.65	0.72	0.115	0.000
				All *loci*				
S2-1	20.5(0.5)	5.2 (1.0)	3.2 (0.6)	0	0.56 (0.16)	0.66 (0.06)	0.176	0.000
S1-2	20.5 (0.5)	3.5 (0.5)	2.1 (0.5)	0	0.46 (0.13)	0.46 (0.11)	0.032	0.000
S1-3	18.7 (0.9)	6.0 (1.0)	4.0 (0.8)	3	0.56 (0.14)	0.72 (0.05)	0.250	0.000
S1-4	22.0 (0.0)	4.7 (1.0)	3.3 (0.5)	2	0.59 (0.08)	0.66 (0.06)	0.132	0.000
S1-1 × S1-3	23.0 (0.0)	5.0 (0.8)	3.8 (0.3)	0	0.65 (0.90)	0.73 (0.02)	0.131	0.000
S1-2 × S1-3	25.7 (0.2)	4.0 (0.8)	2.7 (0.5)	0	0.50 (0.12)	0.56 (0.12)	0.141	0.000

Pvan 0040 locus is that one which has less allelic variants, with only five. Its frequency distribution is unimodal and in 141 size all stocks are represented. Two other loci, Pvan 1758 and Pvan 1815, are unimodal and have private alleles that contribute to the highest quantity of allelic variants. Private alleles correspond to the first generations of the third and fourth introductions. Sizes and frequencies of appeared alleles (within parenthesis) are, for locus 1758: 182 (0.025), 184 (0.250), and 193 (0.025), which appeared in first generation of the third and 210 (0.023) from first generation

of the 4th. To 1815 there is only a private allele from first generation of the 4th: 111 (0.091).

The assignment test results are represented in Table 2. It revealed that 93.3% (125/134) from the shrimp were correctly assigned. S1-3 stock was the most failed assignation with only 80% of right assignation.

F_{st} calculations are shown on Table 3, comparing all stocks among them. It is observed that all comparisons are statistically significant. The biggest values are those that have the presence of the second introduction. On the other hand,

Genetic Characterization of Six Stocks of Litopenaeus vannamei Used in Cuba for Aquaculture by Means of Microsatellite Loci

95

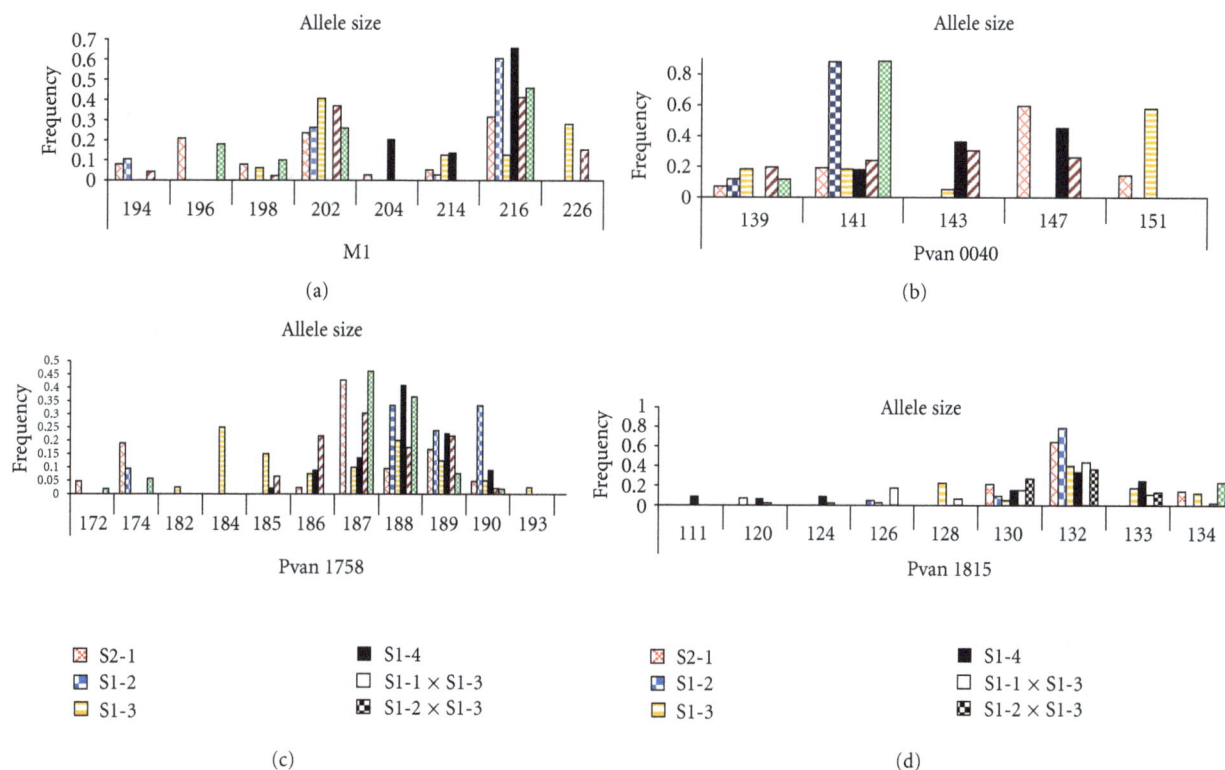

FIGURE 1: Allelic frequencies of loci M-1 [11], Pvan 0040, Pvan 1758 and, Pvan 1815 [12] for six *L. vannamei* stocks used for aquaculture in Cuba.

TABLE 2: Number of assigned individuals to the correspondent stock and/or to any other using the Bayesian assignment test [16] by GeneClass [17]. Simulations were run to determine the probability of assignment, using a probability of rejection set at $P < 0.05$.

Stock	Assigned to the same stock	Assigned to another stock	% of assignment to the same stock
S2-1	19	2	90,50
S1-2	20	1	95,5
S1-3	16	4	80
S1-4	21	1	95,4
S1-1 × S1-3	23	0	100
S1-2 × S1-3	26	1	96
Total	125	9	93.2

TABLE 3: F_{st} values estimated by FSTAT program [13] between *L. vannamei* shrimp stocks used for culture in Cuba. For all comparisons, P value $= 0.00067$; it was obtained after 1500 permutations also by the FSTAT program. Indicative adjusted nominal level (5%) for multiple comparisons is 0.003333.

	S2-1	S1-2	S1-3	S1-4	S1-1 × S1-3
S1-2	0.2163				
S1-3	0.1496	0.2387			
S1-4	0.1192	0.2120	0.1896		
S1-1 × S1-3	0.0667	0.1637	0.0962	0.0663	
S1-2 × S 1–3	0.1639	0.1237	0.2030	0.1922	0.1282

less F_{st} values are observed in comparisons between S1-3, S1-4, and S2-1.

3.3. Relatedness Values. Relatedness coefficients distributions between individual pairs of six stocks are shown in Figure 2, calculated according to [18]. All stocks, except S1-2 and crossing that contains it (S1-2 × S1-3), present unimodal distributions. Three of them present the mode in negative relatedness values (S1-3, S1-4, S1-1 × S1-3), and the only one from a second generation S2–1 presents the mode in zero. In the case of exceptions, for S1-2, one of the modes is near zero, while the second one is near positive values but

lower than 0.5. For crossing (S1-2 × S1-3), the first mode is in negative values and the other one near zero. Mean relatedness coefficients values (Figure 2, right inside) have not significant differences and have standard deviation values. Those are for each stock, S2-1: −0.028; S1-2: 0.044; S1-3: −0.063; S1-4: −0.023; S1-2 × S1-3: −0.019; S1-1 × S1-3: −0.044.

4. Discussion

4.1. Genetic Diversity within L. vannamei Stocks. The four microsatellite loci were polymorphic for all six stocks. The majority of expected and observed heterozigocities are on range according to many authors using microsatellites for penaeid shrimps [2–4, 20–26]. However, they are low and above range for one of the markers, Pvan 0040, which

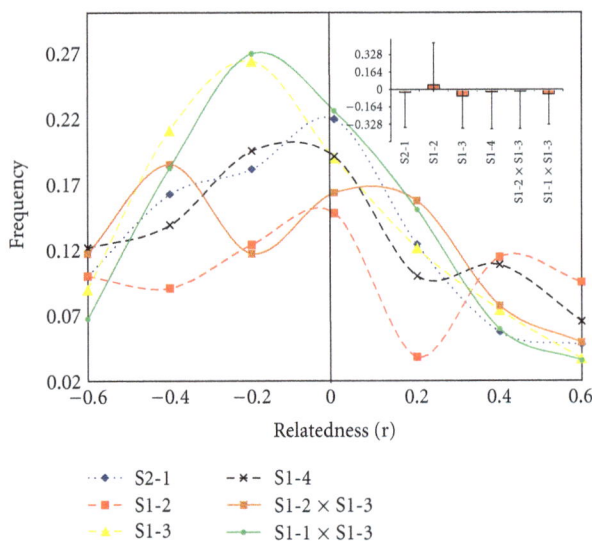

FIGURE 2: Relatedness coefficient distributions between individual pairs from stocks of *L. vannamei* shrimps used for culture in Cuba, calculated according to [18], with 4 microsattelite loci: M-1 [10], Pvan 0040, Pvan 1758, and Pvan 1815 [11]. Rigth inside: mean values and standard deviation of relatedness coefficients for all stocks. *P* value for statistical comparison is 0.2426.

is the less variable loci of all stocks in which the second introduction is involved: in the first generation of the second introduction Ho = 0.24 and He = 0.21 and in crossing of first generation of second introduction with first generation of the first. Previously, [2–4] reported a great contribution of this locus to the decrease of total heterozigocities. Pvan 0040 was monomorphic for the original fourth and fifth introductions [3, 4], and so it could be considered that this locus is sufficiently sensitive to changes in different stocks.

Equilibrium deviations are significant in most loci by stocks, which in accord with other authors' results [3, 4, 20, 22].

Most of allelic frequency graphics are unimodals, with exception of M1 locus. The biggest quantities of allelic variants correspond to locus Pvan 1758, which is at the same time that one in which there are private alleles from first generation of the third. This is in agreement with genetic variability reported for that locus in previous works [20, 22, 25].

4.2. Differentiation between Stocks. Indeed all F_{st} comparisons were significant. They indicate big differences between stocks. The biggest values are those in which there is presence of the second introduction. In a similar way, other authors also obtained significant differences in natural populations as well as in cultured stocks [20, 22]. Using allozymes, the authors in [27, 28] also obtained a genetic differentiation in wild population of *L. shmitti* in Cuba and in wild and cultured *L. vannamei* stocks in Norwest Mexico, respectively. Those differences are in agreement with the assigning exercise, in which only first generation of the third stock had low percentages of assigning. It could be because there are

components of those stocks in others obtained by meanings of crossing with this third introduction.

Although studies with *L. vannamei* in Cuba indicate a decrease in genetic variability, existing stocks at the moment of this genetic characterization, even with little variations, seem, however, significantly different between them. The reason could be that in their origins they constitute genetic lines well diverse in their origins in respect to the main objective for production (e.g., illness resistance, increasing growth, etc.). However, once the selection pressure is increased, genetic variation is decreased as it could be seen in different genetic variability reports for several introductions [4, 29].

4.3. Relatedness Coefficients. Borrell et al. [2] pointed out the risk that could implicate using individuals from the second introduced stock for crossing with others without any methodology such as the following, of relatedness coefficient, that offers information about inbreeding levels. For the second stock, they obtained a mean relatedness coefficient indicative of more related individuals than those of the first one, similar to the results of this work, in which the most consanguineous stock is descendants of this second introduced stock. The highest value, indicative of a great relatedness, was obtained for the fifth introduced stock of *L. vannamei* in Cuba [4], and that is why the use of this molecular tool is of crucial importance for designing and following crossings. The rest of the stocks show means and distributions in agreement with individuals presumed not related, similar to that obtained by [30] for turbot, near zero. The second introduction and its descendants, as well as crossings that contain it, the same as the fifth [4], should be followed with this methodology, and also estimation of heritability for interested characters for culture would be profitable.

4.4. General Remarks. A progressive decrease in genetic variability is observed in successive introductions of *L. vannamei* into Cuba since 2003 and henceforth. It means that the first broodstock should be maintained and combined with others or new stock. Relatedness coefficients support the use of this stock and highlight the fact that the second stock should be carefully used in crossings because of their tendency to positive values.

However, the sustainability of the culture has been achieved, as well as maintenance of no viral diseases that have caused considerable losses in Latin America [31]. In that way, if surveillance program [32], starts with the premise of introducing SPF animals (Shrimp Pathogens Free) for health maintenance in cultures, a strategy of genetic management with microsatellite markers will warranty an adequate productive performance that does not drive to an irreversible endogamy. In agreement with [2], the maintenance of a common origin, besides of health warranty ever crossing stocks genetically different, avoids phenomena as outbreeding depression [33–35] that could damage productive yields in a short time.

Genetic Characterization of Six Stocks of Litopenaeus vannamei Used in Cuba for Aquaculture by Means of
Microsatellite Loci

97

Acknowledgments

Many thanks are due to workers and technicians from the Postlarvae Production Hatchery "YAGUACAM," in Cienfuegos, Cuba, for technical support in maintenance of lines, especially to its director, Ms. Angela Moreno. Special thanks are due to Dr. Vicente Berovides for his critical revision of the text and also to MSc. Román Machado for his comments. The authors also thank the anonymous referees for improving the document with their comments and remarks. They also want to express their gratitude to Ms. Mercedes Escobar for editing and improving English from the original manuscript.

References

[1] R. Tizol, B. Jaime, R. Laria et al., "Introduction of Pacific White Shrimp L. vannamei in Cuba. Quarantine I step (In Spanish). Ocean Docs," http://hdl.handle.net/1834/3588, 2004.

[2] Y. J. Borrell, G. Espinosa, E. Vazquez, J. A. Sánchez, and G. Blanco, "Genetic variability of microsatellite loci in the two first stocks of Litopenaeus vannamei introduced to Cuba for aquaculture," Revista de Investigaciones Marinas, vol. 27, no. 3, pp. 237–244, 2006 (Spanish).

[3] R. Machado, Assessment of genetic variability in two lots of white shrimp, Litopenaeus vannamei (Boone, 1931) introduced to Cuba, M.S. thesis, International Fisheries Management. Department of Aquatic Biosciences. Norwegian College of Fishery Science. University of Tronso, Norway, Oslo, 2006.

[4] A. Artiles, I. Rodríguez, A. Pérez, L. Pérez, and G. Espinosa, "Low genetic variability in the fifth introduction of Litopenaeus vannamei in Cuba, as estimated with microsatellite markers," Biotecnología Aplicada, vol. 28, pp. 142–146, 2011.

[5] D. K. García, M. A. Faggart, L. Rhoades et al., "Genetic diversity of cultured Penaeus vannamei shrimp using three molecular genetic techniques," Molecular Marine Biology and Biotechnology, vol. 3, no. 5, pp. 270–280, 1994.

[6] S. Sunden and K. Davis, "Evaluation of genetic variation in a domestic population of Penaeus vannamei (Boone): a comparison with three natural populations," Aquaculture, vol. 97, no. 2-3, pp. 131–142, 1991.

[7] R. A. Dunham, K. Majumdar, E. Hallerman et al., "Review of the status of aquaculture genetics," in Aquaculture in the Third Millennium. Technical Proceedings of the Conference on Aquaculture in the Third Millennium (February 2000, Bangkok, Thailand), R. P. Subasinghe, P. Bueno, M. J. Phillips, C. Hough, S. E. McGladdery, and J. R. Arthur, Eds., pp. 137–166, NACA/FAO, 2001.

[8] G. Hulata, "Genetic manipulations in aquaculture: a review of stock improvement by classical and modern technologies," Genetica, vol. 111, no. 1-3, pp. 155–173, 2001.

[9] Z. J. Liu and J. F. Cordes, "DNA marker technologies and their applications in aquaculture genetics," Aquaculture, vol. 238, no. 1–4, pp. 1–37, 2004.

[10] G. M. Wolfus, G. K. García, and A. Alcivar-Warren, "Application of the microsatellite technique for analyzing genetic diversity in shrimp breeding programs," Aquaculture, vol. 152, no. 1–4, pp. 35–47, 1997.

[11] P. Cruz, C. H. Mejía-Ruiz, R. Pérez-Enriquez, and A. M. Ibarra, "Isolation and characterization of microsatellites in Pacific white shrimp Penaeus (Litopenaeus) vannamei," Molecular Ecology Notes, vol. 2, no. 3, pp. 239–241, 2002.

[12] R. Peakall and P. E. Smouse, "GENALEX 6: genetic analysis in Excel. Population genetic software for teaching and research," Molecular Ecology Notes, vol. 6, no. 1, pp. 288–295, 2006.

[13] J. Goudet, "FSTAT, a program to estimate and test gene diversities and fixation indexes. (version 2.9.3)," Journal of Heredity, vol. 86, pp. 485–486, 2002.

[14] S. Wright, "The interpretation of population structure by F-statistics with special regards to systems of mating," Evolution, vol. 19, pp. 395–420, 1965.

[15] F. Rousset, Genepop 4.1 for Windows/Linux/Mac OS X, 2008.

[16] B. Rannala and J. L. Mountain, "Detecting immigration by using multilocus genotypes," Proceedings of the National Academy of Sciences of the United States of America, vol. 94, no. 17, pp. 9197–9201, 1997.

[17] S. Piry, A. Alapetite, J. M. Cornuet, D. Paetkau, L. Baudouin, and A. Estoup, "GeneClass 2: a software for genetic assignment and first-generation migrant detection," Journal of Heredity, vol. 95, no. 6, pp. 536–539, 2004.

[18] D. C. Quelle and K. F. Goodnight, "Estimating relatedness using genetic markers," Evolution, vol. 43, no. 2, pp. 258–275, 1989.

[19] GraphPad InStat version 5.04 for Windows, GraphPad Software, La Jolla California USA, http://www.graphpad.com .

[20] R. Valles-Jiménez, P. Cruz, and R. Pérez-Enriquez, "Population genetic structure of Pacific white shrimp (Litopenaeus vannamei) from Mexico to Panama: microsatellite DNA variation," Marine Biotechnology, vol. 6, no. 5, pp. 475–484, 2005.

[21] P. Cruz, A. M. Ibarra, H. Mejia-Ruiz, P. M. Gaffney, and R. Pérez-Enríquez, "Genetic variability assessed by microsatellites in a breeding program of pacific white shrimp (Litopenaeus vannamei)," Marine Biotechnology, vol. 6, no. 2, pp. 157–164, 2004.

[22] R. Pérez-Enríquez, F. Hernández-Martínez, and P. Cruz, "Genetic diversity status of White shrimp Penaeus (Litopenaeus) vannamei broodstock in Mexico," Aquaculture, vol. 297, no. 1–4, pp. 44–50, 2009.

[23] N. Bierne, I. Beuzart, V. Vonau, F. Bonhomme, and E. Bedier, "Microsatellite—associated heterosis in hatchery—propagated stokcs of the srimp Penaeus stylirostris," Aquaculture, vol. 184, pp. 203–219, 2000.

[24] Z. Xu, J. H. Primavera, L. D. de la Pena et al., "Genetic diversity of white and cultured black tiguer shrimp (Penaeus monodon) in the Philippines using microsatellites," Aquaculture, vol. 199, pp. 13–40, 2001.

[25] E. Luvesuto, P. Dominguez de Freitas, and P. M. Galetti Junior, "Genetic variation in a closed line of the white shrimp Litopenaeus vannamei (Penaeidae)," Genetics and Molecular Biology, vol. 30, no. 4, pp. 1156–1160, 2007.

[26] J. A. H. Benzie, "Population genetic structure in penaeid prawns," Aquaculture Research, vol. 31, no. 1, pp. 95–119, 2000.

[27] G. Espinosa López, R. Díaz Fernández, U. Becker Zúñiga et al., "Population analysis of Cuban white shrimp Litopenaeus schmitti using allozymes as genetic markers," Revista de Investigaciones Marinas, vol. 24, no. 1, pp. 11–16, 2003 (Spanish).

[28] J. Soto-Hernández and J. M. Grijalva-Chon, "Genetic differentiation in hatchery strains and wild white shrimp Penaeus (Litopenaeus) vannamei (Boone, 1931) from northwest Mexico," Aquaculture International, vol. 12, no. 6, pp. 593–601, 2004.

[29] V. Sbordoni, E. de Matthaeis, M. Cobolli Sbordoni, G. La Rosa, and M. Mattoccia, "Bottleneck effects and the depression of

genetic variability in hatchery stocks of *Penaeus japonicus* (*Crustacea, Decapoda*)," *Aquaculture*, vol. 57, no. 1–4, pp. 239–251, 1986.

[30] Y. J. Borrell, J. Álvarez, E. Vázquez et al., "Applying microsatellites to the management of farmed turbot stocks (*Scophthalmus maximus L.*) in hatcheries," *Aquaculture*, vol. 241, no. 1–4, pp. 133–150, 2004.

[31] A. Artiles, M. Rubio, E. Gonzalez, R. Laria, and R. Silveira, "Crustacean virus of obligatory declaration by the OIE performance in cultured Litopenaeus vannamei in Cuba from 2003 to 2009," *Ocean Docs Digital Repository*, vol. 28, no. 1, pp. 12–18, 2011 (Spanish).

[32] R. Silveira-Coffigny, "Aquaculture health in Cuba," in *Proceedings of the 1st Meeting of Interamerican OIE Comité for Aquatic Animals*, Panama, Panama City, January 2006.

[33] S. Edmands, "Heterosis and outbreeding depression in interpopulation crosses spanning a wide range of divergence," *Evolution*, vol. 53, no. 6, pp. 1757–1768, 1999.

[34] M. Keller, J. Kollmann, and P. J. Edwards, "Genetic introgression from distant provenances reduces fitness in local weed populations," *Journal of Applied Ecology*, vol. 37, no. 4, pp. 647–659, 2000.

[35] S. Granier, C. Audet, and L. Bernatchez, "Heterosis and outbreeding depression between strains of young-of the-year brook trout (Salvelinus fontinalis)," *Canadian Journal of Zoology*, vol. 89, pp. 190–198, 2011.

12

Density-Dependent Habitat Selection in a Growing Threespine Stickleback Population

Ulrika Candolin and Marita Selin

Department of Biosciences, University of Helsinki, P.O. Box 65, 00014 Helsinki, Finland

Correspondence should be addressed to Ulrika Candolin, ulrika.candolin@helsinki.fi

Academic Editor: Thomas Iliffe

Human-induced eutrophication has increased offspring production in a population of threespine stickleback *Gasterosteus aculeatus* in the Baltic Sea. Here, we experimentally investigated the effects of an increased density of juveniles on behaviours that influence survival and dispersal, and, hence, population growth—habitat choice, risk taking, and foraging rate. Juveniles were allowed to choose between two habitats that differed in structural complexity, in the absence and presence of predators and conspecific juveniles. In the absence of predators or conspecifics, juveniles preferred the more complex habitat. The preference was further enhanced in the presence of a natural predator, a perch *Perca fluviatilis* (behind a transparent Plexiglas wall). However, an increased density of conspecifics relaxed the predator-enhanced preference for the complex habitat and increased the use of the open, more predator-exposed habitat. Foraging rate was reduced under increased perceived predation risk. These results suggest that density-dependent behaviours can cause individuals to choose suboptimal habitats where predation risk is high and foraging rate low. This could contribute to the regulation of population growth in eutrophicated areas where offspring production is high.

1. Introduction

Individuals are usually forced to balance costs against benefits when choosing a habitat [1, 2]. Structurally complex habitats are often more favourable than open habitats, as they provide more resources and better refuges against predators [3–5]. However, complex habitats can also be costly if they harbour more predators and competitors than open habitats [6, 7]. In addition, the costs and benefits of choosing a habitat depend on what other individuals in the population are doing [8–10]. The profitability of a habitat decreases the more individuals that occupy it, because of a reduction in the amount of resources available per individual. As high quality habitats become saturated, more and more individuals are forced to occupy poor-quality habitats [11, 12]. Interactions among individuals will then influence habitat selection, and, thus, density-dependent processes will affect individual fitness and the temporal and spatial distribution of the population.

Human activities are currently altering habitats at an unprecedented rate and scale around the world. The consequences that this will have for the populations that were well adapted to the past conditions are poorly known. An environment that is changing rapidly because of human activities is the Baltic Sea. Increased input of nutrients is enhancing primary production, which is altering habitat structure and water turbidity [13, 14]. This has been found to improve the reproductive success of threespine sticklebacks *Gasterosteus aculeatus* breeding in shallow coastal areas. Enhanced growth of filamentous algae and phytoplankton reduces visibility, which allows more stickleback males to establish breeding territories within an area [15–17] and also enhances their hatching success [18]. Males care alone for their offspring, and reduced visibility lowers the risk of predation and reduces intrusions by competing males, which allows males to allocate time from offspring defence to offspring care, that is, to fan more oxygen-rich water into the nest [18]. These positive effects of eutrophication on offspring production could favour population growth [19]. Increased primary production could also have negative effects on population growth by reducing oxygen levels at night time, favouring the growth of toxic algae species, and increasing the sedimentation of dead organic material. However, recent work indicates that stickleback populations in the Baltic Sea are growing [20].

The ultimate effect of an increased offspring production on population growth depends on offspring survival and dispersal. Interactions among individuals and density-dependent processes are expected to influence these demographic components and, hence, regulate population growth. We investigated if habitat choice, risk taking, and foraging rate of offspring depend on the density of individuals, and, thus, if density-dependent processes could potentially contribute to the regulation of population growth. Three-spine sticklebacks migrate in spring from deeper water to shallow coastal waters to breed [21]. The adults die at the end of the breeding season, but the hatched juveniles stay in the shallow waters until late autumn when they migrate to deeper waters (personal observation). The highest mortality occurs in the youngest age classes, as juveniles are highly vulnerable to predation and starvation [22, 23]. Thus, density-dependent juvenile behavior could have profound effects on mortality, fecundity, and dispersal of individuals and, hence, be an important component in the regulation of population growth [24].

To determine the effect of juvenile density on behaviours that influence population growth—habitat choice, risk taking, and foraging rate—we first allowed juveniles to choose between two habitats differing in structural complexity in the presence and absence of predators and conspecifics. The structural complexity of the habitats of the Baltic Sea is currently changing because of an increased growth of filamentous algae [14]. We then determined if the choice of a habitat influences the activity levels and foraging efficiency of the juveniles.

2. Materials and Methods

First-generation offspring 0+ from a threespine stickle-back population originating from the Baltic Sea close to Tvärminne Zoological station were raised in the laboratory (30 families). The juveniles were housed in mixed families (containing individuals from several families) in large flow through containers that contained both open and vegetated areas, on 6 : 18 light : dark cycle at 12°C. They were fed with *Artemia* nauplii and chironomid larvae at least once a day. The juveniles were 5 months old (L_S: 3.8 ± 0.4 cm; mean ± SD) when used in the experiment. The predators, perch *Perca fluviatilis*, were caught from lake Tuusula (close to Helsinki) with a hand trawl and transported to the University of Helsinki. They were kept in large holding tanks at a density of 4 fish/100-litre for a maximum of 4 weeks, at a temperature of 14°C. They were fed chironomid larvae twice a week.

On the day before the experiment started, the juveniles were transferred from the containers to holding tanks (100 × 40 × 30 cm). The tanks had a temperature of 14°C and a density of 10 fish/100-litre. At the end of the experiment,

the juveniles were returned to the flow through containers, and the perch returned to the lake.

2.1. Experiment 1: Habitat Selection. Juvenile sticklebacks were allowed to choose between two habitats that differed in algae cover, in the presence and absence of a predator (behind a Plexiglas wall) and conspecifics. An aquarium (70 × 40 × 30 cm) was divided into a predator section (70 × 16 cm) and a stickleback section (70 × 24 cm), using a transparent Plexiglas partition (Figure 1). Olfactory contact between the predator and the stickleback sections was allowed through small holes in the partitioning sheet. The bottom was covered with 2 cm of sand. In the stickleback section, an open and a vegetated habitat was created by adding eight bunches of artificial algae to one half of the section (35 × 24 cm) while leaving the other half open (Figure 1). Each bunch of algae consisted of 32 thin, green, 20 cm long polypropylene strings. The artificial algae mimicked filamentous algae, and the density represented the density of algae found in nature. Artificial algae was used to ensure that only the structural complexity of the habitat was manipulated and no other potential effects of algae occurred. The temperature of the water was 14°C.

A stickleback juvenile was placed in a transparent box (19 × 13 × 11 cm) with an open bottom in the middle of the stickleback section. After 10 min of acclimatization, the juvenile was carefully released by lifting the transparent box, without disturbing the water, and its choice of habitat was recorded 15, 30, 60, 120, 180, and 1380 min (23 h) after release. The observations were performed behind a blind with observation holes to minimize disturbances. Each juvenile was exposed to four treatments, in a predetermined order that differed among replicates to prevent an effect of previous experience (all possible orders were used) alone, the presence of a predator, the presence of four conspecific juveniles, and the presence of both a predator and four conspecifics. The four treatments were performed on four consecutive days. The water was changed between treatments to prevent an effect of previous treatments through olfactory cues. The juvenile was marked by cutting the edge of one of the dorsal spines. In total, 28 juveniles were tested.

In the predator treatment, a live perch *Perca fluviatilis*, 15–17 cm standard length, was placed into the predator section one day before the experiment was performed. Perch is the main predator in the region from which the sticklebacks were caught (personal observation). Live perch predators had to be used, as sticklebacks habituate to model predators. In the high density treatment, four additional juveniles were placed into the aquarium at the same time as the focal juvenile, first enclosed in the transparent box and then released. The total number of juveniles in the vegetated habitat was recorded and divided by five to give the probability of a juvenile choosing the vegetated habitat.

The probability that a juvenile would stay in the vegetated habitat was calculated for each of the four treatments as the percent of the six observations in which the juvenile was observed in the vegetated habitat. For the high-density treatment, the mean of the probabilities of a juvenile staying in the vegetation during the six observations was used. Time

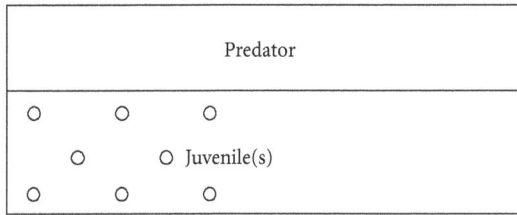

FIGURE 1: Experimental aquarium divided into a predator and a juvenile section using a Plexiglas divider. The vegetated habitat is to the left in the juvenile section.

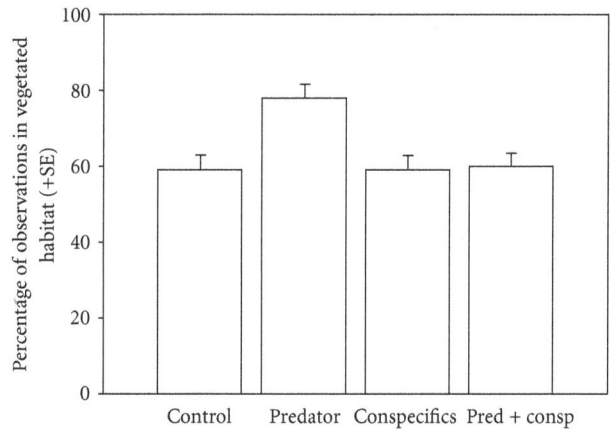

FIGURE 2: Preference of juvenile sticklebacks for the vegetated habitat in the absence and presence of a predator and four conspecifics. $N = 28$ for each treatment.

did not influence the results as qualitatively similar results were gained when the analyses were restricted to only one or a few time periods, and time was hence not further considered in the analyses.

2.2. Experiment 2: Foraging Success and Swimming Activity. Habitat-dependent foraging success and swimming activity was determined by allowing juveniles to feed for 10 min in one of two habitats, open or vegetated, in the presence or absence of a predator. Using the aquarium described above, a vegetated habitat with the same vegetation density as in experiment 1 was created by adding artificial algae to the sticklebacks section (70×24 cm, 16 bunches of artificial algae). The predator section, along the backside of the aquarium (70×16 cm), was left open. In the open habitat treatment, no vegetation was added. All aquaria had 2 cm of sand on the bottom. The stickleback section was divided into three equal sized parts by marking two lines on the front side of the aquarium.

The juveniles were food deprived for 1 day prior to the experiment. Ten chironomid larvae were evenly distributed over the bottom, and a juvenile stickleback was placed in a transparent box in the middle of the aquarium. After 10 min of acclimatization, the juvenile was released and observed for 10 min to determine the number of worms eaten and swimming activity, measured as the number of times the fish crossed the lines marked on the frontside. The number of remaining prey was counted at the end of the experiment. The experiment was done in the presence and absence of a perch in the predator section, using a vegetated and an open habitat, with different juveniles for each treatment. Twenty replicates of each treatment were carried out, each time using different fish.

All data were tested for normality and homogeneity of variances, using Wilk-Shapiro and Bartlett's test for equal variances, respectively. Swimming activity, measured as the number of times the fish crossed the two lines, was square-root transformed before analysis. When assumptions of normality were not met, nonparametric tests were employed.

3. Results

3.1. Habitat Selection. Stickleback juveniles favoured the vegetated habitat over the open habitat in the absence of predators and conspecifics (one sample t-test; $t_{27} = 2.30$, $P = 0.029$, Figure 2). The presence of a predator and/or

conspecifics influenced habitat selection (repeated measures ANOVA, $F_{3,25} = 9.57$, $P < 0.001$). To determine which treatments differed from each other, paired comparisons were carried out. These revealed that the presence of a predator increased the use of the vegetated habitat when the juveniles were held singly (paired t-test for control against predator treatment, $t_{27} = 4.66$, $P < 0.001$) but had no effect on habitat selection when the juveniles were held in a group (paired t-test for conspecifics against predator + conspecifics, $t_{27} = 0.28$, $P = 0.79$, Figure 2). The presence of conspecifics had no effect on habitat selection in the absence of predators (paired t-test for control against conspecifics, $t_{27} = 0.13$, $P = 0.90$), but conspecifics reduced the use of the vegetated habitat in the presence of predators (paired t-test for predator against predator + conspecifics, $t_{27} = 4.24$, $P < 0.001$, Figure 2).

3.2. Foraging Success and Swimming Activity. Vegetation had no effect on foraging success, measured as the number of chironomid larvae eaten (Mann Whitney U-test, $U = 734$, $N1 = N2 = 40$, $P = 0.49$), but the presence of a predator reduced the number of larvae eaten ($U = 590$, $P = 0.029$, Figure 3). The juveniles were less active in the vegetation treatment, measured as number of times they crossed the marks on the frontside of the aquarium (two-way ANOVA; $F_{1,77} = 26.56$, $P < 0.001$, Figure 4). There was a nonsignificant tendency for the presence of a predator to reduce swimming activity ($F_{1,77} = 3.50$, $P = 0.065$, Figure 4), independent of habitat structure ($F_{1,76} = 1.91$, $P = 0.171$).

4. Discussion

Threespine stickleback juveniles preferred the vegetated habitat over the open habitat in the absence of predators and conspecifics. The strength of the preference increased when a juvenile was exposed to a predator behind a Plexiglas wall. This was probably because the vegetated habitat offered better shelter against the predator than the open habitat. A preference for structurally more complex habitats in the face

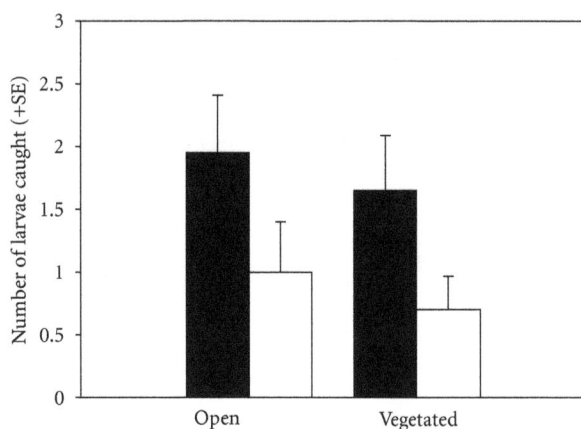

FIGURE 3: Number of chironomid larvae eaten by juvenile stickle-backs in two structurally different habitats, in the absence (black bar) and presence (white bar) of a predator. $N = 20$ for each of the four treatments.

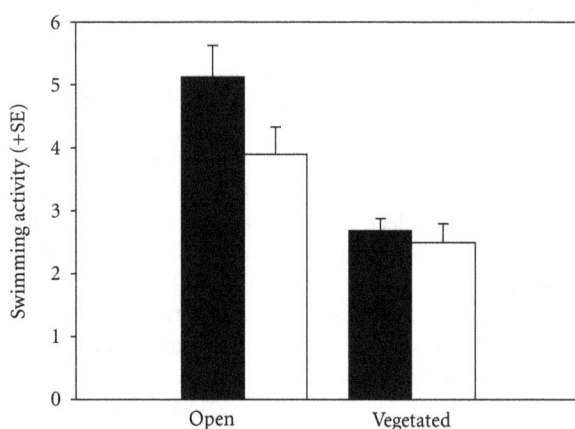

FIGURE 4: Swimming activity of juvenile sticklebacks (square-root transformed + 0.5) + SE in two structurally different habitats in the absence (black bar) and presence (white bar) of a predator. $N = 20$ for each of the four treatments.

of predation is common among fishes, as high structural complexity usually impairs the ability of predators to detect or catch prey [1, 25, 26]. However, increased structural complexity also hinders the detection of predators and can increase their density. Thus, prey has to carefully balance the costs and benefits of staying in each habitat [4, 27]. The preference of the stickleback juveniles for the more complex habitat appears beneficial, as the main predators of sticklebacks in the investigated area are perch and terns, *Sterna hirundo and S. paradisaea* (personal observation), which are visual predators. Perch are efficient piscivores in open areas [28], while birds detect and catch sticklebacks more easily when vegetation is missing or sparse (personal observation).

Interestingly, the predator-induced preference for the more complex habitat was density dependent. An increased density of juveniles reduced the use of the refuge during the predator exposure. The juveniles could have been forced to leave the refuge because of interindividual interactions, or, alternatively, benefitted from leaving the refuge because of the perception of less resources being available per individual in the refuge (food or space, although no food was provided), or a lower per capita mortality risk of venturing out of the refuge when density was higher also outside the refuge. Independent of which scenario, or scenarios, is true, the increased use of the open habitats is expected to increase predation rate on the population. The second experiment on foraging rate and activity found juveniles to reduce their foraging rate when exposed to predators, and they tended to be less active. This suggests that juveniles have to trade foraging for predator avoidance. In the field, juveniles may be forced to use the water column above the vegetation or migrate to areas with sparser vegetation when the density of juveniles increases. This could increase mortality or reduce fecundity, which could contribute to the regulation of the size of the population [29]. Thus, an increased use of predator-exposed areas under high offspring densities could offset the present increase in the reproductive output of the population [15, 18]. However, investigations in the field or under more natural conditions are needed to confirm this.

An increased offspring production because of eutrophication [18] combined with an increased mortality of juveniles because of density-dependent processes could have evolutionary consequences. The strength of natural selection could increase and compensate for a recorded relaxation of sexual selection in eutrophied areas. Reduced visibility due to increased growth of algae has been found to relax sexual selection in sticklebacks by hampering careful mate choice and allowing dishonest sexual signalling [30–32]. As sexual selection is an important evolutionary force that can improve the genetic quality of a population and accelerate adaptation [33], relaxed sexual selection could have negative consequences for population fitness [34–38]. Thus, an increased strength of viability selection when offspring production increases could compensate for relaxed sexual selection. This could accelerate genetic adaptation to the eutrophied conditions.

To determine if density-dependent habitat choice of sticklebacks does regulate population growth under eutrophied conditions, field experiments and monitoring are needed. Moreover, the question of the relative strength of sexual selection, and other components of natural selection in shaping populations during environmental change would deserve more attention [39]. The ability of populations to adapt to changed conditions, such as eutrophication, depends ultimately on genetic changes, which relies on the operation of natural selection. A challenge of future work is to consider the many different factors and pathways that influence population size and viability in changing environments.

Acknowledgments

The authors thank Julia Crombez and Huseyin Sasi for assistance with the experimental procedures, and WANS and the Academy of Finland for financial support. The experimental procedures comply with the laws of the country in which they were performed. They were approved by the Animal Care Committee of the University of Helsinki (no. 133-05).

References

[1] E. E. Werner, G. G. Mittelbach, D. J. Hall, and J. F. Gilliam, "Experimental tests of optimal habitat use in fish: the role of relative habitat profitability," *Ecology*, vol. 64, no. 6, pp. 1525–1539, 1983.

[2] A. I. Houston and J. M. McNamara, *Models of Adaptive Behaviour*, Cambridge University Press, Cambridge, Mass, USA, 1999.

[3] A. J. Kohn and P. J. Leviten, "Effect of habitat complexity on population density and species richness in tropical intertidal predatory gastropod assemblages," *Oecologia*, vol. 25, no. 3, pp. 199–210, 1976.

[4] B. J. Downes, P. S. Lake, E. S. G. Schreiber, and A. Glaister, "Habitat structure and regulation of local species diversity in a stony, upland stream," *Ecological Monographs*, vol. 68, no. 2, pp. 237–257, 1998.

[5] U. Candolin and H. R. Voigt, "Do changes in risk-taking affect habitat shifts of sticklebacks?" *Behavioral Ecology and Sociobiology*, vol. 55, no. 1, pp. 42–49, 2003.

[6] A. Sih, "To hide or not to hide? Refuge use in a fluctuating environment," *Trends in Ecology & Evolution*, vol. 12, pp. 375–376, 1997.

[7] L. Persson and P. Eklöv, "Prey refuges affecting interactions between piscivorous perch and juvenile perch and roach," *Ecology*, vol. 76, no. 1, pp. 70–81, 1995.

[8] P. L. Munday, G. P. Jones, and M. J. Caley, "Interspecific competition and coexistence in a guild of coral-dwelling fishes," *Ecology*, vol. 82, no. 8, pp. 2177–2189, 2001.

[9] S. H. Alonzo, "State-dependent habitat selection games between predators and prey: the importance of behavioural interactions and expected lifetime reproductive success," *Evolutionary Ecology Research*, vol. 4, no. 5, pp. 759–778, 2002.

[10] P. J. Schofield, "Habitat selection of two gobies (*Microgobius gulosus, Gobiosoma robustum*): influence of structural complexity, competitive interactions, and presence of a predator," *Journal of Experimental Marine Biology and Ecology*, vol. 288, no. 1, pp. 125–137, 2003.

[11] E. A. Whiteman and I. M. Côté, "Individual differences in microhabitat use in a Caribbean cleaning goby: a buffer effect in a marine species?" *Journal of Animal Ecology*, vol. 73, no. 5, pp. 831–840, 2004.

[12] J. R. Post, M. R. S. Johannes, and D. J. McQueen, "Evidence of density-dependent cohort splitting in age-0 yellow perch (*Perca flavescens*): potential behavioural mechanisms and population-level consequences," *Canadian Journal of Fisheries and Aquatic Sciences*, vol. 54, no. 4, pp. 867–875, 1997.

[13] S. Råberg, R. Berger-Jönsson, A. Björn, E. Granéli, and L. Kautsky, "Effects of *Pilayella littoralis* on *Fucus vesiculosus* recruitment: implications for community composition," *Marine Ecology Progress Series*, vol. 289, pp. 131–139, 2005.

[14] M. Raateoja, J. Seppälä, H. Kuosa, and K. Myrberg, "Recent changes in trophic state of the Baltic Sea along SW coast of Finland," *Ambio*, vol. 34, no. 3, pp. 188–191, 2005.

[15] U. Candolin, "Effects of algae cover on egg acquisition in male three-spined stickleback," *Behaviour*, vol. 141, no. 11-12, pp. 1389–1399, 2004.

[16] U. Candolin and H. R. Voigt, "Correlation between male size and territory quality: consequence of male competition or predation susceptibility?" *Oikos*, vol. 95, no. 2, pp. 225–230, 2001.

[17] J. Heuschele and U. Candolin, "Reversed parasite-mediated selection in sticklebacks from eutrophied habitats," *Behavioral Ecology and Sociobiology*, vol. 64, no. 8, pp. 1229–1237, 2010.

[18] U. Candolin, J. Engström-Öst, and T. Salesto, "Human-induced eutrophication enhances reproductive success through effects on parenting ability in sticklebacks," *Oikos*, vol. 117, no. 3, pp. 459–465, 2008.

[19] U. Candolin, "Population responses to anthropogenic disturbance: lessons from three-spined sticklebacks Gasterosteus aculeatus in eutrophic habitats," *Journal of Fish Biology*, vol. 75, no. 8, pp. 2108–2121, 2009.

[20] L. Ljunggren, A. Sandström, U. Bergström et al., "Recruitment failure of coastal predatory fish in the Baltic Sea coincident with an offshore ecosystem regime shift," *ICES Journal of Marine Science*, vol. 67, no. 8, pp. 1587–1595, 2010.

[21] U. Candolin and H. R. Voigt, "Size-dependent selection on arrival times in sticklebacks: why small males arrive first," *Evolution*, vol. 57, no. 4, pp. 862–871, 2003.

[22] M. S. Webster, "Role of predators in the early post-settlement demography of coral-reef fishes," *Oecologia*, vol. 131, no. 1, pp. 52–60, 2002.

[23] E. D. Houde, "Mortality," in *Fisheries Science. The Unique Contributions of Early Life Stages*, L. A. Fuiman and R. G. Werner, Eds., pp. 64–87, Blackwell Publishing, Oxford, UK, 2002.

[24] U. Tuomainen and U. Candolin, "Behavioural responses to human-induced environmental change," *Biological Reviews*, vol. 86, no. 3, pp. 640–657, 2011.

[25] J. F. Savino and R. A. Stein, "Behavioural interactions between fish predators and their prey: effects of plant density," *Animal Behaviour*, vol. 37, no. 2, pp. 311–321, 1989.

[26] S. Diehl and P. Eklöv, "Effects of piscivore-mediated habitat use on resources, diet, and growth of perch," *Ecology*, vol. 76, no. 6, pp. 1712–1726, 1995.

[27] D. J. Ferrell and J. D. Bell, "Differences among assemblages of fish associated with *Zostera capricorni* and bare sand over a large spatial scale," *Marine Ecology Progress Series*, vol. 72, no. 1-2, pp. 15–24, 1991.

[28] P. Eklöv and S. Diehl, "Piscivore efficiency and refuging prey: the importance of predator search mode," *Oecologia*, vol. 98, no. 3-4, pp. 344–353, 1994.

[29] W. J. Sutherland and K. Norris, "Behavioural models of population growth rates: implications for conservation and prediction," *Philosophical Transactions of the Royal Society B*, vol. 357, no. 1425, pp. 1273–1284, 2002.

[30] J. Engström-Öst and U. Candolin, "Human-induced water turbidity alters selection on sexual displays in sticklebacks," *Behavioral Ecology*, vol. 18, no. 2, pp. 393–398, 2007.

[31] U. Candolin, T. Salesto, and M. Evers, "Changed environmental conditions weaken sexual selection in sticklebacks," *Journal of Evolutionary Biology*, vol. 20, no. 1, pp. 233–239, 2007.

[32] B. B. M. Wong, U. Candolin, and K. Lindström, "Environmental deterioration compromises socially enforced signals of male quality in three-spined sticklebacks," *American Naturalist*, vol. 170, no. 2, pp. 184–189, 2007.

[33] U. Candolin and J. Heuschele, "Is sexual selection beneficial during adaptation to environmental change?" *Trends in Ecology and Evolution*, vol. 23, no. 8, pp. 446–452, 2008.

[34] A. P. Møller and R. V. Alatalo, "Good-genes effects in sexual selection," *Proceedings of the Royal Society B*, vol. 266, no. 1414, pp. 85–91, 1999.

[35] P. D. Lorch, S. Proulx, L. Rowe, and T. Day, "Condition-dependent sexual selection can accelerate adaptation," *Evolutionary Ecology Research*, vol. 5, no. 6, pp. 867–881, 2003.

[36] M. C. Whitlock, "Fixation of new alleles and the extinction of small populations: drift load, beneficial alleles, and sexual selection," *Evolution*, vol. 54, no. 6, pp. 1855–1861, 2000.

[37] A. F. Agrawal, "Sexual selection and the maintenance of sexual reproduction," *Nature*, vol. 411, no. 6838, pp. 692–695, 2001.

[38] O. Seehausen, "Conservation: losing biodiversity by reverse speciation," *Current Biology*, vol. 16, no. 9, pp. R334–R337, 2006.

[39] U. Candolin and B. B. M. Wong, "Sexual selection in changing environments: consequences for individuals and populations," in *Behavioural Responses to a Changing World: Causes and Consequences*, U. Candolin and B. B. M. Wong, Eds., pp. 201–215, Oxford University Press, Oxford, UK, 2012.

13

Records and Descriptions of Epitoniidae (Orthogastropoda: Epitonioidea) from the Deep Sea off Northeastern Brazil and a Checklist of *Epitonium* and *Opalia* from the Atlantic Coast of South America

Silvio F. B. Lima,[1] Martin L. Christoffersen,[1] José C. N. Barros,[2] and Manuella Folly[3]

[1] *Departamento de Sistemática e Ecologia, Universidade Federal da Paraíba (UFPB), 58059-900 João Pessoa, PB, Brazil*
[2] *Laboratório de Malacologia, Departamento de Pesca e Aquicultura, Universidade Federal Rural de Pernambuco (UFRPE),*
  Avenida Dom Manuel de Medeiros S/N, Dois Irmãos, 52171-030 Recife, PE, Brazil
[3] *Departamento de Zoologia, Instituto de Biologia, Centro de Ciências da Saúde,*
  Universidade Federal do Rio de Janeiro (UFRJ), Ilha do Fundão, 21941-570 Rio de Janeiro, RJ, Brazil

Correspondence should be addressed to Silvio F. B. Lima, sfblima@yahoo.com.br

Academic Editor: Roger P. Croll

A total of six genera and 10 species of marine gastropods belonging to the family Epitoniidae were collected from dredges of the continental slope off Brazil during the development of the REVIZEE (Live Resources of the Economic Exclusive Zone) Program. These species, referable to the genera *Alora, Amaea, Cycloscala, Epitonium, Gregorioiscala,* and *Opalia,* are reported from bathyal depths off northeastern Brazil. *Alora* sp., *Gregorioiscala pimentai* n. sp., and *Opalia revizee* n. sp. are species heretofore unknown to science. A list of the species of *Epitonium* and *Opalia* from the Atlantic coast of South America is presented based primarily on data from the literature. In addition, an overview of the biodiversity and distribution of the genera studied is presented for the Atlantic Ocean.

1. Introduction

Mollusks are a diverse and abundant group, although often inconspicuous in the reef ecosystem. Among gastropods, species of Architectonicidae Gray, 1840, Coralliophilidae Chenu, 1859, Epitoniidae S. S. Berry, 1910, Muricidae Rafinesque, 1815, Nystiellidae Clench and Turner, 1952, Olividae Latreille, 1825, and Ovulidae Fleming, 1822, are known to live on stony and soft corals, hydroids, hydrocorals, discophores, siphonophores, gorgonians, zoanthids, and sea anemones, feeding on living cnidarian tissues [1–5].

Epitoniidae is a cosmopolitan family of carnivorous marine gastropods [1, 6–9] that occurs on a variety of substrata from the intertidal to the abyssal region [1] and feed mainly on cnidarian anthozoans [1–3, 8–17]. These gastropods may also be free-living micropredators [1, 14, 16]

feeding on invertebrates such as annelid worms and nemerteans [14].

The systematics of Epitoniidae remain poorly resolved mainly due to the scarce material collected from the deep sea, which is often represented by one or a few shells [6–9, 14, 16, 18]. Watson [18] reported the first epitoniids for Brazil and Rios [19–22] expanded knowledge on this group in Brazilian waters. However, the alpha taxonomy remains underestimated, and there is fragmented knowledge on the family in the country [22–31].

Brazilian programs of environmental characterization have been very important in the sampling of benthic communities from the continental shelf and deep waters. For example, Miyaji [30] identified five genera and 16 species of Epitoniidae collected in southeastern and southern Brazil during the REVIZEE (Live Resources of the Exclusive Economic Zone) Program. Dr. R. S. Absalão (pers. comm., April 2011) studied new and little known deep-water Epitoniidae (700 to 1950 m) from the Campos Basin off the state of Rio de Janeiro, Brazil. These studies demonstrate the insufficient understanding of the real diversity of Epitoniidae.

This paper presents gastropods of the family Epitoniidae collected from the continental slope off northeastern Brazil during the REVIZEE Program (2000-2001) and lists the species of the genera *Epitonium* Röding, 1798, and *Opalia* Adams and Adams, 1853, reported for the Atlantic coast of South America. A total of 10 species of Epitoniidae were dredged from bathyal depths. Most of the species studied herein are poorly known, have not previously been recorded for the region, or are previously unknown to science. In several cases, only one or a few specimens with damaged shells were collected, identified, and figured. The goal is to provide more alpha taxonomic knowledge on the diversity and geographic ranges of Epitoniidae fauna in Brazil. In addition, an overview of the biodiversity and distribution of the genera studied is presented for the Atlantic Ocean.

2. Material and Methods

This study is based on 13 empty shells collected through dredging in northeastern Brazil by the fishing vessel "Natureza" between depths of 375 and 720 meters. Generic and specific identification is based on comparisons with descriptions and illustrations [6–9, 16, 32, 33]. A checklist of the species of the genera *Epitonium* (Table 1) and *Opalia* (Table 2) known from the Atlantic coast of South America and their geographic and bathymetric distribution is presented based on data from the literature [7, 9, 16, 22–24, 30–32, 34–41], species registered in the online World Register of Marine Species [42], and databases of western Atlantic marine Mollusca [43] and *Conquiliologistas do Brasil* [44].

The supraspecific taxonomy of Epitoniidae is poorly defined and often inconsistent based solely on shell morphology [6, 8, 16]. Thus, the decision was made to assign species only to the genus level until future-changes-based new evidence from anatomical studies complement the taxonomy. The limits of the subgenera (e.g., *Epitonium* and *Opalia*) have not been clearly defined [6, 16].

Each species was photographed under a ZEISS EVO 40 Scanning Electron Microscope at the Management of Biostratigraphy and Applied Paleoecology of the Petrobrás Research Center or under an FEI QUANTA 200F Scanning Electron Microscope at the Center for Technological Research of Northeastern Brazil.

3. Results

3.1. Taxonomic Account.

Epitoniidae Berry, 1910

Epitonium Röding, 1798.

Epitonium fractum Dall, 1927 (Figures 1(a)–1(c)).

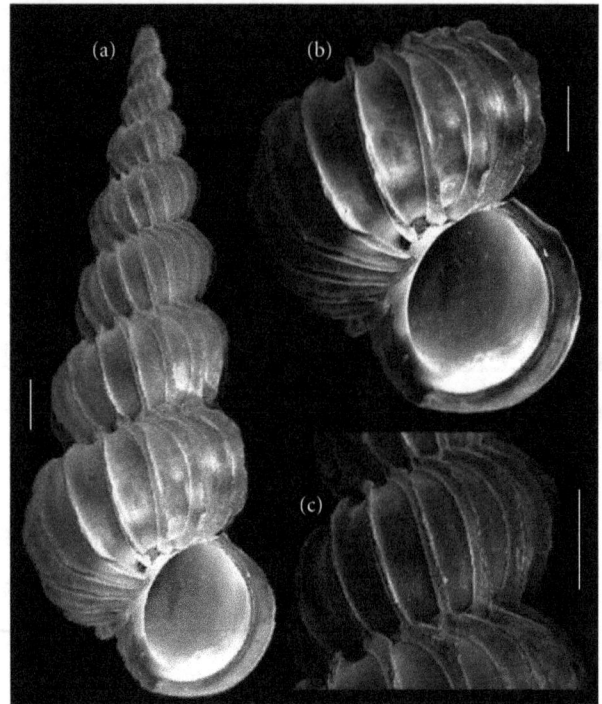

FIGURE 1: (a–c) *Epitonium fractum* (LMUFRPE); (a) ventral view, (b) view of last whorl, and (c) detail of teleoconch ornamentation. Scale bars: (a–c) 500 μm.

Type Material and Locality. Holotype (USNM 108015, not examined)—off Fernandina, Florida [32].

Material Examined. 1 shell (LMUFRPE), Rio Grande do Norte (Brazil, REVIZEE/NE: "Natureza," 04°51′S, 35°06′W, 375 m, 24.xi.2001).

Characterization. Shell elongated, somewhat slender, strongly convex whorls attached by blade-like ribs with well-developed angles at shoulder (Figure 1(a)). Protoconch about 3.5 smooth whorls. Teleoconch with 8 to 16 whorls. About 17 to 20 axial ribs on last whorl. Spire extended (Figure 1(a)). Suture deep (Figures 1(a) and 1(c)). Base sculptured with ribs with well-developed angulation (Figure 1(b)). Aperture subcircular, holostomatous (Figure 1(b)). Umbilicus minute, partially hidden by parietal lip and ribs (Figure 1(b)).

Geographic Distribution. Georgia [43], Florida [22, 32, 37, 45], Gulf of Mexico—off Louisiana [43], West Indies [22], Brazil: Rio Grande do Norte (present study), Espírito Santo [22] to São Paulo [30].

Remarks. The characters of the specimen examined here and those presented by Clench and Turner [32], Abbott [37], and Rios [22] fit the concept of *Epitonium fractum* (see original description in Dall [45]) and figure of type material in Clench and Turner [32]. This species has been collected on the continental shelf and slope [22, 32, 37, 45]. It is similar

Records and Descriptions of Epitoniidae (Orthogastropoda: Epitonioidea) from the Deep Sea off Northeastern Brazil and
a Checklist of Epitonium and Opalia from the Atlantic Coast of South America

107

TABLE 1: Checklist of species of the genus *Epitonium* known for the Atlantic coast of South America with geographic and bathymetric distribution.

Species	Distribution (South America)	Depth (m)
(1) *E. albidum* (d'Orbigny, 1842)	Colombia, Venezuela, Brazil (CE, PE, AL, ES, RJ, SP, PR, SC), Uruguay	0–366
(2) *E. angulatum* (Say, 1831)	Colombia, Brazil (AP, CE, PE, AL, BA, ES, RJ, SP), Uruguay	0–219
(3) *E. apiculatum* (Dall, 1889)	Venezuela	1–90
(4) *E. babylonia* (Dall, 1889)	Colombia, Brazil (AP, CE, PE, SP)	152–1337
(5) *E. candeanum* (d'Orbigny, 1842)	Colombia, Venezuela, Suriname, Brazil (AP, CE, PE, AL, SP), Uruguay, Argentina	0–805
(6) *E.* cf. *candeanum**	Brazil (PI)	0–10
(7) *E. celesti* (Aradas, 1854)	Brazil (RN, CE, RJ, SP, RS)	146–640
(8) *E.* cf. *celesti*	Brazil (RN)	384
(9) *E. dallianum* (Verrill and Smith, 1880)	Brazil (ES, RS)	90–478
(10) *E. denticulatum* (Sowerby II, 1844)	Colombia, Venezuela, Brazil (AP, PA, CE, PE, AL, RJ)	0–1472
(11) *E. fabrizioi* Pastorino and Penchaszadeh, 1998*	Argentina	0–2
(12) *E. foliaceicosta* (d'Orbigny, 1842)	Colombia, Brazil (AL, BA, ES)	0–219
(13) *E. fractum* Dall, 1927	Brazil (RN, ES, RJ, SP)	64–594
(14) *E. frielei* (Dall, 1889)	Brazil (AP, CE, RS)	91–2941
(15) *E. georgettinum* (Kiener, 1838)*	Brazil (BA, ES, SC, RS), Uruguay, Argentina	0–101
(16) *E. hispidulum* (Monterosato, 1874)	Brazil (RJ)	750–800
(17) *E. humphreysii* (Kiener, 1838)	Brazil (CE, PE, RS)	0–95
(18) *E. krebsii* (Mörch, 1875)	Colombia, Venezuela, Brazil (AP, PA, CE, RN, PB, AL, SC)	0–294
(19) *E. lamellosum* (Lamarck, 1822)	Colombia, Venezuela, Brazil (ES)	0–60
(20) *E. magellanicum* (Philippi, 1845)*	Brazil (RS), Uruguay, Argentina, Chile	0–545
(21) *E. matthewsae* Clench and Turner, 1952	Colombia	11–219
(22) *E. multistriatum* (Say, 1826)	Brazil (AL, RS)	2.5–219
(23) *E. novangliae* (Couthouy, 1838)	Colombia, Venezuela, Brazil (CE, PE, AL, ES, RJ, SC)	0–457
(24) *E. occidentale* (Nyst, 1871)	Venezuela, Brazil (CE, PE, AL, BA, RJ, SP)	0–270
(25) *E. polacia* (Dall, 1889)	Brazil (RJ)	65–419
(26) *E. rupicola* (Kurtz, 1860)	Colombia, Suriname (Holocene)	0–65
(27) *E. sericifila* (Dall, 1889)	Colombia	0–7
(28) *E. striatissimum* (Monterosato, 1878)	Argentina	69–183
(29) *E. striatellum* (Nyst, 1871)*	Brazil (SP, RS), Uruguay, Argentina	10–70
(30) *E.* cf. *tiberii*	Brazil (RN)	384
(31) *E. turritellulum* (Mörch, 1875)*	Venezuela, Brazil (AP)	6–40
(32) *E. unifasciatum* (Sowerby II, 1844)	Colombia, Brazil (CE, PE, RS), Uruguay	0–9
(33) *E. venosum* (Sowerby II, 1844)	Venezuela	61
(34) *E. worsfoldi* Robertson, 1994	Brazil (ES)	1
(35) *E. xenicima* (Melvill and Standen, 1903)	Colombia, Brazil (Bahia), Uruguay	0–25

AP: Amapá, PI-Piauí, CE: Ceará, RN: Rio Grande do Norte, PB: Paraíba, PE: Pernambuco, AL: Alagoas, BA: Bahia, ES: Espírito Santo, RJ: Rio de Janeiro, SP: São Paulo, PR: Paraná, SC: Santa Catarina, RS: Rio Grande do Sul; (*) endemic to the Atlantic coast of South America.

to *E. angulatum* (Say, 1831), *E. babylonia* (Dall, 1889) and *E. dallianum* (Verrill and Smith, 1880) in the extended spire and shoulder angulation but differs from these species by the lack of spiral sculpture or spines on the shoulder (Figures 1(a)–1(c)).

Epitonium cf. *celesti* (Aradas, 1854) (Figures 2(a)–2(c)).

Type Material and Locality. off Acitrezza (not examined), Sicily [16].

Material Examined. 1 shell (LMUFRPE), Rio Grande do Norte (Brazil, REVIZEE/NE: "Natureza", 04°51′40″S, 35°08′01″W, 384 m, 24.xi.2001).

Characterization. Shell rather small, broad, slightly globose to turbinate (Figures 2(a) and 2(c)). Protoconch with about 4 whorls (Figure 2(b)). Teleoconch with about 3 almost disjunct whorls attached by blade-like, widely spaced ribs; ribs weakly recurved abaperturally at shoulder (Figures 2(a) and 2(c)). Microscopic sculpture consisting of numerous and

TABLE 2: Checklist of species of the genus *Opalia* known for the Atlantic coast of South America with geographic and bathymetric distribution.

Species	Distribution (South America)	Depth (m)
(1) *O. abbotti* Clench and Turner, 1952	Brazil (PE, RJ, SP)	64–704
(2) *O. aurifila* (Dall, 1889)	Brazil	91–311
(3) *O. burryi* Voss, 1953	Colombia	30–168
(4) *O. crenata* (Linnaeus, 1758)	Venezuela, Brazil (AP, PA, AL to SC)	0.3–82
(5) *O. eolis* Clench and Turner, 1950	Brazil (RN), Southeastern Brazil	60–384
(6) *O. hotessieriana* (d'Orbigny, 1842)	Brazil (AL to SC)	0–165
(7) *O. cf. morchiana**	Southeastern Brazil	—
(8) *O. pumilio* (Mörch, 1875)	Brazil (AP, PI, CE, PE, AL)	0–183
(9) *O. cf. pumilio**	Colombia	—
(10) *O. revizee* n. sp.*	Brazil (AL)	720

AP: Amapá, PA: Pará, PI: Piauí, CE: Ceará, PE: Pernambuco, AL: Alagoas, RJ: Rio de Janeiro, SP: São Paulo, SC: Santa Catarina; (—) depth is not given by Miyaji [30]; (*) endemic to the Atlantic coast of South America.

FIGURE 2: (a–c) *Epitonium* cf. *celesti* (LMUFRPE); (a) ventral view, (b) protoconch, and (c) dorsal view. Scale bars: (a) and (c) 500 μm, (b) 200 μm.

fine spiral threads (Figures 2(a) and 2(c)). Suture deep (Figure 2(c)). Base broad and rounded. Aperture oval, rather thickened and deflected. Umbilicus rather deep, narrow, sometimes slightly covered by reflection of lip, and/or thick extension of columella (Figure 2(a)).

Geographic Distribution. Eastern Atlantic—Mediterranean Sea, Portugal, Madeira, and Azores Islands [16, 22]; Western Atlantic—New Jersey [7, 16, 22], North Carolina, Florida, Bermuda, Bahamas Islands [7], Gulf of Mexico—Yucatan Strait [43], Cuba [7], Virgin Islands [7, 16, 22], Barbados [7], Brazil [21]: Rio Grande do Norte (present study); São Paulo [30]; Rio Grande do Sul [22].

Remarks. This species is similar to *Epitonium krebsii* (Mörch, 1875) in the globose-turbinate shell and blade-like ribs widely spaced and slightly abaperturally recurved at the shoulder but differs in that it has numerous fine spiral threads and

does not have the wide umbilicus characteristic of *E. krebsii*. Clench and Turner [32], Abbott [37], and Rios [21] did not recognize any spiral sculpture on the teleoconch of *E. krebsii*.

The only specimen collected matches *Epitonium celesti* in the conical protoconch with about 4 whorls as well as in the shell shape and in axial and spiral sculpture (see Bouchet and Warén [16]). The prominent axial ribs strongly recurved abaperturally at the shoulder forming a spine in *Epitonium celesti* that seems to be somewhat different from the specimens illustrated herein. However, the only shell collected here is somewhat worn.

Epitonium cf. *tiberii* (de Boury, 1890) (Figures 3(a)–3(c)).

Material Examined. 1 shell (LMUFRPE), Rio Grande do Norte (Brazil, REVIZEE/NE: "Natureza," 04°51′40″S, 35°08′01″W, 384 m, 24.xi.2001).

Characterization. Shell conical-turbinate (Figure 3(a)). Protoconch conical, about 3.5 whorls, sculptured with subsutural spiral threads and opisthocline incremental lines (Figure 3(b)). Teleoconch about 3.5 whorls attached by prosocline blade-like ribs, slightly expanded, angulated, and with weak spine at shoulder (Figures 3(a) and 3(c)). Microscopic sculpture of numerous spiral threads (Figure 3(c)). First, second, and third teleoconch whorl sculptured with about 14, 18, and 22 axial ribs, respectively (Figure 3(a)). Suture deep and very constricted. Base conical, moderately elongated. Aperture oval. Outer and inner lip thin. Umbilicus minute, partially hidden by parietal lip and ribs (Figure 3(a)).

Geographic Distribution. Eastern Atlantic—Bay of Biscay to Cape Verde Islands [46]; western Atlantic—continental slope of Rio Grande do Norte (northeast Brazil: present study).

Remarks. Despite the wear, concretions on the surface of the teleoconch and the juvenile stage of the shell, this specimen approaches *E. algerianum* (Weinkauff, 1866), *E. tiberii* (de

Records and Descriptions of Epitoniidae (Orthogastropoda: Epitonioidea) from the Deep Sea off Northeastern Brazil and a Checklist of Epitonium and Opalia from the Atlantic Coast of South America

109

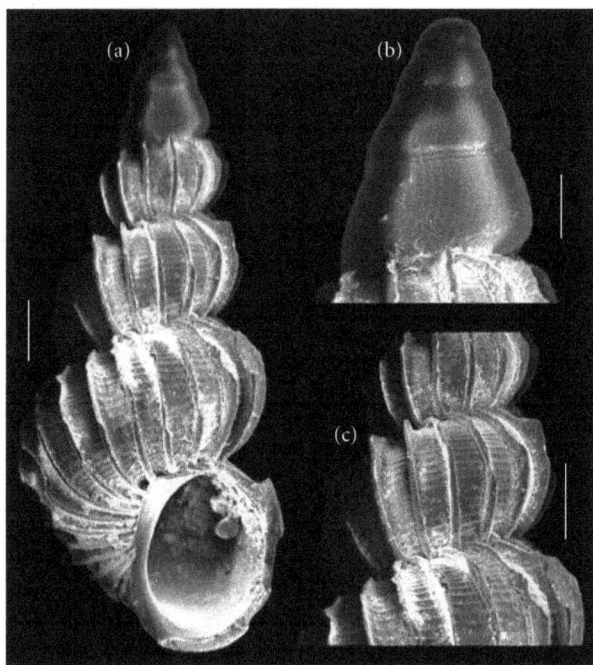

FIGURE 3: (a–c) *Epitonium* cf. *tiberii* (LMUFRPE); (a) ventral view, (b) protoconch and (c) detail of teleoconch ornamentation. Scale bars: (a) and (c) 500 μm, (b) 200 μm.

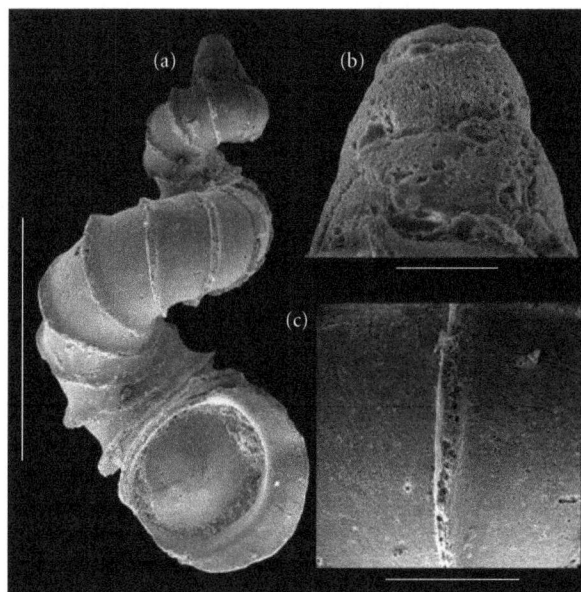

FIGURE 4: (a–c) *Cycloscala echinaticosta* (MNRJ 17.165); (a) ventral view, (b) protoconch view, and (c) detail of teleoconch ornamentation. Scale bars: (a) 1 mm, (b) 100 μm, and (c) 200 μm.

Boury, 1890), and *E. tryoni* (de Boury, 1913) (all from the eastern Atlantic) in shape and pattern of axial and spiral sculpture, including the slightly spinose projections at the shoulder. *Epitonium algerianum* has a robust shell and sinuous axial lamellae; *E. tryoni* has orthocline axial lamellae and an open umbilicus, while *Epitonium* cf. *tiberii* is characterized by a thin shell, prosocline axial ribs, and a minute umbilicus. *Epitonium* cf. *tiberii* approaches *E. tiberii*, although the protoconch of the present specimen is rather different from that shown by Bouchet and Warén [21] for *E. tiberii*. More material is necessary for further comparisons.

Cycloscala Dall, 1889.

Cycloscala echinaticosta (d'Orbigny, 1842) (Figures 4(a)–4(c)).

Type Material and Locality. Holotype (not examined)—probably in NHMUK and St. Thomas—Virgin Islands [32].

Material Examined. 1 shell (MNRJ 17.165), Rio Grande do Norte (Brazil, REVIZEE/NE: "Natureza," 04°51′40″S, 35°08′01″W, 384 m, 24.xi.2001).

Characterization. Shell slightly globose to elongate-turbinate. Protoconch with 3.5 to 4 whorls. First postnuclear whorls attached and remaining disjunct or usually all disjunct whorls, sculptured with widely spaced axial ribs (Figure 4(a)). First postnuclear whorl narrowly coiled (Figure 4(a)). Second postnuclear whorl much more widely coiled (Figure 4(a)). First and second postnuclear whorls

with 8 to 9 scalloped axial costae completely encircling teleoconch (Figure 4(a)). Axial interspaces smooth (Figure 4(c)). Aperture circular (Figure 4(a)). Outer and inner lip thin (Figure 4(a)).

Geographic Distribution. North Carolina [43], Florida [22, 24, 32, 37], Bermuda [22, 24, 32, 33, 43, 47, 48], Bahamas [32, 33, 48, 49], Gulf of Mexico—off Louisiana [43], Belize [33], Costa Rica, Panama [43], Turks and Caicos, Cuba [32], Jamaica [32, 50], Haiti, Dominican Republic, Puerto Rico [43], Virgin Islands [32, 51–53], Anguilla [32], Barbados [32, 37], Bonaire [33, 54], Colombia [43], Brazil [22, 23, 37]: Pará [31], Rio Grande do Norte (present study), Fernando de Noronha Archipelago [22, 24, 48], Espírito Santo [43].

Remarks. Woodring [50] and Clench and Turner [32] classified this species as *Epitonium* (*Cycloscala*). Later, Kilburn [8] considered the subgenus to have characters sufficiently well defined to warrant full generic status. However, subsequent studies continued to recognize *Epitonium echinaticosta* [23, 24, 31, 48, 49, 54], except Garcia [33], who also discussed distinguishing characters in favor of treating *Cycloscala* on the generic level, mainly due to the disjunct postnuclear whorl, scalloped costae (Figure 4(a)), and shell body lacking spiral ornamentation (Figure 4(c)) [8, 33].

Currently, *Cycloscala* is a valid genus with about 13 species in the Pacific Ocean [46] and only *Cycloscala echinaticosta* in the Atlantic Ocean (between depths of 2 and 384 m) [8, 22, 33, 55]. This taxon has considerable conchological variability [32, 33, 48, 49].

Amaea H. and A. Adams, 1853.

Amaea retifera (Dall, 1889) (Figures 5(a)–5(d)).

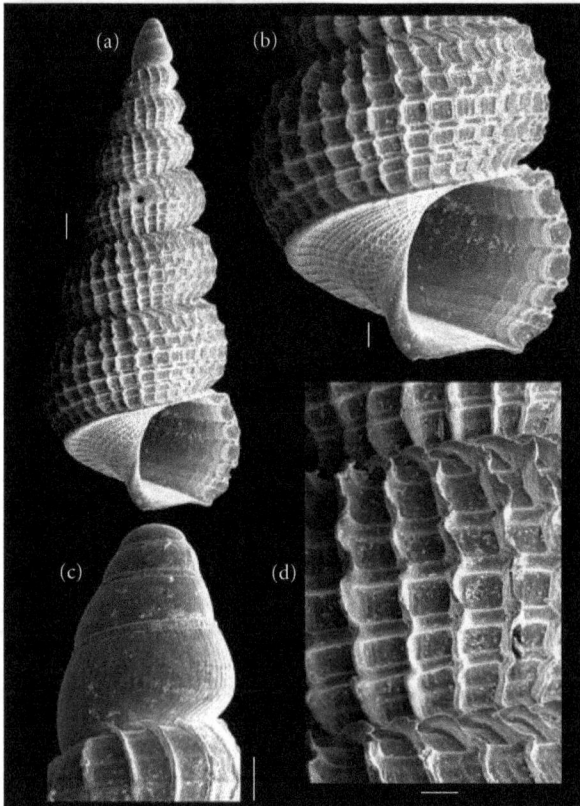

FIGURE 5: (a–d) *Amaea retifera* (LMUFRPE); (a) ventral view, (b) view of last whorl, (c) protoconch, and (d) detail of teleoconch ornamentation. Scale bars: (a), (b) and (d) 500 μm, (c) 100 μm.

Type Material and Locality. Holotype (USNM 83733, not examined)—off Cape Hatteras, North Carolina [6].

Material Examined. 1 shell (LMUFRPE)—Pernambuco (Brazil—REVIZEE/NE: "Natureza," 08°46′00″S, 34°44′ 00″W, 690 m, 18.xi.2000).

Characterization. Shell turriform, elongate, thin. Teleoconch whorls with 6 to 16 whorls strongly convex, constricted (Figures 5(a) and 5(b)), sculptured with blade-like axial ribs (about 32 on last whorl) and spiral threads (7 to 8 on last whorl) with reticulated pattern (Figure 5(d)). Sculpture with rectangular intervals ornamented by axial striae (Figure 5(d)). Suture deep (Figures 5(a), 5(b) and 5(d)). Basal disc well developed, flat, sculptured with axial and spiral threads, delimited by prominent cord (Figure 5(b)). Base imperforate. Aperture oval (Figure 5(b)). Outer lip thickened (Figure 5(b)). Inner lip thin (Figure 5(b)).

Geographic Distribution. North Carolina [6, 37, 53, 56], Florida [6, 37, 56], Mexico [6, 56], Cuba, Puerto Rico [56], Anguilla, St. Barthelemy/St. Bartholomew, Antigua [43], Barbados [6, 56], Colombia, Venezuela, Surinam [43], Brazil [37]: Amapá, Pará [22, 31], and Pernambuco (present study).

Remarks. *Amaea retifera* and *A. mitchelli* Dall, 1896 are the only species reported for the western Atlantic [6, 22, 37, 39, 56, 57]. The conchological characters of *Amaea retifera* are unequivocal and differ from *Amaea mitchelli* by the presence of blade-like axial ribs and strongly convex, constricted whorls. *Amaea retifera* has been recorded by Rios [21, 22] and Oliveira and Rocha-Barreira [31] only in northern Brazil (Amapá and Pará).

Alora H. Adams, 1861.

Alora sp. (Figures 6(a)–6(f)).

Material Examined. 3 shells (MNRJ 17.166) and 2 shells (MZSP 101282)—Pernambuco (Brazil—REVIZEE/NE: "Natureza," 08°46′00″S, 34°44′00″W, 690 m, 18.xi.2000).

Characterization. Shell thin, fragile, small, biconic, white (Figures 6(c) and 6(d)). Protoconch multispiral, conical, with 3.75 slightly convex whorls, sculptured with subsutural band, faint axial, and spiral striae (Figures 6(a) and 6(b)). Transition to teleoconch marked by straight axial edge and faint riblets (Figures 6(a) and 6(b)). Spire low, about 1/4 of total length (Figures 6(c) and 6(d)). Teleoconch with about 3 convex whorls; whorls rapidly increasing in diameter (Figures 6(c) and 6(e)), sculptured with closely spaced and fine axial incremental lines and raised spiral threads, rather irregularly disposed (Figure 6(f)); axial elements undulating as they cross spiral threads (Figure 6(f)). Suture rather deep (Figures 6(d) and 6(e)). Last whorl capacious, with strong peripheral carina, about 2/3 of total length of shell (Figures 6(c)–6(e)); posterior half strongly convex, sculptured with about 15 spiral threads; base subtrigonal, delimited posteriorly by carina; 15 to 20 spiral threads below carina (Figures 6(c), 6(d) and 6(f)). Aperture large, rather ovate (Figure 6(c)). Parietal region straight (Figure 6(c)). Inner lip and columella straight (Figure 6(c)). Umbilicus enlarged, half-moon shaped (Figure 6(c)).

Geographic Distribution. Known from the continental slope off northeastern (Pernambuco: present study) and southeastern Brazil (Dr. R. S. Absalão—pers. comm., April 2011).

Remarks. Only two deep sea epitoniids of the genus *Alora* H. Adams, 1861, are described for the Atlantic Ocean: *A. tenerrima* (Dautzenberg and Fischer, 1896) and *A. retifera* Bouchet and Warén, 1986. *Alora tenerrima* is amphi-Atlantic (USA: Georgia and eastern Atlantic: Azores [16, 45]), while *A. retifera* is described for the northeastern Atlantic.

Alora sp. is an undescribed species from the Atlantic Ocean. Dr. R. S. Absalão (pers. comm., April 2011) identified this species for the Campos Basin (Brazil: Rio de Janeiro) based on dozens of specimens. A formal specific epithet is being provided by this researcher. *Alora* sp. and *A. tenerrima* are similar in the convexity of the teleoconch whorls (except for the last whorl), in the dominating spiral sculpture, in the shape of the aperture, parietal region, and straight inner lip, in the deep suture and in the enlarged umbilicus (half-moon

Records and Descriptions of Epitoniidae (Orthogastropoda: Epitonioidea) from the Deep Sea off Northeastern Brazil and a Checklist of Epitonium and Opalia from the Atlantic Coast of South America

111

FIGURE 6: (a–f) *Alora* sp. (MNRJ 17.166); (a) and (b) protoconch view, (c) ventral view, (d) dorsal view, (e) apical view, and (f) detail of teleoconch ornamentation. Scale bars: (a) 200, (b) 300, (c) 1 mm, (d) 2 mm, (e) 500 μm, and (f) 400 μm.

FIGURE 7: (a–e) *Opalia abbotti* (IBUFRJ 18.828); (a) protoconch, (b) detail of protoconch ornamentation, (c) detail of teleoconch ornamentation, (d) ventral view, and (e) view of last whorl and penultimate whorl. Scale bars: (a) 100 μm, (b) 20 μm, (c) 10 μm, (d) 200 μm, and (e) 100 μm.

shape). *Alora* sp. differs from *A. tenerrima* by the presence of a peripheral carina, more numerous and weaker spiral threads, more closely spaced axial incremental lines, and a subtrigonal, not regularly convex last whorl. *Alora tenerrima* has about 5 spiral threads above the peripheral cord and 7 to 9 stronger threads below the cord; axial incremental lines are well spaced, and the last whorl is globose and regularly convex [16].

Opalia H. and A. Adams, 1853.

Opalia abbotti Clench and Turner, 1952 (Figures 7(a)–7(e)).

Type Material and Locality. Holotype (MCZ 184511, not examined) and off Puerto Tanamo, Cuba [7, 16].

Material Examined. 1 shell (IBUFRJ 18.828), Pernambuco (Brazil, REVIZEE/NE: "Natureza," 08°46'00''S, 34°44'00''O, 690 m, 18.xi.2000).

Characterization. Shell small, conical, whitish. Protoconch with 3.5 to 4 whorls sculptured with subsutural spiral band, slight axial threads, and microscopic pits (Figures 7(a) and 7(b)). Teleoconch with 5 to 7 whorls, regularly convex, constricted, sculptured with strong, thick, high, rounded, widely

spaced, and prosocline axial ribs (about 12 on last whorl) (Figures 7(d) and 7(e)) and numerous microscopic pits (Figure 7(c)). Basal disc delimited by prominent basal ridge (Figures 7(d) and 7(e)). Aperture circular (Figure 7(e)). Peristome thickened (Figure 7(e)). Umbilicus narrow and chinked (Figure 7(e)).

Geographic Distribution. Eastern Atlantic-northeastern Atlantic [16]; western Atlantic—Florida [7, 16, 37], Gulf of Mexico—off Louisiana [43], Cuba [16, 37, 56], Brazil: Pernambuco (present study) and São Paulo [30].

Remarks. This is an amphi-Atlantic species, but with extensive geographic distribution throughout the western Atlantic [7, 16, 30, 37, 56]. There are no previous records of *Opalia abbotti* for northeastern Brazil.

The only shell found here did not exhibit any morphological variation with respect to the specimens illustrated by Clench and Turner [7] and Bouchet and Warén [16]. According to Clench and Turner [7], the axial ribs on the last whorl in this species do not extend to the basal disc. However, these ribs may slightly invade the region (see Bouchet and Warén [16]).

Opalia eolis Clench and Turner, 1950 (Figures 8(a)–8(c)).

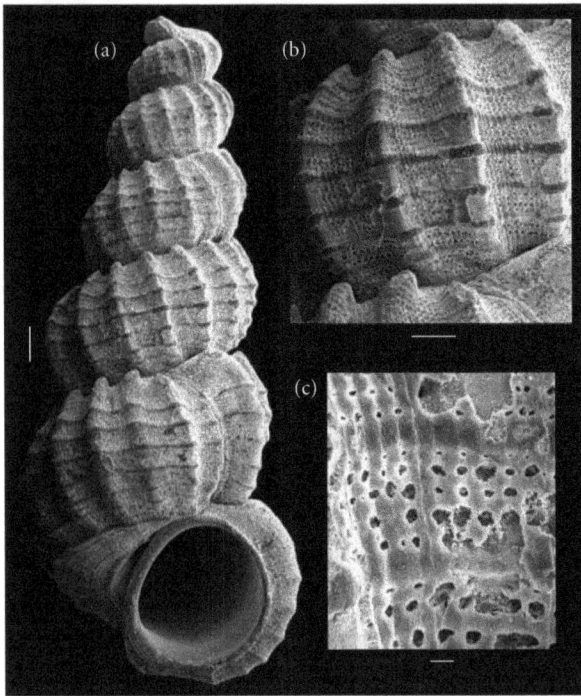

FIGURE 8: (a–c) *Opalia eolis* (MNRJ 17.170); (a) ventral view and (b), (c) detail of teleoconch ornamentation. Scale bars: (a) 500 μm and (b), (c) 100 μm.

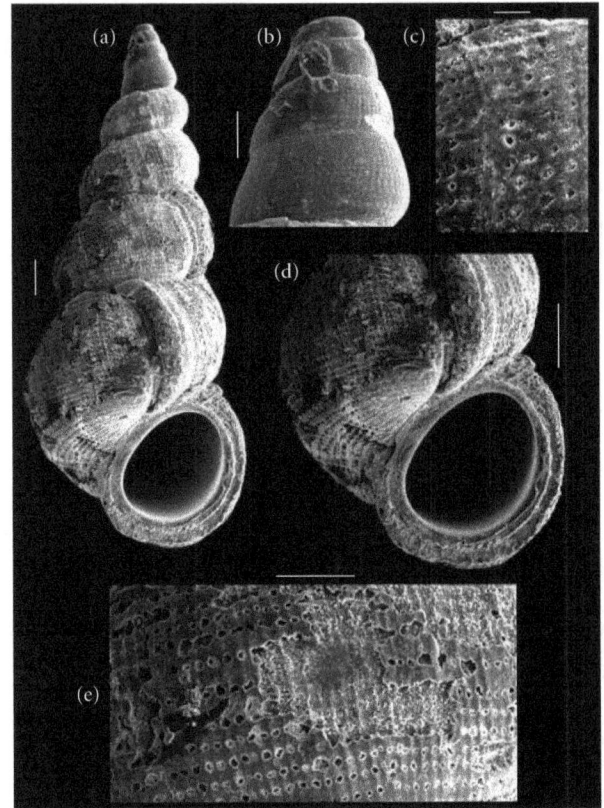

FIGURE 9: (a–e) *Opalia revizee* new species (holotype—MNRJ 18.307); (a) ventral view, (b) protoconch, (c) detail of protoconch ornamentation, (d) last whorl and aperture view, and (e) detail of teleoconch ornamentation. Scale bars: (a) 300 μm, (b) and (e) 100 μm, (c) 20 μm, and (d) 200 μm.

Type Material and Locality. Holotype (MCZ 187110, not examined) and off Looe Key, Lower Florida Keys, Florida (128 to 164 m); paratypes (USNM, not examined) and off Fowey Light, Sand Key, and Palm Beach, Florida [6].

Material Examined. 1 shell (MNRJ 17.170), Rio Grande do Norte (Brazil, REVIZEE/NE: "Natureza," 04°51′40″S, 35°08′01″W, 384 m, 24.xi.2001).

Characterization. Shell conical, elongate. Protoconch smooth, with 3 whorls. Teleoconch with 8 to 11 whorls regularly convex, rounded, constricted, sculptured with 16 to 20 strong, high, prosocline axial ribs, with crenulations at suture, 7 to 10 spiral threads, varix prosocline (Figures 8(a) and 8(b)), surface covered by intritacalx sculptured with microscopic pits in oval pattern (Figure 8(c)). Intersection of ribs and threads forming slight nodules (Figure 8(b)). Suture deeply impressed (Figure 8(b)). Basal disc delimited by slight spiral thread (Figure 8(a)). Aperture subcircular (Figure 8(a)). Outer and inner lip thickened (Figure 8(a)). Umbilicus narrow and chinked (Figure 8(a)).

Geographic Distribution. Florida [6, 22], Mexico, Bahamas [49], Anguilla, Barbados [6, 22], Brazil: Rio Grande do Norte (present study), Rio de Janeiro [22], and São Paulo [30].

Remarks. This species is described in some studies as imperforate [6, 22]. In some specimens, however, there is a small chink-like umbilicus [49]. Although not usually in

the description of this species, varices are seen in the images [6, 22].

Opalia revizee n. sp. (Figures 9(a)–9(e)).

Type Material. Holotype, 1 shell (MNRJ 18.307), Alagoas (Brazil, REVIZEE/NE: "Natureza," 10°06′35″S, 35°46′41″W, 720 m, 16.xii.2001).

Description. Shell small, conical, whitish (Figure 9(a)). Protoconch conical, with 3 slightly convex whorls (Figure 9(b)), sculptured with spiral rows of microscopic pits (Figure 9(c)). Teleoconch with 5 rounded whorls, regularly convex (Figure 9(a)), surface covered by intritacalx densely sculptured with spiral rows of microscopic pits (Figure 9(e)). Varix prosocline and strong on penultimate whorl (Figures 9(a) and 9(d)). Suture well impressed (Figure 9(a)). Base weakly convex, without basal ridge, disc or umbilicus (Figure 9(d)). Aperture ovate (Figure 9(e)); peristome thickened, complete (Figure 9(e)).

Etymology. Named for the REVIZEE Program (Live Resources of the Exclusive Economic Zone).

Type Locality. State of Alagoas ("Natureza," 10°06′35″S, 35°46′41″W) at a depth of 720 meters, muddy bottom, 16.xii.2001.

Geographic Distribution. Known only from the type locality.

Remarks. *Opalia revizee* new species is similar to and may be confused with *O. eolis* and *O. fortunata* Bouchet and Warén, 1986, due to the presence of a varix.

Opalia revizee and *O. eolis* are similar in the presence of a thickened outer lip, numerous microscopic pits, and a varix on the teleoconch. *Opalia revizee* is distinguished from *O. eolis* by the absence of a crenulated suture, spiral sculpture, axial threads, and ribs. *Opalia eolis* displays a crenulated suture, heavy axial ribs, and strong spiral cords [6, 37, 49].

Opalia fortunata (northeastern Atlantic) is the species most closely related to *O. revizee*. Both have a similar outline of the shell, 3 whorls on the protoconch, about 5 whorls on the teleoconch, whorls regularly convex and sculptured with microscopic pits, a prosocline varix, a regularly convex base, a thickened and complete peristome, and the absence of a basal ridge or basal disk. *Opalia revizee* differs from *O. fortunata* by exhibiting an ovate aperture, surface covered with intritacalx densely sculptured with spiral rows of microscopic pits and no axial ribs or crenulated suture. *Opalia fortunata* has a rounded aperture, whorls covered by a smooth, finely pitted intritacalx, weak axial ribs, and suture weakly crenulated [16].

Gregorioiscala Cossmann, 1912.

Gregorioiscala pimentai n. sp. (Figures 10(a)–10(d)).

Type Material. Holotype, 1 shell (MNRJ 18.306), Alagoas (Brazil, REVIZEE/NE: "Natureza," 10°06′35″S, 35°46′41″W, 720 m, 16.xii.2001).

Description. Shell whitish, small, strong, thick, conical (Figure 10(a)). Protoconch with about 2 moderately convex whorls, sculptured with microscopic pits (Figure 10(c)). Spire moderately high (Figure 10(a)). Suture moderately deep (Figures 10(a)–10(c)). Teleoconch with about 6 constricted whorls; whorls strongly convex, irregular in outline (Figure 10(a)); surface covered by intritacalx densely sculptured with microscopic pits in square pattern (Figure 10(b)). Axial sculpture with strong, thick, high, rounded, prosocline, widely spaced ribs that do not form crenulations in subsutural region (Figures 10(a) and 10(b)). Last whorl sculptured with 10 to 14 axial ribs, faint on basal ridge, weakly invading basal disc (Figure 10(d)). Base delimited posteriorly by prominent ridge keel (Figure 10(d)). Basal disc strongly flattened, large, sculptured with microscopic pits, weak ribs; prominent spiral rib at periphery of inner lip, with nodules in intersection of axial ornamentation (Figure 10(d)). Aperture rounded (Figures 10(a) and 10(d)); peristome thickened (Figure 10(d)); umbilicus very narrow and chinked (Figure 10(d)).

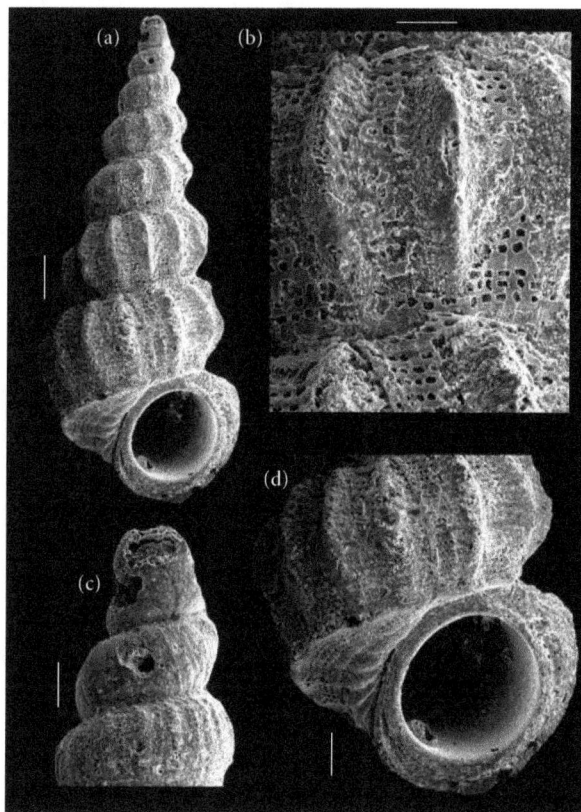

FIGURE 10: (a–d) *Gregorioiscala pimentai* new species (holotype—MNRJ 18.306); (a) ventral view, (b) detail of teleoconch ornamentation, (c) protoconch, and (d) view of last whorl. Scale bars: (a) 300 μm, (b), (c) 100 μm, and (d) 200 μm.

Etymology. In homage to Dr. Alexandre Dias Pimenta (MNRJ) for his initiative in reviewing the epitoniids and nystiellids of the Brazilian coast.

Type Locality. State of Alagoas ("Natureza," 10°06′35″S, 35°46′41″W), at a depth of 720 meters, muddy bottom, 16.xii.2001.

Geographic Distribution. Known only from the type locality.

Remarks. The taxon *Gregorioiscala* Cossmann, 1912, was erected to include deep-water epitoniid species with non-crenulated sutures, a relatively wide basal disk and a strongly defined basal ridge, thickened outer lip, thick pitted intritacalx, and strong axial ribs, some of which may form varices [16, 58]. About 14 *Gregorioiscala* species are currently known in seas worldwide [16, 55, 58]. *Gregorioiscala pachya* (Locard, 1897) is the only deep sea congener reported for the western Atlantic (Gulf of Mexico) [55].

Gregorioiscala pimentai resembles *G. pachya* in presenting a teleoconch with constricted whorls; whorls irregular in outline; strong, thick, prosocline, widely spaced axial ribs, axial ribs of subsequent whorls not aligned in a row, suture moderately deep, and peristome thickened. *Gregorioiscala pimentai* is distinguished from *G. pachya* by the presence of a conical, less solid shell, teleoconch with 6 whorls, columellar axis not curved, teleoconch whorls not developing a shoulder, lacking varices, basal disc sculptured with weak axial ribs and a spiral rib present at the periphery of inner lip, forming nodules in the intersections with axial ribs. *Gregorioiscala pachya* is recognized by the turriculate, very solid, heavy shell, teleoconch with about 12 whorls, curved columellar axis, shouldered teleoconch whorls, some axial ribs forming varices and basal disc not sculptured with ribs [16].

4. Discussion

The present paper permits an extended analysis of the biodiversity and distribution of the genera studied in the Atlantic Ocean, with an emphasis on the taxa reported for South America. A review of the literature and together with the database offered by the present study resulted in records of three *Alora*, including *A.* sp. [9, 16], about four *Amaea* [6, 9, 22, 57, 59], one *Cycloscala* [33, 49, 55], about four *Gregorioiscala*, including *G. pimentai* n. sp. [9, 16], fourteen *Opalia*, including *O. revizee* n. sp. [6, 9, 16, 22, 37, 39, 42, 49], and about eighty *Epitonium* species [7, 9, 16, 22, 32, 37, 39, 42, 43, 49, 54] reported for the Atlantic Ocean. Information on the geographic and bathymetric distribution of *Alora* and *Cycloscala* was presented previously.

The genera *Amaea* and *Gregorioiscala* are among the most poorly represented epitoniids on both Atlantic coasts and species richness values therefore have little comparative meaning between regions. The two *Amaea* species from the western Atlantic are spread over a broad geographic area [6, 9, 22, 37, 39, 55, 57], but only *A. retifera* has been recognized as significantly expanding the area of occurrence of the group to the Atlantic coast of South America and apparently beyond the continental shelf [22]. Other species have been found in the eastern Atlantic off West Africa (*A. africana* Bouchet and Tillier, 1978; *A. guineensis* Bouchet and Tillier, 1978) [9, 42, 59].

At least eleven *Opalia* species have been reported for the western Atlantic, five of which (*O. crenata* (Linnaeus, 1758), *O. hotessieriana* (d'Orbigny, 1842), *O. pumilio* (Mörch, 1875), *O. eolis*, and *O. abbotti*) have distributions shared between the Gulf of Mexico, West Indies, and the Atlantic coast of South America (Table 2) [6, 9, 16, 22, 39, 42, 49, 55], extending the southern limit of distribution (except *O. pumilio*) to southeastern/southern Brazil [22]. *Opalia leeana* (Verrill, 1882) and *O. tortilis* (Watson, 1883) are only known from their type localities [43]. This genus is still very poorly studied in the eastern Atlantic, with four species known for Europe [9, 16, 42]. Despite the occurrence of *O. abbotti* on both sides of the Atlantic [7, 16], there are wide gaps between the reported localities, especially between the West Indies and the northeastern portion of South America (see [7, 22, 37]). Most known species of *Opalia* in the Atlantic Ocean occur on the continental shelf [6, 7, 9, 22, 37, 39, 49, 56], with the exception of *O. abbotti*, *O. aurifila* (Dall, 1889), *O. fortunata*

Bouchet and Warén, 1986, *O. leeana*, *O. tortilis*, and *O. revizee* n. sp., which have been recorded in deep waters (between depths of 200 and 713 m) [6, 9, 16, 18, 22, 37, 53, 60].

The genus *Epitonium* has the highest number of described epitoniids [7–9, 11, 14, 22, 23, 32, 37, 39], with about forty species on both sides of the Atlantic [42, 43]. Currently, the *Epitonium* fauna of the western Atlantic may be divided into six categories on the basis of their distribution: (A) amphi-Atlantic species (4.79%); (B) species widely distributed from the United States to the Caribbean Sea and the Atlantic coast of South America (38.09%); (C) species distributed from the United States to the Caribbean Sea (16.66%); (D) species known only for the coast of the United States (16.66%); (E) species known only for the Caribbean Sea (11.90%); (F) species restricted to the Atlantic coast of South America (11.90%).

The data reviewed here reveal that *Epitonium* fauna with the greatest similarity occur between the Caribbean Sea and the Brazilian coast, as nearly half the species have records in both regions [7, 22, 32, 37, 41, 43, 49]. In the western Atlantic, about 20 species (47.61%) are found at depths of less than 200 m, while the other half has also been collected from deep waters [43]. In South America, the vertical range of *Epitonium* as a whole extends from the sublittoral zone (15 species; 42.8%) to the bathyal zone (20 species; 57.2%) (Table 1).

The only studies documenting *Epitonium* species in subregions of the Atlantic coast of South America were carried out by Diaz and Puyana [39] for Colombia (at least 16 species) and Rios [19–22] for Brazil. Not surprisingly, the Brazilian coast has the greatest richness of *Epitonium* in South America, with at least 25 species (Table 1) [22, 23, 31, 32, 37, 41, 44]. However, this total is far from being considered satisfactory due to the vast areas with scarce or no information on Epitoniidae.

Abbreviations

IBUFRJ:	Laboratório de Malacologia, Instituto de Biologia, Universidade Federal do Rio de Janeiro, Rio de Janeiro, Brazil;
LMUFRPE:	Laboratório de Malacologia, Universidade Federal Rural de Pernambuco, Recife, Brazil;
MCZ:	Museum of Comparative Zoology, Cambridge, USA;
MNRJ:	Setor de Malacologia, Museu Nacional, Universidade Federal do Rio de Janeiro, Rio de Janeiro, Brazil;
MZSP:	Museu de Zoologia, Universidade de São Paulo, São Paulo, Brazil;
NHMUK:	The Natural History Museum, London, Great Britain;
SEM:	Scanning Electronic Microscope;
USNM:	National Museum of Natural History, Washington, USA.

Records and Descriptions of Epitoniidae (Orthogastropoda: Epitonioidea) from the Deep Sea off Northeastern Brazil and
a Checklist of Epitonium and Opalia from the Atlantic Coast of South America

115

Acknowledgments

The authors are grateful to the Research and Management Center of Fishing Resources of the Northeastern Coast—CEPENE/IBAMA for the collection of sediment and donation of material to the Malacology Laboratory of the Universidade Federal Rural de Pernambuco (UFRPE), Brazil; Dr. Fábio H. V. Hazin (DEPAq/UFRPE) for his constant support of our research; Drs. Emilio F. García (USA), R. N. Kilburn (Natal Museum, Pietermaritzburg), Alexandre D. Pimenta (MNRJ), Paulo M. S. Costa (MNRJ), Cristina A. Rocha-Barreira (Universidade Federal do Ceará), and M Sc Bruno G. Andrade (MNRJ) for their generous assistance in obtaining the literature; Drs. Priscila A. Grohmann (IBUFRJ), Carlos H. S. Caetano (Universidade Federal do Estado do Rio de Janeiro), and M Sc Raquel M. A. Figueira (IBUFRJ) for their suggestions for improving part of the study; Dr. Emilio F. García (USA) and M. Sc Bruno G. Andrade for their important contribution to the revision and corrections of the paper, confirmation of the identification of Epitoniidae, and for showing us the correct identification of *Gregorioiscala*; Dr. Roger P. Croll (Dalhousie University, Canada-Academic Editor of the International Journal of Zoology) and anonymous referees for their suggestions on the paper; Dr. Ricardo S. Absalão (IBUFRJ) for the supervision of the study of Manuella Folly; Mr. Richard Boike for the English revision of the paper.

References

[1] R. Robertson, "Wentletraps (Epitoniidae) feeding on sea anemones and corals," *Proceedings of the Malacological Society of London*, vol. 35, no. 2-3, pp. 51–63, 1963.

[2] R. Robertson, "Review of the predators and parasites of stony corals, with special reference to symbiotic prosobranch gastropods," *Pacific Science*, vol. 24, pp. 43–54, 1970.

[3] R. Robertson, "*Epitonium millecostatum* and *Coralliophila clathrata*: two prosobranch gastropods symbiotic with indopacific *Palythoa* (Coelenterata: Zoanthidae)," *Pacific Science*, vol. 34, no. 1, pp. 1–17, 1980.

[4] M. C. Hadfield, "Molluscs associated with living tropical corals," *Micronesica*, vol. 12, pp. 133–148, 1976.

[5] W. F. Ponder, D. J. Colgan, J. M. Healy, A. Nützel, L. R. L. Simone, and E. Strong, "Caenogastropoda," in *Phylogeny and Evolution of the Mollusca*, W. F. Ponder and D. R. Lindberg, Eds., University of California Press, Berkeley, Calif, USA, 2008.

[6] W. J. Clench and R. D. Turner, "The genera *Sthenorytis, Cirsotrema, Acirsa, Opalia* and *Amaea* in the Western Atlantic," *Johnsonia*, vol. 2, no. 29, pp. 221–246, 1950.

[7] W. J. Clench and R. D. Turner, "The genus *Epitonium* (Part II), *Depressiscala, Cylindriscala, Nystiella* and *Solutiscala* in the Western Atlantic," *Johnsonia*, vol. 2, no. 31, pp. 289–356, 1952.

[8] R. N. Kilburn, "The family Epitoniidae (Mollusca: Gastropoda) in southern Africa and Mozambique," *Annals of the Natal Museum*, vol. 27, pp. 239–337, 1985.

[9] A. Weil, L. Brown, and B. Neville, *The Wentletrap Book: A Guide to the Tecent Epitoniidae of the World*, Evolver, Rome, Italy, 1999.

[10] V. Fretter and A. Graham, *British Prosobranch Molluscs*, Ray Society, London, UK, 1962.

[11] V. Fretter and A. Graham, "The prosobranch molluscs of Britain and Denmark. Part 7. Heterogastropoda (Cerithiopsacea, Triphoracea, Epitoniacea, Eulimacea)," *Journal of Molluscan Studies*, supplement 11, pp. 363–434, 1982.

[12] H. F. Bosh, "A Gastropod parasite of solitary Corals in Hawaii," *Pacific Science*, vol. 19, no. 2, pp. 267–268, 1965.

[13] F. Perron, "The habitat and feeding behavior of the wentletrap *Epitonium greenlandicum*," *Malacologia*, vol. 17, no. 1, pp. 63–72, 1978.

[14] H. DuShane, "The family Epitoniidae (Mollusca: Gastropoda) in the northeastern Pacific," *The Veliger*, vol. 22, no. 2, pp. 91–134, 1979.

[15] H. Dushane, "Geographical distribution of some Epitoniidae (Mollusca: Gastropoda) associated with fungiid corals," *The Nautilus*, vol. 102, no. 1, pp. 30–35, 1988.

[16] P. Bouchet and A. Warén, "Revision of the Northeast Atlantic bathyal and abyssal Aclididae, Eulimidae, Epitoniidae (Mollusca, Gastropoda)," *Bollettino Malacologico*, supplement 2, pp. 299–576, 1986.

[17] B. Kokshoorn, "Epitoniid parasites (Gastropoda, Caenogastropoda, Epitoniidae) and their host sea anemones (Cnidaria, Actiniaria, Ceriantharia) in the Spermonde archipelago, Sulawesi, Indonesia," *Basteria*, vol. 71, pp. 33–56, 2007.

[18] R. B. Watson, "Report on the Scaphopoda and Gasteropoda collected by H.M.S. "Challenger" during the years 1873–1876," *Reports on the Scientific Results of the Challenger Expedition, Zoology*, vol. 42, pp. 1–756, 1886.

[19] E. C. Rios, *Brazilian Marine Mollusks Iconography*, Fundação Universidade do Rio Grande, Rio Grande, Brazil, 1975.

[20] E. C. Rios, *Seashells of Brazil*, Fundação Cidade do Rio Grande/ Museu Oceanográfico, Rio Grande, Brazil, 1985.

[21] E. C. Rios, *Seashells of Brazil*, Fundação Cidade do Rio Grande/ Museu Oceanográfico, Rio Grande, Brazil, 1994.

[22] E. C. Rios, *Compendium of Brazilian Sea Shells*, Evangraf, Rio Grande, Brazil, 2009.

[23] E. C. Rios and R. S. Absalão, "Contribución al conocimiento de la familia Epitoniidae S.S. Berry, 1910 en el Brasil," *Comunicaciones de la Sociedad Malacologica del Uruguay*, vol. 6, pp. 367–370, 1986.

[24] J. H. Leal, *Marine Prosobranch Gastropods from Oceanic Islands off Brazil*, Universal Books Services, Olgstgeet, The Netherlands, 1991.

[25] R. L. S. Mello, "Moluscos do Brasil. I. Gastropoda, Bivalvia e Scaphopoda, coletados durante as viagens do navio oceanográfico "Almirante Saldanha". Comissão Sul I. Considerações biogeográficas," *Boletim do Museu de Malacologia da UFRPE*, vol. 1, pp. 31–49, 1993.

[26] O. Tenório, J. C. N. Barros, and R. L. S. Mello, "Gastrópodes da Margem Leste e Sul não citados para o Brasil," *Trabalhos Oceanográficos da Universidade Federal de PE*, vol. 22, pp. 305–323, 1993.

[27] J. C. N. Barros and E. A. Oliveira, "Comentários sobre três gastrópodes raros descritos por R. B. Watson, entre 1879 e 1885," *Boletim do Museu de Malacologia*, vol. 2, pp. 135–146, 1994.

[28] R. S. Absalão, A. D. Pimenta, and P. M. S. Costa, "Novas ocorrências de gastrópodes no litoral do Rio de Janeiro (Brasil)," *Nerítica*, vol. 10, pp. 57–68, 1996.

[29] J. C. N. Barros, F. N. Santos, M. C. F. Santos, E. Cabral, and F. D. Acioli, "Redescoberta de moluscos obtidos durante a "Challenger" Expedition (1873—1876): micromoluscos de águas profundas," *Boletim Técnico-Científico do CEPENE*, vol. 9, pp. 9–24, 2001.

[30] C. Miyaji, "Classe gastropoda," in *Biodiversidade Bentônica da Região Sudeste-Sul do Brasil—Plataforma Externa e Talude Superior*, A. C. Z. Amaral and C. L. B. Rossi-Wongtschowski, Eds., Séries documentos Revizee: Score Sul, Instituto Oceanográfico, São Paulo, Brazil, 2004.

[31] F. M. R. Oliveira and C. A. Rocha-Barreira, "A família Epitoniidae (Mollusca: Gastropoda) do norte e nordeste do Brasil," *Arquivos de Ciências do Mar*, vol. 42, no. 1, pp. 121–127, 2009.

[32] W. J. Clench and R. D. Turner, "The genus Epitonium in the Western Atlantic. Part I," *Johnsonia*, vol. 2, no. 30, pp. 249–288, 1951.

[33] E. F. Garcia, "On the genus Cycloscala Dall, 1889 (Gastropoda: Epitoniidae) in the Indo-Pacific, with comments on the type species, new records of known species, and the description of three new species," *Novapex*, vol. 5, no. 2-3, pp. 57–68, 2004.

[34] A. d'Orbigny, "Mollusques," *Voyage dans l'Amérique Méridionale*, vol. 5, pp. 59–69, 1839.

[35] P. Dautzenberg, "Croiseres du yatch "Chazalie" dans l' Atlantique, Mollusques," *Mémoires de la Société Zoologique de France*, vol. 13, pp. 145–265, 1900.

[36] B. Tursch and J. Pierret, "New species of mollusks from the coast of Brazil," *The Veliger*, vol. 7, pp. 35–37, 1964.

[37] R. T. Abbott, *American Seashells*, Van Nostrand Reinhold, New York, NY, USA, 2nd edition, 1974.

[38] C. O. van R. Altena, "The marine Mollusca of Suriname (Dutch Guiana), Holocene and recent III, Gastropoda and Cephalopoda," *Zoologisches Verhandlinger*, vol. 139, pp. 1–104, 1975.

[39] M. J. M. Diaz and H. M. Puyana, *Moluscos del Caribe Colombiano*, Colciencias, Fundación Natura Colômbia, Bogotá, Colombia, 1994.

[40] G. Pastorino and P. Penchaszadeh, "*Epitonium fabrizioi* (Gastropoda: Epitoniidae), a new species from Patagonia, Argentina," *Nautilus*, vol. 112, no. 2, pp. 63–68, 1998.

[41] B. G. Andrade, P. M. S. Costa, and A. D. Pimenta, "Revisão taxonômica do gênero Epitonium no Brasil (Gastropoda, Epitoniidae), exceto subgênero Asperiscala," in *Proceedings of the Libro de Resúmenes del 8th Congreso Latinoamericano de Malacología (CLAMA '11)*, G. Bigatti and S. V. Molen, Eds., Universidad Tecnológica Nacional, Universidad Nacional de la Patagonia San Juan Bosco, Buenos Aires, Agrnetina, Consejo Nacional Investigaciones Científicas Técnicas—CONICET, 2011.

[42] S. Gofas, *Epitoniidae*, World Register of Marine Species, 2011, http://www.marinespecies.org.

[43] G. Rosenberg, *Malacolog 4.1.1: A database of western Atlantic marine Mollusca*, 2009, http://www.malacolog.org/.

[44] Conquiliologistas do Brasil, 2011, http://www.conchasbrasil.org.br/.

[45] W. H. Dall, "Small shells from dredgings off the southeast coast of the United States by the United States Fisheries Steamer 'Albatross' in 1885 and 1886," *Proceedings of the United States National Museum*, vol. 70, pp. 1–134, 1927.

[46] S. Gofas, J. Le Renard, and P. Bouchet Mollusca, *European Register of Marine Species: A Check-List of the Marine Species in Europe and a Bibliography of Guides to their Identification*, vol. 50 of *Collection Patrimoines Naturels, edited by M. J. Costello, C. S. Emblow and R. J. White*, 2001.

[47] A. E. Verrill and K. J. Bush, "Additions to the marine Mollusca of the Bermudas," *Transactions of the Connecticut Academy of Arts and Sciences*, vol. 10, pp. 513–544, 1900.

[48] R. Robertson, "Protoconch size variation along depth gradients in a planktotrophic *Epitonium*," *The Nautilus*, vol. 107, no. 4, pp. 107–112, 1994.

[49] C. Redfern, *Bahamian Seashells: A Thousand Species from Abaco, Bahamas*, Bahamianseashells, Boca Raton, Fla, USA, 2001.

[50] W. P. Woodring, "Miocene Mollusks from Bowden, Jamaica. Part II. Gastropods and discussion of results," *Carnegie Institute of Washington Publication*, vol. 385, pp. 1–564, 1928.

[51] A. d'Orbigny, "Mollusques," *Histoire Physique, Politique et Naturelle de lile de Cuba*, vol. 2, pp. 1–112, 1842.

[52] O. A. L. Mörch, "Synopsis familiae Scalidarum Indiarum occidentalium. Oversigt over Vestindiens Scalarier," *Videnskabelige Meddelelser fra den Naturhistoriske Forening i Kjöbenhavn*, pp. 250–268, 1875.

[53] W. H. Dall, "Reports on the results of dredgings, under the supervision of Alexander Agassiz, in the Gulf of Mexico (1877-78) and in the Caribbean Sea (1879-80), by the U. S. Coast Survey Steamer 'Blake'," *Bulletin of the Museum of Comparative Zoology*, vol. 18, pp. 1–492, 1889.

[54] K. M. De Jong and H. E. Coomans, *Marine Gastropods from Curaçao, Aruba and Bonaire*, E.J. Brill, Leiden, The Netherlands, 1988.

[55] G. Rosenberg, F. Moretzsohn, and E. F. García, "Gastropoda (Mollusca) of the Gulf of Mexico," in *Gulf of Mexico–Origins, Waters, and Biota. Biodiversity*, D. L. Felder and D. K. Camp, Eds., Texas A and M Press, College Station, Tex, USA, 2009.

[56] J. Espinosa and R. Fernández-Garcés, "La família Epitoniidae (Mollusca: Gastropoda) en Cuba," *Poeyana*, vol. 404, pp. 1–13, 1990.

[57] W. H. Dall, "On some new species of Scala," *The Nautilus*, vol. 9, pp. 111–112, 1896.

[58] E. F. Garcia, "New records of Opalia-like mollusks (Gastropoda: Epitoniidae) from the Indo-Pacific, with the description of fourteen new species," *Novapex*, vol. 5, no. 1, pp. 1–18, 2004.

[59] P. Bouchet and S. Tiller, "Two new giant Epitoniids (Mollusca: Gastropoda) from West Africa," *The Veliger*, vol. 20, no. 4, pp. 345–348, 1978.

[60] A. E. Verrill, "Catalogue of marine Mollusca added to the fauna of the New England region, during the past ten years," *Transactions of the Connecticut Academy of Arts and Sciences*, vol. 5, pp. 451–587, 1882.

14

Growth, Mineral Deposition, and Physiological Responses of Broiler Chickens Offered Honey in Drinking Water during Hot-Dry Season

Monsuru Oladimeji Abioja, Kabir Babatunde Ogundimu, Titilayo Esther Akibo, Kayode Ezekiel Odukoya, Oluwatosin Olawanle Ajiboye, John Adesanya Abiona, Tolulope Julius Williams, Emmanuel Oyegunle Oke, and Olusegun Ayodeji Osinowo

Department of Animal Physiology, College of Animal Science and Livestock Production, University of Agriculture, Abeokuta PMB 2240, Nigeria

Correspondence should be addressed to Monsuru Oladimeji Abioja, dimejiabioja@yahoo.com

Academic Editor: Eugene S. Morton

Growing broilers were offered either 0 (0H), 10 (10H), 20 mL (20H) honey, or 0.5 g vitamin C/litre water (AA) during hot-dry season. Honey had no significant ($P > 0.05$) effect on feed intake (FI), weight gain (WG), feed conversion ratio (FCR), water intake (WI), survival (SURV), dressed percentage (DRE), breas tmeat (BRE), gizzard (GIZ), drumstick (DRU), shank (SHA), thigh (THI), tibia volume (VOL), and magnesium (MAG). Effect of honey was significant ($P < 0.05$) on tibial weight (WEI), density (DEN), calcium (CAL), and phosphorus (PHO). WEI and DEN increased with increasing level of honey. 20H broilers had higher CAL than 0H and 10H groups. Broilers offered honey had significantly lower PHO than AA group but the difference between honey groups was not significant. Honey significantly affected PR ($P < 0.001$) and HR ($P < 0.001$) but not RT ($P > 0.005$). Higher dose of honey lowered PR and HR. Honey significantly ($P < 0.05$) increased THY but LIV, KID, LUN, SPL, BUR, and HEA were not significantly ($P > 0.05$) affected. 20H broilers had higher THY than 0H and 10H groups. In conclusion, honey did not affect growth but might improve broilers' welfare when offered up to 20H during hot periods.

1. Introduction

The fast-growing commercial broilers are more sensitive to high temperature during the growing-finishing phase than starter phase. This may be adduced to the inferior development of the cardiovascular, respiratory, and thermoregulatory systems compared with their high rate of growth [1]. They do not possess sweat glands [2, 3]. Environmental temperature in South-Western Nigeria is often higher than 18–21°C [4] recommended for optimal productivity of growing broiler chickens [5]. Often, growth and welfare of the birds are compromised [6–9] and survival lowered [10] because of the birds responses to the stressor. Post [11] reported that high environmental temperature leads to excretion of some minerals like Ca, Fe, Zn, and results in decreased bone strength. HS showed deleterious effects by decreasing length and width of tibia, ash and its strength [12]. Fast-growing broilers as a result typically have significant skeletal problems and may suffer lameness. Thus optimal broiler production in the hot season therefore requires an adequate and appropriate management system that can reduce the effects of HS to the minimum. Abioja and colleagues [13] had earlier reported that addition of vitamin C to broilers' drinking water could reduce rectal temperature and panting rate during afternoon in open-sided poultry house during hot-dry season. This agrees with the reports of previous works on effects of vitamin C in heat-stressed birds [14–18]. Vitamin C helps in inhibiting the secretion and release of corticosterone, which may be cytocytic at high concentration during stress episodes.

There are various plant materials in nature which had been suggested that could be of help in reducing the effects of HS in farm animals. Ramnath et al. [19] used Brahma Rasayana, a nontoxic poly herbal preparation while others used extract from propolis, a plant gum-like waxy resin collected by honey bees to ameliorate the effects of HS in broilers. The authors found these to be efficacious. Chen et al. [20] reported that pretreating of cyclists with propolis extract reversed or reduced the hyperthermia-induced effects, including reduced cell death, inhibition of the overproduction of superoxide, and an attenuation of the depletion of glutathione in the exercising cyclists. Wang et al. [21] worked and reported on the use of bee pollen. Diets supplemented with 1.5% bee pollen could boost the early development of thymus and Fabrici bursa, retard the bursa degeneration, and promote the immune response of spleen chickens. These include ginger root [22] tried ginger root in heat-stressed chickens. The proposal of all these plant materials to ameliorate the effects of HS was based on their content of substances that have antioxidant properties. Honey is a good example of natural substance that contains phytochemicals such as vitamin C, thiamine, riboflavin, pyridoxine, pantothenic acid, nicotinic acid, phenolic compounds, and enzymes glucose oxidase, catalase, and peroxidase. In human beings, honey has been reported to have antioxidant [23] and antibacterial activities [24, 25]. Adebiyi et al. [26] had earlier reported that Nigerian honey was quite rich in minerals. Twelve elements—K, Ca, Ti, Cr, Mn, Fe, Ni, Cu, Zn, Se, Br and Rb—were detected in honey sampled in South-West and South-East Nigeria. Reports on the use of honey in chickens are not readily available. This study is therefore aimed at determining the efficacy of honey in improving the growth rate and physiological welfare of broilers reared during hot-dry season.

2. Materials and Methods

2.1. Experimental Location and Meteorological Observations. The research was carried out at the Poultry Unit of the Teaching and Research Farms, University of Agriculture, Abeokuta, Nigeria (latitude 7° 13′ 49.46″ N; longitude 3° 26′ 11.98″ E and altitude 76 metre above sea level). The climate is humid and located in the rain forest vegetation zone of western Nigeria. The daily minimum, maximum, and mean ambient temperatures, relative humidity and wet- and dry-bulb temperatures at the level of the birds in the pen at 0800 h and 1400 h were monitored throughout the experimental period. The temperature-humidity index was determined from relative humidity and wet- and dry-bulb temperature data.

2.2. Experimental Animals and Management. The birds at d21 were randomly allotted to 12 groups of 11 birds each and allowed adjust to the pens for a week. Each 3 groups received one of 4 treatments: either 0 mL (0H; negative control), 10 mL (10H), 20 mL (20H) honey or 0.5 g vitamin C (AA; positive control) per litre water from d28 to d56. Water and standard finisher mash were given *ad libitum*. The composition of the diet is shown in Table 1.

2.3. Data Collection

2.3.1. Growth Performance. Liveweight of the birds in each replicate was monitored with a sensitive scale every week during the experiment. Records on daily weight gain (WG), feed intake (FI), water intake (WI), and survivability (SURV) were kept. Feed conversion ratio (FCR) was calculated as the ratio of gain to feed. At d56, three birds per replicate were sacrificed by exsanguinations. The birds were processed by scalding and weight of the cut parts measured. Cut parts considered include breast meat (BRE), gizzard (GIZ), drumstick (DRU), shank (SHA), and thigh (THI).

2.3.2. Physiological Responses. Daily rectal temperature (RT) was measured with aid of *Jorita* digital thermometer (model: ECT-5) with ±0.1°C accuracy, inserted into the rectum of the birds and held till the thermometer beeped. Panting rate (PR) was taken as the number of flank movement per minute. A count was taken as a cycle of in and out movement. A stethoscope was placed on the chest region to monitor the heart rate (HR) per minute.

2.3.3. Mineral Deposition in Tibiae and Relative Weight of Organs. Right and left tibiae were removed and boiled to detach all flesh. The weight and volume of the fresh bones were taken. The bones were oven-dried to a constant weight before analyzed to determine for calcium, phosphorus and magnesium contents. Weight of liver (LIV), kidney (KID), lungs (LUN), spleen (SPL), bursa of *Fabricius* (BUR), thymus (THY), and heart (HEA) was taken and relative weight was determined as a percent of liveweight of the birds.

2.4. Statistical Design and Analysis. Data on feed intake (FI), weight gain (WG), feed conversion ratio (FCR), water intake (WI), survivability (SURV), percent dressed weight (DRE), percent breast meat (BRE), percent gizzard (GIZ), percent drumstick (DRU), percent shank (SHA), percent thigh (THI), tibial volume (VOL), tibial weight (WEI), tibial density (DEN), calcium (CAL), phosphorus (PHO), and magnesium (MAG) content of tibia bones, rectal temperature (RT), panting rate (PR), heart rate (HR), relative weight of liver (LIV), kidney (KID), lungs (LUN), spleen (SPL), bursa of *Fabricius* (BUR), thymus (THY), and heart (HEA) were subjected to one-way analysis of variance using GLM method of SYSTAT statistical package [27], with level of significance taken as $P \leq 0.05$. The statistical model is $Y_{ij} = \mu + M_i + \Sigma_{ij}$: where Y_{ij} = yield; μ = population mean, $M_i = i$th effect due to treatment ($i = 1, 2, 3, 4$), and Σ_{ij} = residual error. Means observed to be significantly different were subjected to mean separation using Tukey procedure of SYSTAT [27].

3. Results

Summary of the climatic measurement is presented in Table 2. The morning temperature and temperature-humidity index were lower in value than the values for the afternoon. Opposite was the case for relative humidity. Table 3 shows the result of the effect of honey on growth

Growth, Mineral Deposition, and Physiological Responses of Broiler Chickens Offered Honey in Drinking Water during Hot-Dry Season

119

TABLE 1: Composition of diets for broiler (starter and finisher phases).

Ingredients	Starter phase (%)	Finisher phase (%)
Maize	46.00	50.00
Soybean meal	18.50	12.00
Groundnut cake	15.00	11.00
Fishmeal	2.00	2.00
Wheat offal	12.45	19.05
Bone meal	2.00	2.00
Oyster shell	3.00	3.00
Salt	0.25	0.25
*Premix	0.25	0.25
Methionine	0.30	0.25
Lysine	0.25	0.20
	100	100
Calculated:		
Crude protein (%)	23.05	19.91
ME (MJ/Kg)	11.73	11.71
Ether extract (%)	3.93	3.89
Crude fibre (%)	3.67	3.79
Calcium (%)	1.75	1.74
Phosphorus (%)	0.43	0.41

*1 kg of premix contains: vitamin A: 10,000,000 IU; vitamin D3: 2,000,000 IU; vitamin E: 20,000 IU; vitamin K: 2,250 mg; thiamine B1: 1,750 mg; riboflavin B2: 5,000 mg; pyridoxine B6: 2,750 mg; niacin: 27,500 mg; vitamin B12: 15 mg; pantothenic acid: 7,500 mg; folic acid: 7,500 mg; biotin: 50 mg; choline chloride: 400 g; antioxidant: 125 g; magnesium: 80 g; zinc: 50 g; iron: 20 g; copper: 5 g; iodine: 1.2 g; selenium: 200 mg; cobalt: 200 mg.

TABLE 2: Average daily values for meteorological parameters observed during the experimental period.

Parameter	08.00 h	16.00 h	Average
Dry-bulb temperature (°C)	26.4	34.1	30.2
Wet-bulb temperature (°C)	25.3	30.9	28.1
Relative humidity (%)	94.2	83.8	89.0
Temperature-humidity index	85.2	131.1	108.1

TABLE 3: Effect of honey on growth performance of broiler chickens during hot-dry season.

Parameter	0H	10H	20H	AA	sem
Feed intake (Kg/day/bird)	0.101	0.100	0.099	0.099	0.0050
Weight gain (Kg/day/bird)	0.029	0.028	0.029	0.025	0.0030
Feed conversion ratio	4.73	3.64	3.84	4.13	0.547
Water intake (L/day)	2.34	2.27	2.37	2.38	0.079
Survivability (%)	100.0	99.0	99.0	100.0	0.71
Slaughtered weight (%)	92.4	93.8	88.6	93.8	2.23
Dressed weight (%)	86.8	89.2	84.8	88.2	2.06
Breast meat (%)	15.4	15.4	16.6	16.6	0.78
Gizzard (%)	2.6	2.7	2.5	2.9	0.20
Drumstick (%)	10.8	11.2	10.3	10.9	0.39
Shank (%)	5.0	5.1	4.5	4.8	0.29
Thigh (%)	10.3	10.8	10.2	10.4	0.61

TABLE 4: Effect of honey on mineral deposition in tibiae of broiler chickens during hot-dry season.

Parameter	0H	10H	20H	AA	Sem
Volume of tibiae (mL)	6.4	8.2	8.2	6.8	0.62
Weight of tibiae (g)	8.8[b]	10.0[a]	10.2[a]	7.8[c]	0.17
Density of tibiae (g/mL)	1.4[a]	1.2[ab]	1.2[ab]	1.1[b]	0.09
Calcium (mg/g)	118.5[b]	117.6[b]	120.2[a]	120.3[a]	0.32
Phosphorus (mg/g)	77.9[ab]	76.5[b]	76.2[b]	79.9[a]	0.51
Magnesium (mg/g)	4.9	4.7	4.9	4.8	0.05

[a,b,c] Means on the same row with different superscripts differ significantly ($P < 0.05$).

performance of broiler chickens during hot-dry season. The treatment had no significant ($P > 0.05$) effect on FI, WG, FCR, WI, SURV, DRE, BRE, GIZ, DRU, SHA, and THI. The values recorded were similar for the treatment group and the control. Effect of honey on bone strength and mineral deposition in tibiae of broiler chickens was presented in Table 4. There were no significant ($P > 0.05$) differences in tibial volume from broilers offered either honey or no honey. Magnesium content of tibiae was not ($P > 0.05$) different for the four treatment groups. However, treatment had significant ($P < 0.05$) effect on weight and density of tibiae. Broilers in 20H group had the heaviest tibiae (10.2 g), while the lightest were from 0H group. The weight of tibiae

from broilers in 10H was similar to that of 20H. In the same vein, calcium, and phosphorus deposition in bone was significantly ($P < 0.05$) affected by treatment. Broilers in 20H group had similar calcium content in bone to AA group. These were significantly higher than 0H and 10H groups. Broilers offered honey had significantly lower PHO content in tibiae than AA group, though the PHO content in OH group was not different from AA group.

Table 5 shows the effect of honey on physiological response and relative weight of organs in broiler chickens during hot-dry season. RT was not ($P > 0.05$) significantly affected by treatment. However, the effect of the treatment on PR ($P < 0.001$) and HR ($P < 0.01$) was highly significant. Addition of 20H honey to drinking water lowered PR (32.3 breaths per minute) in broiler chickens compared to 0H (39.0 breaths per minute) though this was not different from 10H (35.4 breaths per minute). HR decreased linearly as honey treatment increases (from 0H to 20H) Broilers offered 20H had lower HR compared to 0H and AA groups. The PR decreased gradually as concentration of honey in water increased. The relative weight of LIV, KID, LUN, SPL, BUR, and HEA was not ($P > 0.05$) affected by treatment. However, the relative weight of THY was significantly ($P < 0.05$) increased by honey. 0H and 10H groups (0.07 and

TABLE 5: Effect of honey on physiological responses and relative weight of organs of broiler chickens during hot-dry season.

Parameter	Mean				
	0H	10H	20H	AA	sem
Rectal temperature (°C)	40.5	40.8	40.7	40.6	0.09
Panting rate (breaths/minute)	39.0[a]	35.4[ab]	32.3[b]	39.2[a]	1.52
Heart rate (beats/minute)	108.7[a]	76.4[bc]	72.5[c]	95.2[ab]	6.06
Liver (%)	2.5	2.7	2.2	2.5	0.15
Kidney (%)	0.02	0.02	0.02	0.03	0.003
Lung (%)	0.60	0.81	0.61	0.63	0.066
Spleen (%)	0.10	0.16	0.11	0.10	0.018
Bursa (%)	0.04	0.06	0.05	0.06	0.015
Thymus (%)	0.07[b]	0.06[b]	0.10[a]	0.11[a]	0.012
Heart (%)	0.45	0.41	0.44	0.42	0.026

[a,b,c] Means with different superscripts in the same row differ significantly ($P < 0.05$).

0.06%, resp.) recorded lighter thymus than 20H (0.10%). The weight at 20H was similar to that of AA group.

4. Discussion

HS causes reduction in feed consumption and growth rate of broiler chickens and increases the death rate [8, 9]. Singleton [28] reported that about 75% of the metabolizable energy consumed by the bird will be converted to body heat and required to be lost to the environment. Thus, reduction in feed intake is an important physiological adaptive mechanism to reduce HS. Reduction in feed intake consequentially results in deficiency in the essential nutrients. Howlider and Rose [29] and May et al. [30] reported that growth rate is reduced in broiler birds when environmental temperature rises because energy obtained from the small feed consumed is expended in panting. The result is that birds had lower final body weight. Efficacious intervention that is capable of improving these traits therefore must be put in place to optimize productivity in broiler production. In the present study, neither honey (10H and 20H) nor AA (0.5 g vitamin C/L water) was effective in improving the weight gain, feed intake, feed conversion ratio, and survival in broilers under hot-dry condition. Miraei-Ashtiani et al. [31] had earlier given a report similar to this. The authors found that there was no difference in weight gain, feed intake, feed conversion ratio and final liveweight offered feed supplemented with or without 200 ppm vitamin C. The outcome of this present study is contrary to the report of [32] that dietary vitamin C improves growth performance in broilers. It is possible that the concentrations of honey and vitamin C used in this study were not sufficient.

In this study, the volume of tibia bones for broilers offered either honey or not remained the same. However, weight, density, and calcium content of the bone were improved by addition of honey to drinking water of heat-stressed broiler chickens at high dose (20H). This might be adduced to improvement in calcium metabolism of the birds.

Mineral metabolism and deposition in bones are affected by HS in poultry [33]. Impaired growth of cartilage and bone was one of the effects of HS mentioned by [34] in a review on managing stress in broiler breeders. Corticosteroid hormones, the hormones released during episode of stress have strongly been implicated in mammalian osteoporosis [35]. These steroids have multitudes of effects on cells such as slowing cell division and differentiation. In mature animals, corticosteroids can affect remodeling, perhaps by preventing the recruitment of osteoblasts and causing bone to weaken by preventing normal bone formation. However, [36] reported that both male and female White Leghorns can maintain bone Na, K, Ca, Mg, P, Cu, Zn, Fe, Mn, and percentage ash in ambient temperature ranging between 21.1 and 35.0°C. HS causes depletion of vitamin C in animals. Vitamin C is required for conversion of 25-hydroxy vitamin D_3 produced by the liver into the hormone calcitriol in the kidney. Calcitriol is essential in the regulation of calcium metabolism [33, 37]. Therefore, the depletion of vitamin C during HS results in obstruction of calcium metabolism. Chung et al. [38] reported that hens fed Vitamin C supplemented feed had higher tibial bone breaking strength than those fed control diet though tibial bone weight and length were not affected by dietary treatments.

The panting and heart rate of broilers were reduced by honey, especially at 20 mL per litre water compared with other groups. This corroborates the work of [39] that panting rate was reduced in heat-stressed birds given vitamin C. Some constituents of honey might be effective at adrenal cortex level where they may either inhibit secretion or release of corticosterone, the main stress hormone in chickens. Vitamin C had been reported to carry out the same functions in birds under stress conditions. Further studies should be carried out on the use of honey in broilers, especially on the constituents that are bioactive. It is possible that it contains some phyto-hormones that have control on contraction and relaxation of muscles of the heart and lungs.

Honey had effect only on the size of thymus out of all organs sampled. The relative weight of thymus to the live bodyweight was increased by 20H honey. The weight was similar to that of AA group. HS has been reported to inhibit immune functions in chickens [40–42]. Surgical removal of thymus (thymectomy) has been used to demonstrate its immunologic role [43]. Efficacy of Sb-Asper-C, a combined ascorbic acid and acetylsalicylic acid treatment in reducing the effects of HS was tested in broilers by [44]. The authors reported that the treatment increased the ratio of thymus to body weight. Actually, the thymus of heat-stressed chickens not supplemented with Sb-Asper-C was atrophied. Honey up to 20 mL per litre water reduced the effect of HS on thymus.

5. Conclusion

Addition of honey to drinking water of broiler chickens in this experiment neither affected the growth nor reduced the body temperature. However, at a dosage of 20 mL honey per litre water, the panting and heart rates were reduced in birds during hot spell. Honey may improve bone strength and immunity in heat-stressed broiler chickens.

Growth, Mineral Deposition, and Physiological Responses of Broiler Chickens Offered Honey in Drinking Water during Hot-Dry Season

121

References

[1] S. Yahav, "Domestic fowl: strategies to confront environmental conditions," *Avian and Poultry Biology Reviews*, vol. 11, no. 2, pp. 81–95, 2000.

[2] D. Grieve, *Heat Stress in Commercial Layers and Breeders.*, vol. 19, Iowa HLST, Technical Bulletin Hy-Line International, 2003.

[3] N. Chaiyabutr, "Physiological reactions of poultry to heat stress and methods to reduce its effects on poultry production," *Thai Journal of Veterinary Medicine*, vol. 32, no. 2, pp. 17–30, 2004.

[4] D. R. Charles, "Responses to the thermal environment," in *Poultry Environment Problems, A Guide To Solutions*, D. A. Charles and A. W. Walker, Eds., pp. 1–16, Nottingham University Press, Nottingham, UK, 2002.

[5] M. O. Abioja, *Temperature-humidity effects on egg fertility and evaluation of vitamin C and cold water on broiler growth in hot season [Ph.D. thesis]*, Department of Animal Physiology, University of Agriculture, Abeokuta, Nigeria, 2010.

[6] S. M. Shane, "Factors influence health and performance of poultry in hot climates," *Poultry Biology*, vol. 1, pp. 247–269, 1988.

[7] O. Altan, A. Altan, I. Oguz, A. PabuAscuoglu, and S. Konyalioğlu, "Effects of heat stress on growth, some blood variables and lipid oxidation in broilers exposed to high temperature at an early age," *British Poultry Science*, vol. 41, no. 4, pp. 489–493, 2000.

[8] L. Hai, D. Rong, and Z. Y. Zhang, "The effect of thermal environment on the digestion of broilers," *Journal of Animal Physiology and Animal Nutrition*, vol. 83, no. 2, pp. 57–64, 2000.

[9] Z. H. M. Abu-Dieyeh, "Effect of chronic heat stress and long-term feed restriction on broiler performance," *International Journal of Poultry Science*, vol. 5, pp. 185–190, 2006.

[10] J. T. Brake, *Stress and Modern Poultry Management: Animal Production Highlights*, Hoffman-La Roche, Basle, Switzerland, 1987.

[11] J. Post, J. M. J. Rebel, and A. A. H. M. Ter Huurne, "Physiological effects of elevated plasma corticosterone concentrations in broiler chickens. An alternative means by which to assess the physiological effects of stress," *Poultry Science*, vol. 82, no. 8, pp. 1313–1318, 2003.

[12] R. Vakili, A. A. Rashidi, and S. Sobhanirad, "Effects of dietary fat, vitamin E and zinc supplementation on tibia breaking strength in female broilers under heat stress," *African Journal of Agricultural Research*, vol. 5, no. 23, pp. 3151–3156, 2010.

[13] M. O. Abioja, O. A. Osinowo, O. F. Smith, D. Eruvbetine, and J. A. Abiona, "Evaluation of cold water and vitamin C on broiler growth during hot-dry season in south-western Nigeria," *Archivos De Zootecnia*, vol. 60, pp. 1095–1103, 2011.

[14] S. L. Pardue and J. P. Thaxton, "Ascorbic acid in poultry: a review," *World's Poultry Science Journal*, vol. 42, pp. 107–123, 1986.

[15] K. Sahin, N. Sahin, and O. Kucuk, "Effects of chromium, and ascorbic acid supplementation on growth, carcass traits, serum metabolites, and antioxidant status of broiler chickens reared at a high ambient temperature (32°C)," *Nutrition Research*, vol. 23, no. 2, pp. 225–238, 2003.

[16] K. Z. Mahmoud, F. W. Edens, E. J. Eisen, and G. B. Havenstein, "Ascorbic acid decreases heat shock protein 70 and plasma corticosterone response in broilers (Gallus gallus domesticus)

subjected to cyclic heat stress," *Comparative Biochemistry and Physiology B*, vol. 137, no. 1, pp. 35–42, 2004.

[17] R. A. Sobayo, *Effect of ascorbic acid supplementation on the performance and haematological profile of pullets and layers [Ph.D. thesis]*, Department of Animal Nutrition, University of Agriculture, Abeokuta, Nigeria, 2005.

[18] C. A. Mbajiorgu, J. W. Ng'ambi, and D. Norris, "Effect of time of initiation of feeding after hatching and influence of dietary ascorbic acid supplementation on productivity, mortality and carcass characteristics of ross 308 broiler chickens in South Africa," *International Journal of Poultry Science*, vol. 6, no. 8, pp. 583–591, 2007.

[19] V. Ramnath, P. S. Rekha, and K. S. Sujatha, "Amelioration of heat stress induced disturbances of antioxidant defense system in chicken by Brahma Rasayana," *Evidence-Based Complementary and Alternative Medicine*, vol. 5, no. 1, pp. 77–84, 2008.

[20] Y. J. Chen, A. C. Huang, H. H. Chang et al., "Caffeic acid phenethyl ester, an antioxidant from propolis, protects peripheral blood mononuclear cells of competitive cyclists against hyperthermal stress," *Journal of Food Science*, vol. 74, no. 6, pp. H162–H167, 2009.

[21] J. Wang, G. M. Jin, Y. M. Zheng, S. H. Li, and H. Wang, "Effect of bee pollen on development of immune organ of animal," *Zhongguo Zhong Yao Za Zhi*, vol. 30, pp. 1532–1536, 2005.

[22] G. F. Zhang, Z. B. Yang, Y. Wang, W. R. Yang, S. Z. Jiang, and G. S. Gai, "Effects of ginger root (Zingiber officinale) processed to different particle sizes on growth performance, antioxidant status, and serum metabolites of broiler chickens," *Poultry Science*, vol. 88, no. 10, pp. 2159–2166, 2009.

[23] M. A. Akanmu, C. Echeverry, F. Rivera, and F. Dajas, "Antioxidant and neuroprotective effects of Nigerian honey," in *Proceedings of the Nueroscience Meeting Planner*, Washington, DC, USA, 2009.

[24] F. O. Adetuyi, T. A. Ibrahim, Jude-Ojei, and G. A. Ogundahunsi, "Total phenol, tocopherol and antibacterial quality of honey Apis mellifera sold in owo community, Ondo State, Nigeria," *Electronic Journal of Environmental, Agricultural and Food Chemistry*, vol. 8, no. 8, pp. 596–601, 2009.

[25] B. O. Omafuvbe and O. O. Akanbi, "Microbiological and physico-chemical properties of some commercial Nigerian honey," *African Journal of Microbiology Research*, vol. 3, no. 12, pp. 891–896, 2009.

[26] F. M. Adebiyi, I. Akpan, E. I. Obiajunwa, and H. B. Olaniyi, "Chemical/physical characterization of Nigerian honey," *Pakistan Journal of Nutrition*, vol. 3, pp. 278–281, 2004.

[27] SYSTAT, *Systat Analytical Computer Package (Version 5. 0)*, Systat Inc., 1992.

[28] R. Singleton, "Hot weather broiler and breeder management," *Asian Poultry Magazine*, pp. 26–29, 2004.

[29] M. A. R. Howlider and S. P. Rose, "Temperature and the growth of broilers," *World's Poultry Science Journal*, vol. 43, pp. 228–237, 1987.

[30] J. D. May, B. D. Lott, and J. D. Simmons, "The effect of air velocity on broiler performance and feed and water consumption," *Poultry Science*, vol. 79, no. 10, pp. 1396–1400, 2000.

[31] S. R. Miraei-Ashtiani, P. Zamani, M. Shirazad, and A. Zareshahned, "Comparison of the effect of different diets on acute heat stressed broilers," in *Proceeding of the 22nd World Poultry Congress*, p. 552, Istanbul, Turkey, 2004.

[32] W. B. Gross, "Effect of ascorbic acid on the mortality of Leghorn-type chickens due to overheating," *Avian Diseases*, vol. 32, no. 3, pp. 561–562, 1988.

[33] J. Brake, *Stress and Modern Poultry Management. Animal Production Highlights 2/87*, Hoffmann-La Roche, Basle, Switzerland, 1988.

[34] A. G. Rosales, "Managing stress in broiler breeders: a review," *Journal of Apllied Poultry Research*, vol. 3, pp. 199–207, 1994.

[35] I. R. Reid, "Glucocorticoid-induced osteoporosis: assessment and treatment," *Journal of Clinical Densitometry*, vol. 1, no. 1, pp. 65–73, 1998.

[36] K. V. Vo, M. A. Boone, and A. K. Torrence, "Electrolyte content of blood and bone in chickens subjected to heat stress," *Poultry science*, vol. 57, no. 2, pp. 542–544, 1978.

[37] H. Weiser, M. Schlachter, and R. Fenster, "The importance of vitamin C for the metabolism of vitamin D3 in poultry," in *Proceedings of the 18th World Poultry Congress*, p. 831, Nagoya, Japan, 1988.

[38] M. K. Chung, J. H. Choi, Y. K. Chung, and K. M. Chee, "Effects of dietary vitamins C and E on egg shell quality of broiler breeder hens exposed to heat stress," *Asian-Australasian Journal of Animal Sciences*, vol. 18, no. 4, pp. 545–551, 2005.

[39] H. R. Kutlu and J. M. Forbes, "Changes in growth and blood parameters in heat-stressed broiler chicks in response to dietary ascorbic acid," *Livestock Production Science*, vol. 36, no. 4, pp. 335–350, 1993.

[40] D. Curca, V. Andronie, and I. C. Andronie, "The effect of ascorbic acid on poultry under thermal stress," in *Proceedings of the 3rd International Congress of Pathophysiology*, Lahti, Finland, June 1998.

[41] D. Curca, M. D. Codreanu, A. Pop, I. Codreanu, and C. P. Constantinescu, "Ascorbic acid level like redox system in parasitic disease in animal blood, liver and adrenal glands," *Simpozionul de Fiziopatologie, Craiova*, vol. 4-5, pp. 38–41, 2003.

[42] M. M. Mashaly, G. L. Hendricks, M. A. Kalama, A. E. Gehad, A. O. Abbas, and P. H. Patterson, "Effect of heat stress on production parameters and immune responses of commercial laying hens," *Poultry Science*, vol. 83, no. 6, pp. 889–894, 2004.

[43] D. Panigraphi, G. L. Waxler, and V. H. Mallman, "The thymus in the chicken and its anatomical relationship to the thyroid," *Journal of Immunology*, vol. 107, no. 1, pp. 289–292, 1971.

[44] B. Anwar, S. A. Khan, A. Maqbool, and K. A. Khan, "Effects of ascorbic acid and acetylsalicylic acid supplementation on the performance of broiler chicks exposed to heat stress," *Pakistan Veterinary Journal*, vol. 24, pp. 109–111, 2004.

15

Colony Development and Density-Dependent Processes in Breeding Grey Herons

Takeshi Shirai

Department of Biology, Tokyo Metropolitan University, Minamiosawa 1-1, Hachioji, Tokyo 192-0397, Japan

Correspondence should be addressed to Takeshi Shirai; ardea@k00.itscom.net

Academic Editor: Eugene S. Morton

The density-dependent processes that limit the colony size of colonially breeding birds such as herons and egrets remain unclear, because it is difficult to monitor colonies from the first year of their establishment, and the most previous studies have considered mixed-species colonies. In the present study, single-species colonies of the Grey Heron (*Ardea cinerea*) were observed from the first year of their establishment for 16 years in suburban Tokyo. Colony size increased after establishment, illustrating a saturation curve. The breeding duration (days from nest building to fledging by a pair) increased, but the number of fledglings per nest decreased, with colony size. The reproductive season in each year began earlier, and there was greater variation in the timing of individual breeding when the colony size was larger. The prolonged duration until nestling feeding by early breeders of the colony suggests that herons at the beginning of the new breeding season exist in an unsteady state with one another, likely owing to interactions with immigrant individuals. Such density-dependent interference may affect reproductive success and limit the colony size of Grey Herons.

1. Introduction

Wading birds such as herons and egrets nest in colonies in relatively limited areas and at high density, often reaching hundreds or thousands of pairs [1]. Herons and egrets are usually quite large birds, and the dynamics of their population attract much attention, as estimating the number of breeding pairs is easier in colony-breeding birds than in noncolony breeders. Population fluctuations may be caused by density-dependent and density-independent processes, although the two are not mutually exclusive.

Density dependence is a negative feedback between population growth rate and population density [2]. In a White Heron (*Egretta alba*) colony, the mean number of fledglings per nest decreased as the number of nests in the colony increased, suggesting a density-dependent response [3]. The first egg-laying date was delayed when colony size was large in a Purple Heron (*Ardea purpurea*) colony [4]. However, little unequivocal evidence shows that density-dependent factors regulate population fluctuations of wading birds, because intraspecific competition for breeding and

colony sites appears to be unimportant in determining the reproductive success or survival in most species [5, 6]. Instead of tracking yearly changes in a single colony, intercolonial comparisons in the same year are useful to detect density-dependent processes, and such comparisons have been conducted in herons and egrets. Results have shown that final breeding success is negatively correlated with colony size in Grey Herons (*Ardea cinerea*) in Belgium [7] and Great Blue Herons (*Ardea herodias*) in Canada [8]. However, in Grey Herons in northern Poland, breeding success increases with colony size [9]. Different conditions within each colony habitat, such as the distance to feeding sites and predation risks, may cause opposing results for the relationship between colony size and breeding success.

Density-independent factors have also been suggested to affect the population fluctuations of herons and egrets. Unpredictable factors such as cold weather, high winds, disease, food shortage, and human impacts can explain reduced breeding success and increased mortality in some regions without invoking density-dependent mechanisms

[5, 10, 11]. In particular, winter temperature is known to affect populations of herons and egrets via increased mortality [12–15]. Winter temperature also affects the onset of the breeding season in Grey Herons [16–19].

Short-term studies are suboptimal for detecting the effects of density on the dynamics of wading bird populations, because birds are long-lived and breed annually [1]. Even in the several long-term studies conducted to date, census data were not recorded beginning with colony establishment but with mature colonies (e.g., [10–12, 20–23]). The tendency of wading birds to form mixed-species groups has posed another difficulty for detecting density-dependent processes in colony size regulation; in these cases, both intra- and interspecific interactions affect individual behavior and survival [24, 25].

In the present study, to simplify the factors affecting density-dependent processes of colony breeders, a single-species colony of Grey Herons was selected and monitored for 16 years beginning at its establishment. I investigated the long-term population fluctuation patterns in this colony and examined density-dependent phenomena based on individual breeding behavior and reproductive successes in the colony. With respect to density-independent factors, I examined how winter temperatures affect colony size and individual breeding behavior.

2. Methods

2.1. Study Animal and Study Sites. The breeding range of Grey Herons covers most of the Old World south of the Arctic Circle, including Europe, Africa, Asia, and the East Indies islands to Wallace's Line [1]. After the breeding season, generally when the young can fly, they disperse in all directions [1]. Therefore, the nonbreeding range expands outside of the breeding range. These birds are sedentary and widely distributed in Japan [26]. They feed on a variety of aquatic animals such as fish, amphibians, crustaceans, insects, and worms [1]. Feeding sites include ponds, lakes, rivers, marshes, and seashores (shallow tidal bays), and they often use artificial water environments such as rice fields, fish farms, parks, and dams [27].

In 1996, a breeding colony of Grey Herons became established at a hillside at Renkoji in suburban Tokyo (35°39'N, 139°28'E); this was selected as a long-term monitoring site. This site is located near the Oguri-gawa River where it just joins the Tama-gawa River. The colony (250 m along the major axis, 80 m along the minor axis) was located in a small area of woods surrounded by houses in a newly built town. The woods mainly consisted of *Quercus serrata* and *Q. acutissima* and some *Pinus densiflora*. In 2000, however, the colony abandoned this site and relocated to a new breeding site on a hilltop in Tama Zoological Park (35°38'N, 139°24'E), located about 5 km west of Renkoji along the Tama-gawa River. In the zoo, woods are patchily distributed, and the Grey Herons use an area (290 m along the major axis, 220 m along the minor axis) consisting of *Quercus serrata*, *Q. acutissima*, *Carpinus tschonoskii*, *Prunus jamasakura*, and *Pinus densiflora*.

2.2. Observations. After colony establishment was first noted at Renkoji in mid-1996 [28], the herons were observed systematically beginning from the next year (1997) using binoculars (×10) and a spotting scope (×20–60; TSN-1, Kowa, Nagoya). Herons that stayed in their nests were identified to species. I monitored the typical behavior of breeding individuals (nest building, incubation, or guarding) and the number and body size of chicks (small, medium, or large), if present, at intervals of 3–7 days in 1997 and every 7–10 days in 1998–2000. In 2000, the 49 breeding pairs of this colony had disappeared by late April, and no herons were observed afterwards. However, 14 pairs were observed making nests since late March in 2000. Therefore, I continued to census this consecutive colony at intervals of 7–10 days in 2001, but at intervals of 3-4 days in 2002–2008 and 7 days in 2009–2011. From the census data, the breeding and nestling feeding durations of each nest were calculated as the period from the day that nest-building behavior was first recorded until fledging was completed (all fledglings disappeared in the nest) and the period from ending incubation (judged from continuous standing of parents) to complete fledging, respectively. If predation occurred in the field, the predators were identified and recorded.

To examine the effects of winter temperature on population fluctuations and the timing of breeding, air temperatures were used from public data recorded by the automated meteorological data acquisition system of the Japan Meteorological Agency at Fuchu, Tokyo (35°41'N, 139°29'E), located 8.4 km northeast from Tama Zoological Park. The minimum value among the mean 10-day temperatures recorded from December 1 to January 31 was used as the coldness value starting the breeding season in each year.

Some unintended effects of observations on breeding heron behavior have been identified [29, 30], particularly researcher visits to colonies during nest building and early incubation, leading to abandonment of nests [31]. Artificial effects on heron behavior may be negligible in this study, because census observations were made carefully and far from the nests to avoid disturbing the herons, and because the nests were so high in the trees that any eggs in the nests were not examined.

3. Results

In 2001 and 2002, eight and two pairs of Black-Crowned Night Herons (*Nycticorax nycticorax*) were found, respectively, in the Tama Zoological Park Grey Heron colony. The park staff also reported that this colony included at least 13 nests of Black-Crowned Night Herons in 2000 [32]. In all other years, only Grey Herons nested within the colony. The number of nests used by the pairs increased annually after colony establishment at Renkoji but tended to become saturated after about 10 years (Figure 1). The mean duration of breeding by each pair ranged from 99.1 to 129.0 days, and the mean duration of nestling feeding ranged from 53.6 to 67.8 days (Table 1). The proportion of successful nests that fledged was 62.6%–85.3%, excluding all that were unsuccessful by the abandonment of the site in 2000 (Table 1). The maximum

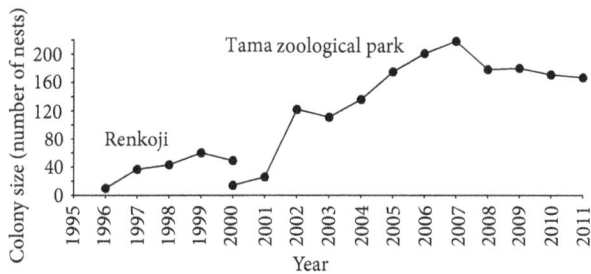

FIGURE 1: Yearly changes in the number of nests in the Renkoji and Tama Zoological Park colonies. In 2000, the Renkoji site was abandoned, and a new colony was established at the Tama Zoological Park close to Renkoji.

number of fledglings was four, and the mean number of fledglings per successful nests ranged from 1.8 to 3.1 (Table 1). The date that the first pair(s) began to breed in the colony differed from year to year, ranging from January 4 to March 3 (Table 1). The mean start date of breeding by pairs also differed yearly, and the coefficient of variation (CV) in the start date of colony members ranged from 20.8% to 55.6% (Table 1).

For the effect of colony size on breeding parameters (Figure 2), breeding duration increased with colony size, but nestling feeding duration was not correlated with colony size. Colony size did not affect the proportion of successful nests, but the mean number of fledglings per successful nest decreased with colony size. As the colony size increased, the date of the first breeding became earlier, and variation in the timing of the onset of breeding by individual pairs was higher.

In the relationships between the date that each pair began breeding and the duration of breeding or nestling feeding (Figure 3, Table 2), the breeding duration was usually longer in early breeders than in late breeders. However, the nestling feeding duration did not differ greatly between earlier and later breeding pairs, suggesting that the duration until nestling feeding was elongated in earlier breeders. In most years, the number of fledglings decreased as the date of the onset of breeding was delayed (Figure 4, Table 2).

Air temperatures were warmer (4.0–5.4°C in mean temperature for a period of 10 days of the coldest period) at the beginning of breeding in 2000, 2002, 2004, 2007, and 2009, but were colder (1.9–2.9°C) in 1998, 2001, 2003, 2006, and 2011, compared to temperatures in 1997, 1999, 2005, 2008, and 2010 (3.6–3.9°C). However, winter temperature was unlikely to explain yearly changes in colony size (Figure 1). Moreover, the two groups of warmer and colder years did not significantly differ in the date of the onset of breeding of the colony (Student's t-test; $t = 0.70$, df = 8, $P = 0.70$), the CV in individual breeding date ($t = 1.14$, df = 7, $P = 0.29$), or other reproductive traits (breeding duration: $t = 0.04$, df = 7, $P = 0.97$; nestling feeding duration: $t = 0.22$, df = 7, $P = 0.83$; number of successful nests: $t = 0.51$, df = 7, $P = 0.63$; number of fledglings per successful nest: $t = 0.43$, df = 7, $P = 0.68$).

Predation rarely occurred during the censuses. The observed predators of eggs and chicks were the Japanese rat snake *Elaphe climacophora* ($N = 1$) and the jungle crow *Corvus macrorhynchos* ($N = 14$).

4. Discussion

4.1. Density Dependence. The Renkoji site was abandoned 4 years after establishment and the colony relocated to a new site in Tama Zoological Park close to Renkoji along the Tama-gawa River. Higher rates of nest mortality lead to significant decreases in colony size, and breeding colony distributions often shift in association with these reductions in colony size [33, 34]. For example, a colony of Great Blue Herons at Pender Harbour, British Columbia, abandoned their site after 2 years of high predation by Ravens (*Corvus corax*), Bald Eagles (*Haliaeetus leucocephalus*), and probably raccoons (*Procyon lotor*) [8, 35]. Heavy predation may thus force colony abandonment by herons. In the study population, however, predation was rarely observed. The snake *Elaphe climacophora* and the crow *Corvus macrorhynchos* were the only predators of eggs and chicks recorded in the present study. Colony nesting of birds sometimes damages vegetation, mainly by altering soil nutrient concentration [36, 37]. Vegetation degradation may cause shifting the colony site [38]. In this study, however, obvious change in vegetation occurred for 16 years even at the central part of colonies. Thus, the cause of the relocation of the colony remains unknown.

In the present study, the colony size increased yearly after establishment and later became saturated (Figure 1), similar to the patterns reported for Grey Herons in two Spanish colonies [39, 40]. Density-dependent processes may be important factors causing such a demographic pattern. The mean clutch size and fledglings per successful nest have been reported to be 3.3–4.2 and 2.1–3.9, respectively, for Grey Herons [41]. In the present study, fledglings per successful nest ranged from 1.8 to 3.1 with a mean of 2.2 (SD = 0.3, $N = 14$), but the fledgling number decreased with colony size (Figure 2(d)), suggesting a density-dependent response. In a study on White Herons in New Zealand, the mean number of fledglings per nest also decreased as the number of nests in the colony increased [3]. In the present study, the timing of breeding may have also been density dependent, because the breeding duration became longer (Figure 2(a)), the reproductive season occurred earlier (Figure 2(e)), and variation among individuals increased (Figure 2(f)) as the colony size increased. The tendency to delay the initial egg-laying date when colony size is large has also been reported for Purple Herons in southern France [4].

The mechanisms of density dependence in demography involve both extrinsic (food) and intrinsic (social behavior) factors [42]. Increased feeding visits by parents (i.e., higher food availability to chicks) decrease the chick mortality of Grey Herons [9]. The nestling feeding duration of Grey Herons, particularly when feeding small chicks, becomes longer when the frequency of returns to the nest decreases, likely because of less food availability at hunting sites [43]. In a field experiment using food supplements, Great Blue Herons showed increased clutch size and fledging success but no difference in the seasonal timing of nesting [44]. Thus,

TABLE 1: Mean values of the breeding duration, nestling feeding duration, number of fledglings per successful nest, and date starting breeding in each nest. The proportion of successful nests to fledge and the date of the first breeding in the colony are also shown.

Colony	Year	Number of nests used	Breeding duration (days)			Nestling feeding duration (days)			Successful nests		Number of fledglings/ successful nest			Date of the first breeding in the colony[b]	Date of the beginning of each nest breeding[b]			
			Mean	SD	N	Mean	SD	N	Percentage	N	Mean	SD	N		Mean	SD	CV (%)	N
Renkoji	1996	10																
Renkoji	1997	37	102.5	6.9	19	67.3	17.1	15	85.3	34	3.1	0.7	29	48	78.1	26.4	33.8	34
Renkoji	1998	43	106.7	14.7	24	58.6	16.3	24	92.1	38	2.5	0.9	35	62	74.9	18.1	24.1	33
Renkoji	1999	60	106.6	16.0	20	64.6	17.0	20	64.4	45	2.2	0.7	29	62	79.6	16.5	20.8	58
Renkoji	2000[a]	49			0			0	0.0	49			0	47				0
Tama Zoo	2000	14																
Tama Zoo	2001	26	114.6	15.0	15	65.9	16.8	7	85.0	20	1.9	0.7	17	45	83.8	18.0	21.5	25
Tama Zoo	2002	122	99.1	16.2	56	57.3	12.1	43	70.7	92	2.1	0.9	65	49	75.5	22.1	29.3	78
Tama Zoo	2003	111	113.4	15.5	82	64.6	11.0	76	82.5	103	2.5	0.8	85	42	75.9	26.0	34.3	106
Tama Zoo	2004	136	111.1	19.2	79	60.6	17.6	77	79.5	122	2.4	0.9	97	26	72.1	29.2	40.6	135
Tama Zoo	2005	175	120.9	22.1	112	60.4	13.4	95	81.5	157	2.1	0.8	128	4	64.3	35.8	55.6	156
Tama Zoo	2006	201	123.8	21.1	102	61.7	14.5	89	62.6	182	2.0	0.8	114	12	68.6	27.4	39.9	195
Tama Zoo	2007	218	126.6	21.3	125	63.4	17.4	106	79.9	174	2.1	0.8	139	35	69.4	29.2	42.0	208
Tama Zoo	2008	178	114.3	16.9	102	59.3	11.6	96	75.8	149	2.0	0.8	113	27	67.4	21.2	31.4	166
Tama Zoo	2009	180	129.0	21.1	105	67.8	16.7	85	79.1	153	1.8	0.6	121	23	57.0	24.4	42.8	183
Tama Zoo	2010	171	125.9	21.3	117	53.6	17.4	113	77.7	157	1.9	0.6	122	31	67.9	20.9	30.8	155
Tama Zoo	2011	167	125.2	24.7	93	63.3	18.5	99	79.3	150	2.0	0.8	119	38	73.6	31.2	42.4	130

[a] All nests were abandoned in this year; [b] date is presented by days since the 1st of January in each year.

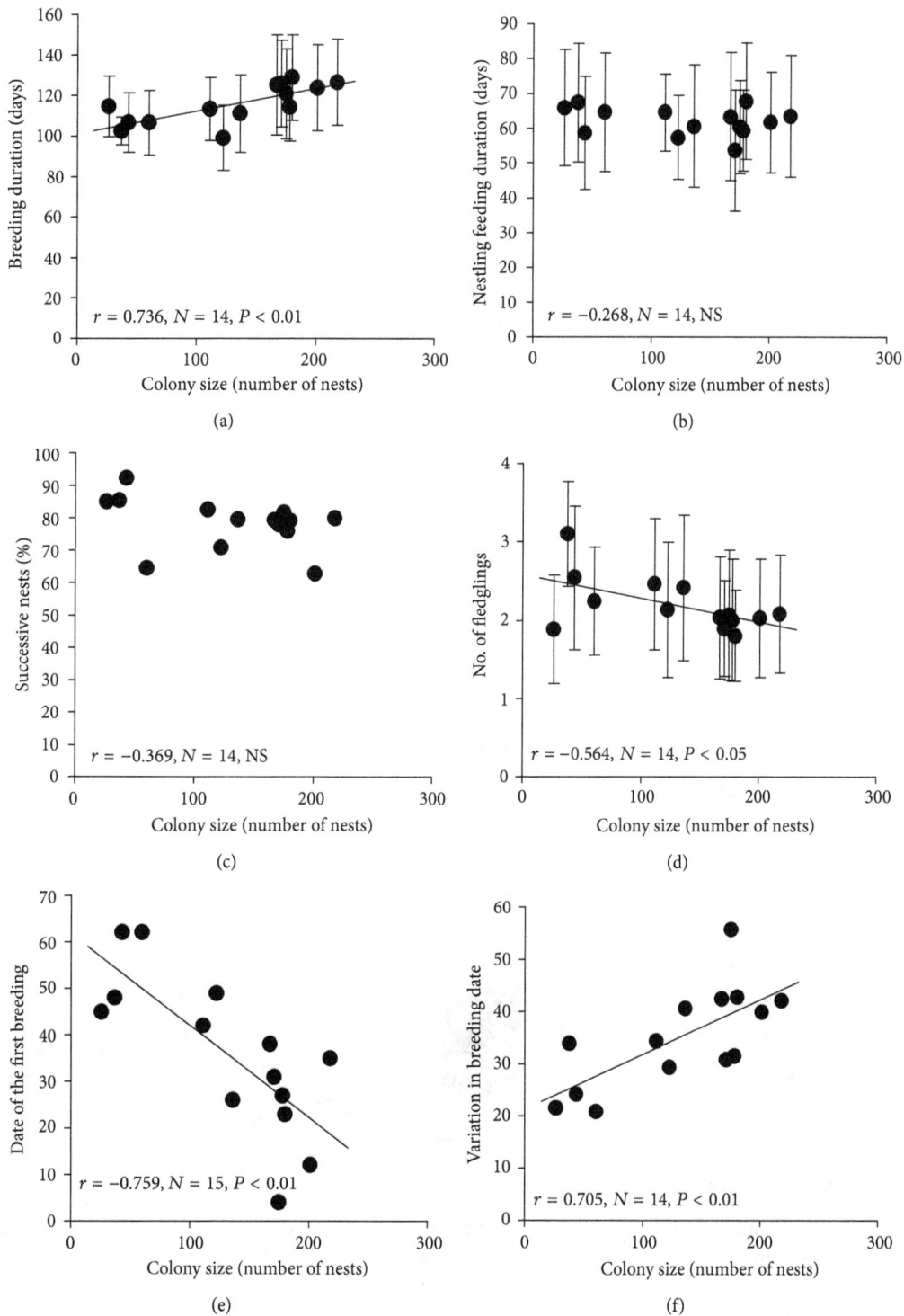

FIGURE 2: Relationships between colony size and (a) breeding duration, (b) nestling feeding duration, (c) proportion of successfully fledged nests, (d) number of fledglings per successful nest, (e) date of the first breeding in the colony, and (f) coefficient of variation in the date of the onset of breeding by individuals. Bars indicate ± SD.

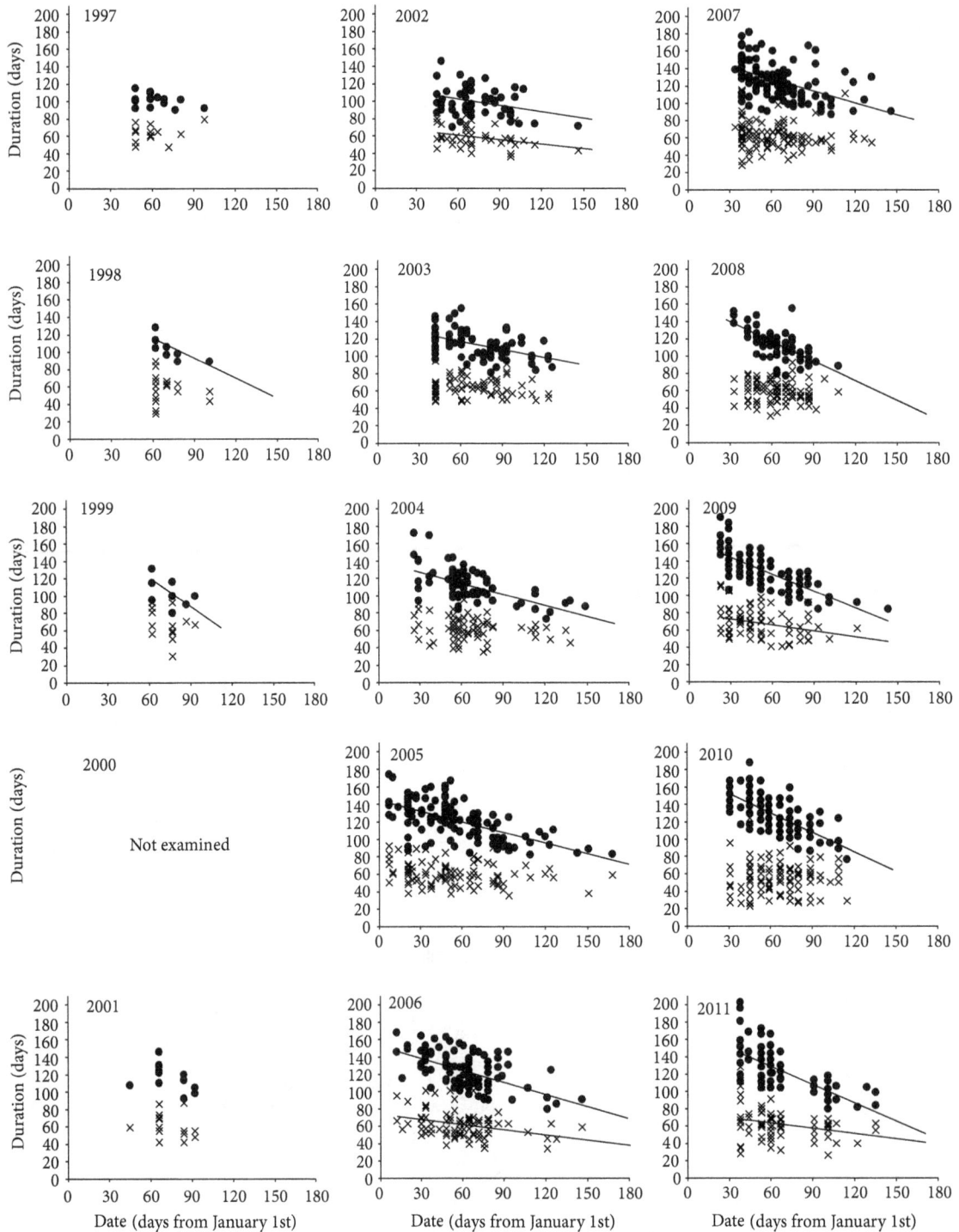

FIGURE 3: Breeding (•) or nestling feeding (×) durations of each pair of Grey Herons in relation to the date of the beginning of breeding in each year. Regression lines are shown when the correlation coefficients are statistically significant (see Table 2).

competition for limited food is a possible source of density dependence. However, the effect of food availability on colony size is usually difficult to detect, because food searching areas used by colony members cannot be tracked completely. In Portugal, Cattle Egret (*Bubulcus ibis*) colonies reportedly depend on the areas of dry pasture and crops within a 5 km radius of the colony center, and Little Egrets (*Egretta garzetta*)

depend on the presence of freshwater and saltwater habitats within a 5 km area [45]. Estimating the amount of available food over such a wide range of habitat would be difficult.

Social factors involved in density dependence may be direct behavioral interference among colony members. In herons and egrets, intraspecific interactions may be critical in courtship and nesting [1, 46]. In the present study,

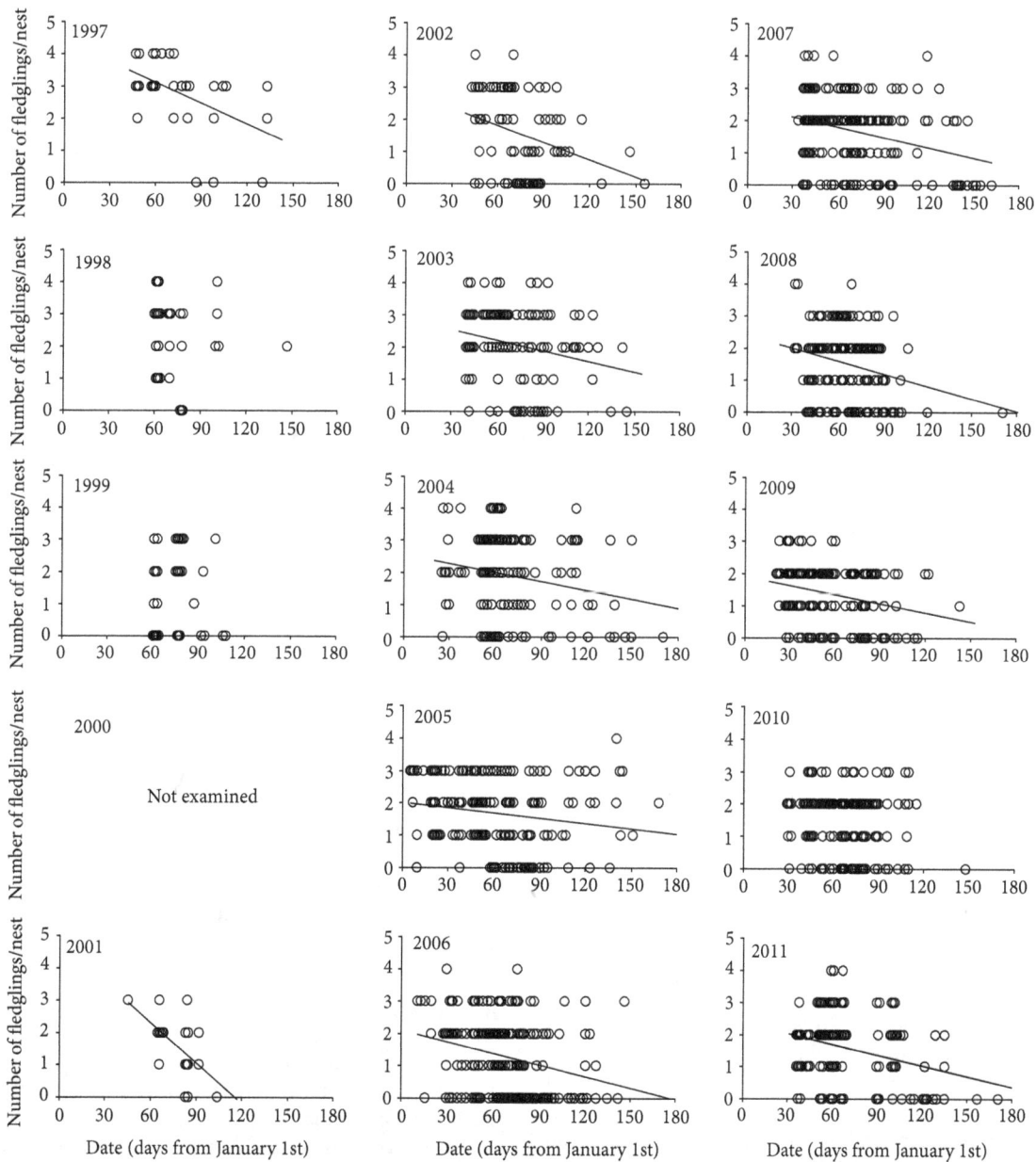

FIGURE 4: Number of fledglings per nest in relation to the date of the beginning of breeding in each nest. Regression lines are shown when the correlation coefficients are statistically significant (see Table 2).

earlier breeders took a longer period of time until nestling feeding (i.e., the period from nest building to completed egg incubation) than later breeders (Figure 3). The prolonged prenestling feeding duration may have been caused by unstable states of colony members at the beginning of the breeding season. During this period, interference is expected to be frequent and intense, because the colony includes many birds that have arrived and joined the colony that year. To confirm this interference hypothesis, quantitative data examining the relationship between interference frequency and colony size are needed.

Despite these possible intense interactions in the early breeding season, the number of fledged chicks was greater for earlier breeders than for later breeders (Figure 4). A similar decline in the number of fledglings with a later date of the onset of breeding was observed for Grey Herons in Poland, although it was not observed every year [19]. Brood size is often larger in earlier breeders within a reproductive season in birds [47], as observed in Grey Herons [7, 48, 49] and other heron and egret species [50–52]. On the other hand, mean brood size increases with the age of parents [53, 54]. Older individuals may be more successful in acquiring nest sites of good quality and have more experience in parental care [54, 55]. In the present study, these age effects could not be assessed. In the future, to discriminate the effects of the age of parents and the timing of breeding on reproductive success,

TABLE 2: Correlation coefficients between the date of starting breeding and the breeding duration, the nestling feeding duration, or the number of fledglings in each nest.

Colony	Year	Breeding duration			Nestling feeding duration			Number of fledglings		
		r	N	P	r	N	P	r	N	P
Renkoji	1996									
Renkoji	1997	−0.376	19	NS	0.285	16	NS	−0.495	30	**
Renkoji	1998	−0.751	24	***	−0.135	24	NS	−0.048	30	NS
Renkoji	1999	−0.647	20	**	−0.350	19	NS	−0.013	37	NS
Renkoji	2000									
Tama Zoo	2000									
Tama Zoo	2001	−0.422	15	NS	−0.233	14	NS	−0.591	19	**
Tama Zoo	2002	−0.329	54	*	−0.323	40	*	−0.327	78	**
Tama Zoo	2003	−0.475	79	***	−0.121	74	NS	−0.233	100	*
Tama Zoo	2004	−0.589	78	***	−0.213	76	NS	−0.218	117	*
Tama Zoo	2005	−0.614	111	***	−0.203	94	NS	−0.178	150	*
Tama Zoo	2006	−0.548	102	***	−0.342	89	**	−0.287	180	**
Tama Zoo	2007	−0.500	125	***	−0.118	106	NS	−0.311	170	***
Tama Zoo	2008	−0.743	101	***	−0.136	95	NS	−0.252	143	**
Tama Zoo	2009	−0.751	105	***	−0.311	84	**	−0.267	137	**
Tama Zoo	2010	−0.709	117	***	0.032	110	NS	−0.110	150	NS
Tama Zoo	2011	−0.702	93	***	−0.255	87	*	−0.307	120	*

NS, $P > 0.05$; *$P < 0.05$; **$P < 0.01$; ***$P < 0.001$.

the age of the Grey Herons must be determined by marking individuals.

4.2. Density Independence. Unpredictable factors can explain reduced breeding success and increased mortality in some regions without invoking density-dependent mechanisms [5, 10, 11]. For example, in a study conducted in the Yucatan Peninsula, most herons and egrets failed to reproduce because of food shortages caused by exceptionally heavy rains and flooding of their lowland habitats [10].

Another well-known density-independent process that can reduce population size in herons and egrets is unpredicted severely low temperatures in winter. The Grey Heron population in England and Wales usually numbers 4500–4800, but after severe winters, it decreases to around 3000 [12]. In particular, winter (January to March) temperature strongly affects the survival rate of first-year Grey Herons in England, as estimated by the annual recovery of banded nestlings from 1955 to 1974 [13]. In those studies, in a year when temperatures reached about 1°C, few young birds survived, whereas during years with winter temperatures above 6°C, more than half survived. A large number of Grey Herons also died during a cold spell in January and February 1976 in The Netherlands [14]. In northwest Italy, Grey Herons increased following an increase in winter temperature [15]. In France, however, the survival of yearling Little Egrets did not appear to be affected by winter severity [56].

In the present 16-year study, winter temperature was unlikely to explain yearly variation in reproductive parameters. In Poland, Grey Herons in inland colonies begin to breed later than those in coastal colonies, because spring air temperatures are lower and ice cover is present on feeding sites for a longer duration in inland locations [19]. Thus, the accessibility of feeding sites (lack of ice cover) in spring is an important factor affecting the onset of breeding. However, in suburban Tokyo, winter temperatures are relatively mild and rarely fall below 0°C. Therefore, annual differences in the onset of breeding are not likely related to temperature, as in the case of Little Egrets in France [56].

Acknowledgments

The author thanks Fumio Hayashi, Tamotsu Kusano, and Tadashi Suzuki for their help with many aspects of this study and particularly F. Hayashi for the improvement of an earlier version of this paper. He also thanks Kazuyoshi Ito, Etsuo Narushima, Tomoko Yabe, Yasumasa Tomita, and Heizo Sugita for allowing him to study at Tama Zoological Park and Fumio Nakamura for his help in initiating a study there.

References

[1] J. A. Kushlan and J. A. Hancock, *The Herons*, Oxford University Press, New York, NY, USA, 2005.

[2] I. Newton, *Population Limitation in Birds*, Academic Press, London, UK, 1998.

[3] C. Miller, "Long-term monitoring of a breeding colony of white herons (*Egretta alba*) on the Waitangiroto river, South Westland, New Zealand," *Notornis*, vol. 48, no. 3, pp. 157–163, 2001.

[4] C. Barbraud, M. Lepley, V. Lemoine, and H. Hafner, "Recent changes in the diet and breeding parameters of the Purple Heron *Ardea purpurea* in Southern France," *Bird Study*, vol. 48, no. 3, pp. 308–316, 2001.

[5] R. W. Butler, "Population regulation of wading ciconiiform birds," *Colonial Waterbirds*, vol. 17, no. 2, pp. 189–199, 1994.

[6] R. Lande, S. Engen, and B. E. Sæther, "Estimating density dependence in time-series of age-structured populations," *Philosophical Transactions of the Royal Society B*, vol. 357, no. 1425, pp. 1179–1184, 2002.

[7] J. van Vessem and D. Draulans, "The adaptive significance of colonial breeding in the Grey Heron *Ardea cinerea*: inter- and intra-colony variability in breeding success," *Ornis Scandinavica*, vol. 17, no. 4, pp. 356–362, 1986.

[8] R. W. Butler, P. E. Whitehead, A. M. Breault, and I. E. Moul, "Colony effects on fledging success of Great Blue Herons (*Ardea herodias*) in British Columbia," *Colonial Waterbirds*, vol. 18, no. 2, pp. 159–165, 1995.

[9] D. Jakubas, "Factors affecting the breeding success of the Grey Heron (*Ardea cinerea*) in Northern Poland," *Journal of Ornithology*, vol. 146, no. 1, pp. 27–33, 2005.

[10] A. Lopez-Ornat and C. Ramo, "Colonial waterbird populations in the Sian Ka'an Biosphere Reserve (Quintana Roo, Mexico)," *The Wilson Bulletin*, vol. 104, no. 3, pp. 501–515, 1992.

[11] E. M. Kirsch, B. Ickes, and D. A. Olsen, "Assessing habitat use by breeding Great Blue Herons (*Ardea herodias*) on the upper Mississippi River, USA," *Waterbirds*, vol. 31, no. 2, pp. 252–267, 2008.

[12] J. Stafford, "The heron population of England and Wales, 1928–1970," *Bird Study*, vol. 18, no. 4, pp. 218–221, 1971.

[13] P. M. North and B. J. T. Morgan, "Modelling heron survival using weather data," *Biometrics*, vol. 35, no. 3, pp. 667–681, 1979.

[14] E. J. van der Molen, A. A. Blok, and G. J. de Graaf, "Winter starvation and mercury intoxication in Grey Herons (*Ardea cinerea*) in The Netherlands," *Ardea*, vol. 70, no. 2, pp. 173–184, 1982.

[15] M. Fasola, D. Rubolini, E. Merli, E. Boncompagni, and U. Bressan, "Long-term trends of heron and egret populations in Italy, and the effects of climate, human-induced mortality, and habitat on population dynamics," *Population Ecology*, vol. 52, no. 1, pp. 59–72, 2010.

[16] J. Walmsley, "The development of a breeding population of Grey Herons (*Ardea cinerea*) in the Camargue," *La Terre et la Vie*, vol. 29, pp. 89–99, 1975.

[17] L. Marion, "Territorial feeding and colonial breeding are not mutually exclusive: the case of the Grey Heron (*Ardea cinerea*)," *Journal of Animal Ecology*, vol. 58, no. 2, pp. 693–710, 1989.

[18] F. Campos and M. Fernández-Cruz, "The breeding biology of the Grey Heron (*Ardea cinerea*) in the Duero river basin in Spain," *Colonial Waterbirds*, vol. 14, pp. 57–60, 1991.

[19] D. Jakubas, "The influence of climate conditions on breeding phenology of the Grey Heron *Ardea cinerea* L. in Northern Poland," *Polish Journal of Ecology*, vol. 59, no. 1, pp. 179–192, 2011.

[20] J. A. Kushlan and H. Hafner, *Heron Conservation*, Academic Press, London, UK, 2000.

[21] H. Hafner, R. E. Bennetts, and Y. Kayser, "Changes in clutch size, brood size and numbers of nesting Squacco Herons *Ardeola ralloides* over a 32-year period in the Camargue, Southern France," *Ibis*, vol. 143, no. 1, pp. 11–16, 2001.

[22] J. P. Kelly, K. Etienne, C. Strong, M. McCaustland, and M. L. Parkes, "Status, trends, and implications for the conservation of heron and egret nesting colonies in the San Francisco Bay area," *Waterbirds*, vol. 30, no. 4, pp. 455–478, 2007.

[23] A. Ismail and F. Rahman, "Population dynamics of colonial waterbirds in upper Bisa, Putrajaya Wetlands, Malaysia," *Acta Biologica Malaysiana*, vol. 1, no. 1, pp. 36–40, 2012.

[24] R. E. Bennetts, M. Fasola, H. Hafner, and Y. Kayser, "Influence of environmental and density-dependent factors on reproduction of Little Egrets," *The Auk*, vol. 117, no. 3, pp. 634–639, 2000.

[25] L. Dami, R. E. Bennetts, and H. Hafner, "Do Cattle Egrets exclude Little Egrets from settling at higher quality sites within mixed-species colonies?" *Waterbirds*, vol. 29, no. 2, pp. 154–162, 2006.

[26] Committee for Check-list of Japanese Birds, *Check-List of Japanese Birds*, Ornithological Society of Japan, Tokyo, Japan, 6th edition, 2000.

[27] Y. Sawara, N. Azuma, K. Hino, K. Fukui, G. Demachi, and M. Sakuyama, "Feeding activity of the Grey Heron *Ardea cinerea* in tidal and non-tidal environments," *Japanese Journal of Ornithology*, vol. 39, no. 2, pp. 45–52, 1990.

[28] T. Shirai, "Breeding biology of Grey Herons *Ardea cinerea* in the Tama River," *Strix*, vol. 17, pp. 85–91, 1999 (Japanese).

[29] D. K. Goering and R. Cherry, "Nestling mortality in a Texas heronry," *The Wilson Bulletin*, vol. 83, no. 3, pp. 303–305, 1971.

[30] P. C. Frederick and M. W. Collopy, "Researcher disturbance in colonies of wading birds: effects of frequency of visit and egg-marking on reproductive parameters," *Colonial Waterbirds*, vol. 12, no. 2, pp. 152–157, 1989.

[31] J. Tremblay and L. N. Ellison, "Effects of human disturbance on breeding of black-crowned night herons," *The Auk*, vol. 96, no. 2, pp. 364–369, 1979.

[32] H. Sugita, "Wild Grey Herons nesting in the zoo. The bird watching group of Tama Zoo," *Animals and Zoos*, vol. 53, no. 9, p. 21, 2001 (Japanese).

[33] T. Boulinier, "On breeding performance, colony growth and habitat selection in Buff-Necked Ibis," *Condor*, vol. 98, no. 2, pp. 440–441, 1996.

[34] E. Danchin, T. Boulinier, and M. Massot, "Conspecific reproductive success and breeding habitat selection: implications for the study of coloniality," *Ecology*, vol. 79, no. 7, pp. 2415–2428, 1998.

[35] K. Simpson, J. N. M. Smith, and J. P. Kelsall, "Correlates and consequences of coloniality in great blue herons," *Canadian Journal of Zoology*, vol. 65, no. 3, pp. 572–577, 1987.

[36] A. Ishida, "Changes of soil properties in the colonies of the common cormorant, *Phalacrocorax carbo*," *Journal of Forest Research*, vol. 1, no. 1, pp. 31–35, 1996.

[37] H. T. Mun, "Effects of colony nesting of *Adrea cinerea* and *Egretta alba modesta* on soil properties and herb layer composition in a *Pinus densiflora* forest," *Plant and Soil*, vol. 197, no. 1, pp. 55–59, 1997.

[38] K. Ogasawara, K. Abe, and T. Naito, "Ecological study of Grey Heron in Oga Peninsula, Akita Prefecture," *Journal of Yamashina Institute for Ornithology*, vol. 14, no. 2-3, pp. 232–245, 1982.

[39] M. Fernández-Cruz and F. Campos, "The breeding of Grey Herons (*Ardea cinerea*) in Western Spain: the influence age," *Colonial Waterbirds*, vol. 16, no. 1, pp. 53–58, 1993.

[40] J. Prosper and H. Hafner, "Breeding aspects of the colonial ardeidae in the Albufera de Valencia, Spain: population changes, phenology, and reproductive success of the three most abundant species," *Colonial Waterbirds*, vol. 19, no. 1, pp. 98–107, 1996.

[41] J. Kim and T.-H. Koo, "Nest site characteristics and reproductive parameters of Grey Herons *Ardea cinerea* in Korea," *Zoological Studies*, vol. 48, pp. 657–664, 2009.

[42] V. Bretagnolle, F. Mougeot, and J.-C. Thibault, "Density dependence in a recovering osprey population: demographic and behavioural processes," *Journal of Animal Ecology*, vol. 77, no. 5, pp. 998–1007, 2008.

[43] J. van Vessem and D. Draulans, "Factors affecting the length of the breeding cycle and the frequency of nest attendance by Grey Herons *Ardea cinerea*," *Bird Study*, vol. 33, no. 2, pp. 98–104, 1986.

[44] G. V. N. Powell, "Food availability and reproduction by Great White Herons, *Ardea herodias*: a food addition study," *Colonial Waterbirds*, vol. 6, pp. 139–147, 1983.

[45] J. C. Farinha and D. Leitão, "The size of heron colonies in Portugal in relation to foraging habitat," *Colonial Waterbirds*, vol. 19, no. 1, pp. 108–114, 1996.

[46] D. A. McCrimmon Jr., "Nest site characteristics among five species of herons on the North Carolina coast," *The Auk*, vol. 95, pp. 267–280, 1978.

[47] D. Lack, *The Natural Regulation of Animal Numbers*, Clarendon Press, Oxford, UK, 1954.

[48] G. Creutz, "Zur Brutbiologie des Graureihers (*Ardea cinerea* L.) in der Oberlausitz," *Beiträge zur Vogelkunde*, vol. 21, pp. 161–171, 1975.

[49] J. van Vessem, "Timing of egg-laying, clutch size and breeding success of the Grey Heron, *Ardea cinerea*, in the North of Belgium," *Le Gerfaut*, vol. 81, pp. 177–193, 1991.

[50] J. A. Rodgers Jr., "Breeding ecology of the Little Blue Heron (*Florida caerulea*) on the West coast of Florida, USA," *The Condor*, vol. 82, no. 2, pp. 164–169, 1980.

[51] J. A. Rodgers Jr., "Breeding chronology and reproductive success of Cattle Egrets and Little Blue Herons on the West coast of Florida, USA," *Colonial Waterbirds*, vol. 10, no. 1, pp. 38–44, 1987.

[52] G. S. Ranglack, R. A. Angus, and K. R. Marion, "Physical and temporal factors influencing breeding success of Cattle Egrets (*Bubulcus ibis*) in a West Alabama colony," *Colonial Waterbirds*, vol. 14, no. 2, pp. 140–149, 1991.

[53] G. S. Baxter, "The influence of synchronous breeding, natal tree position and rainfall on egret nesting success," *Colonial Waterbirds*, vol. 17, no. 2, pp. 120–129, 1994.

[54] F. Thomas, Y. Kayser, and H. Hafner, "Nestling size rank in the little egret (*Egretta garzetta*) influences subsequent breeding success of offspring," *Behavioral Ecology and Sociobiology*, vol. 45, no. 6, pp. 466–470, 1999.

[55] F. Thomas, F. Renaud, T. De Meeus, and F. Cézilly, "Parasites, age and the Hamilton-Zuk hypothesis: inferential fallacy?" *Oikos*, vol. 74, no. 2, pp. 305–309, 1995.

[56] H. Hafner, Y. Kayser, V. Boy et al., "Local survival, natal dispersal, and recruitment in Little Egrets *Egretta garzetta*," *Journal of Avian Biology*, vol. 29, no. 3, pp. 216–227, 1998.

16

An Evaluation of Ad Hoc Presence-Only Data in Explaining Patterns of Distribution: Cetacean Sightings from Whale-Watching Vessels

Louisa K. Higby,[1,2] **Richard Stafford,**[3] **and Chiara G. Bertulli**[2,4]

[1] *School of Ocean Sciences, Bangor University, Menai Bridge, Anglesey LL59 5AB, UK*
[2] *School of Engineering and Natural Sciences, Faculty of Life and Enviromental Sciences, Elding Whale-Watching, Ægisgata 7, 101 Reykjavik, Iceland*
[3] *Division of Science, Institute of Biomedical and Environmental Science and Technology, University of Bedfordshire, Luton LU1 3JU, UK*
[4] *School of Engineering and Natural Sciences, Faculty of Life and Enviromental Sciences, University of Iceland, Sturlugata 7, 101 Reykjavik, Iceland*

Correspondence should be addressed to Louisa K. Higby, louisahigby@gmail.com

Academic Editor: Anne Goodenough

The analysis of presence-only data is a problem in determining species distributions and accurately determining population sizes. The collection of such data is common from unequal or nonrandomised effort surveys, such as those surveys conducted by citizen scientists. However, causative regression-based methods have been less well examined using presence-only data. In this study, we examine a range of predictive factors which might influence Cetacean sightings (specifically minke whale sightings) from whale-watching vessels in Faxaflói Bay in Iceland. In this case, environmental variables were collected regularly regardless of whether sightings were recorded. Including absences as well as presence in the analysis resulted in a multiple-generalised linear regression model with significantly more explanatory power than when data were presence only. However, by including extra information on the sightings of the whales, in this case, their observed behaviour when the sighting occurred resulted in a significantly improved model over the presence-only data model. While there are limitations of conducting nonrandomised surveys for the use of predictive models such as regression, presence-only data should not be considered as worthless, and the scope of collection of these data by citizen scientists using modern technology should not be underestimated.

1. Introduction

Presence-only data, data where presence of a species or individual is recorded, but where absences are not, are frequent in many *ad hoc* scientific surveys, such as those datasets collected by volunteers or citizen scientists [1–3]. While presence-only data have been shown to produce good maps of species ranges in some occasions (e.g., [1]), using such data to infer changes in distribution, population sizes, and other ecological parameters can be difficult [2]. Reasons for this largely relate to unequal sampling effort [1–4]. Low- or zero-sampling effort could easily miss the presence of a low-density species in certain areas, but the amount of effort applied is largely unknown, and hence lack of presence of

a species could relate to a real absence, or simply a lack of effort. Misidentification of species, or misreporting of locations, can confound such studies, although such issues can also occur in any volunteer programme, regardless of sampling strategy employed [1, 2].

Volunteers, and citizen scientists recruited through "crowd-sourcing" events, however, are a cheap method of collecting data over a wide spatial or temporal scale [1, 5]. As such, presence-only data are becoming common, and an evaluation of their use in scientific research is timely. While this study does not strictly use citizen science data, the sampling regime used is, by necessity, not of equal effort in space or time, and some aspects of the dataset collected are, again by necessity, presence only. However, since absence data were also recorded, the dataset provides an ideal opportunity to test the use of presence-only data in causative regression models.

Multiple linear regression and associated linear model reduction methods are a common tool in addressing habitat or environmental variables related to habitat selection by organisms [6–8]. For example, in a stepwise model reduction approach, a large number of explanatory factors can be used to predict presence (or number) of individuals in each location, and through examination of each factor's relative importance, the most important factors involved in explaining the majority of the variation in the dependent variable can be found [9].

Such a multiple regression approach to predicting sightings of minke whales (*Balaenoptera acutorostrata*) was used in the current study. Data on sightings of these species were collected from whale-watching vessels, and a range of environmental factors, such as weather conditions, sea state, and temperature, were also recorded or obtained from existing public data sources. These factors were used in conjunction with sightings of Cetaceans and observations on their behaviour to identify the role of environmental factors in the occurrence of whale sightings. Since data were collected regularly throughout the cruise, many factors were known when sightings did not occur, and removing these "absence" data points in subsequent analyses allowed for an evaluation of presence-only data in causative regression models.

2. Methods

Boat-based surveys were carried out from Faxaflói Bay (64.20871° N, 22.19869° W), on the south west coast of Iceland during whale-watching trips. Each survey session was broken down into 15 min observation intervals where environmental variables were recorded (Table 1). In addition, minke whale (*Balaenoptera acutorostrata*) encounters were recorded, along with the numbers seen at each sighting, and this figure was used as a dependent variable in the analysis (with zero indicating absence).

A total of 634 data points were used in the analysis, of which 133 recorded sightings of minke whales (15-minute periods when 1 or more minke whales were observed). These surveys were carried out over 104 days, between April and July 2010 (days when the sea state exceeded Beaufort scale 3 were discarded from the analysis—following prior recommendations, e.g., [10, 11]).

A multiple linear model was constructed with all environmental factors listed in Table 1 as possible explanatory factors for the number of minke whales seen in any 15-minute time period (with the exception of behaviour—which could not be included in models where absence data was used). A stepwise model reduction process was then undertaken (using forward and backward processes) as described in [9], using AIC as a model reduction method. Next, any 15-minute period where no minke whales were seen was removed to create a presence-only dataset, and the stepwise model process ran again. Next, the presence-only dataset was analysed again using all the previous

environmental factors along with behaviour of the cetacean when it was seen.

Given that data are count data and residuals from standard linear models were not normally distributed, generalised linear models were used for analysis (as per [12, 13]). The full dataset showed variance far exceeded the mean for the counts of both minke whales and white-beaked dolphins, and GLMs based on the negative binomial distribution were used [13]. When zero counts were excluded, data were still not normally distributed, but followed the assumptions of Poisson distributions (mean~variance), and these GLMs were based on this distribution [13]. Given GLM does not provide a goodness-of-fit statistic (such as R^2), we use (1) an evaluation of the sum of the residuals as a measure of goodness of fit to compare presence and absence data models, where lower values indicate better fit, (2) manually calculated R^2 as a model comparison mechanism, but not as a true measure of the predictive power of the model, by subtracting the quotient of the residual and model deviance from one, and (3) ran the analysis using traditional linear models, despite limitations applied to count data, to provide a more comparative study of the proportion of variability explained. In this case, all data for dependent variables (minke whale sighting number) were $\log_{10} + 1$ transformed, since this normalised the residuals of the presence-only dataset.

3. Results

Analysis of the full dataset, excluding behaviour but including absences of minke whales, produced a reduced model with five significant ($P < 0.05$) explanatory factors (Table 2), a mean squared residual value of 0.528, and an estimated $R^2 = 0.394$. In comparison, traditional linear modelling (despite nonnormal residuals) resulted in a highly significant final model ($F_{9, 624} = 16.22$; $P < 0.001$) and an adjusted $R^2 = 0.178$.

The reduced dataset—only using data when minke whales were present—produced a model with considerably the worst explanatory power (although significance testing was not possible due to the different sizes of datasets), with two significant explanatory factors (Table 2), with a mean-squared residual value of 0.907 and an estimated $R^2 = 0.140$. Similar decreases in fit were obtained by traditional linear models ($F_{5, 127} = 2.91$; $P = 0.005$; adjusted $R^2 = 0.086$).

Inclusion of the behaviour of the minke whale as an observation using the presence-only dataset gave an improved fit regression with two significant explanatory factors (Table 2), and mean squared residual value of 0.804 and an R^2 of 0.240. In this case, feeding behaviour occurred during significantly more sightings than the other behaviours. Again, similar trends were found with traditional linear models ($F_{7, 125} = 4.67$; $P < 0.001$; adjusted $R^2 = 0.163$); this model was significantly better than the presence-only model not including behaviour (ANOVA test on two fitted models, $P = 0.0015$).

From an analysis of explanatory variables in the reduced models (Table 2), it can be seen that cetacean sightings were affected by similar factors in most models—sea and

An Evaluation of Ad Hoc Presence-Only Data in Explaining Patterns of Distribution: Cetacean Sightings from
Whale-Watching Vessels

135

TABLE 1: Explanatory factors initially included in the linear model. Use of continuous and discrete (category) variables follows suggestions of
[14], where category variables are categorised in as few as possible meaningful categories, and ordinal variables are assumed to be continuous.
Logistic variables are categorised as discrete with $n = 2$ levels.

Factor	Variable type	Notes
Date	Continuous	From day 1 to day 158
Boat	Category ($n = 2$)	Differences in height of sighting platform; possible differences in acoustics of boat engines
Time of day	Continuous	That is, 4:30 pm = 16.5
Behaviour of cetacean	Category ($n = 3$)	Surfacing, feeding, and other
Number of humpback whales	Continuous	
Number of killer whale	Continuous	
Number of white-beaked dolphins	Continuous	
Sea state	Continuous	
Percentage of cloud cover	Continuous	
Weather conditions	Category ($n = 3$)	Sun, Cloud, and rain
Wind direction	Category ($n = 5$)	N, E, S, W, and no wind
Tidal conditions	Category ($n = 2$)	Flood or ebb
Swell height	Continuous	
Visibility	Continuous	
Sea surface temperature	Continuous	
Observer	Category ($n = 3$)	Three different observers recorded results

weather conditions were important. Sea temperature was also important in some cases, and the behaviour being performed by the cetacean was important, when included, for explaining the observed sightings of minke whales.

4. Discussion

In general, we demonstrate that presence-only data limit the explanatory properties of models such as multiple linear regression. However, the inclusion of explanatory factors, which can only be included in presence-only models (i.e., they relate to the actual sightings), can increase the power of such modelling approaches on presence-only data.

One important consideration is that datasets containing multiple zero counts require more complex models than those which do not [12, 13]. As such, presence-only data, by its nature, excludes all zero counts, and means models built using standard linear model techniques can be used for analysis. Such models give a much better and intuitive understanding of model fit using the familiar R^2 variable. While in this study, presence-only data were analysed using generalised linear models for comparison with absence data, a simple log transform of the dependent variable normalised the residuals of the linear model approach.

Given that the current dataset was not collected by citizen scientists, we need to consider how the results might apply to citizen-science-collected data, and whether the technique is valuable if applied to the collection of data using citizen science methods. Firstly, what is clearly important is the amount of data available. A successful citizen science programme could greatly increase the number of records and may result in high-quality predictive models being built on presence-only data. Furthermore, it should also be noted

that if larger numbers of participants are taking part in such surveys, the chance of missing an actual sighting of a cetacean on a whale-watching trip will be reduced. As such, if the number of "returns" or submission of data is high for each whale-watching trip, then it could be considered that most actual sightings will have been recorded, hence it is more likely that where no data are present, there were no Cetaceans, rather than this being a false absence.

Explanatory models frequently showed factors such as cloud cover, visibility, and sea state to be important. While these factors could influence actual distribution of Cetaceans, it is more likely that they influence the observer's ability to detect them [10, 15]. While such issues may cause some concern for conversion to citizen-science-collected data, the extra volumes of data collected may be able to be standardised for "detection" conditions, by subsetting the data prior to analysis (i.e., into rough or calm conditions, or into sunny versus overcast conditions).

When behaviour was included in the predictive model, there was a significant increase in its explanatory power. In this study, behaviour helped to explain the number of minke whales present at a particular sighting, with more minke whales present when feeding was occurring than for the other behaviours, likely as minke whales may be more likely to be or remain in the presence of a boat when there is significant food available in the area. However, the use of such an approach is only possible when using presence data, since behaviour cannot be recorded if no individual is seen. However, recent research suggests that untrained volunteers are not good at recording behaviour accurately [16]. Despite this, given that data submission could be via photograph or video, this recognition of behaviour could be verified by researchers (e.g., [1]), as could other accuracy-of-data issues,

TABLE 2: Factors present in the stepwise-reduced model for each dataset where significant multiple linear regressions were obtained. P indicates that the factor was present in the reduced model, and + or − indicates whether there was a positive or negative relationship of the factor compared to the number of Cetaceans. Where no + or − is present, the factor was a category variable. N/A indicates that this variable was not used as an explanatory factor in the analysis.

Factor	Full dataSet	Presence only	Presence only and behaviour
Date	P+		
Boat	P		
Time of day			
Behaviour of cetacean	N/A	N/A	P
Presence of humpback whales			
Presence of killer whale			
Presence of white-beaked dolphin			
Sea state		P−	
Percentage of cloud cover	P+		
Weather conditions			
Wind direction			
Tidal conditions			
Swell height			
Visibility	P+	P+	
Sea surface temperature	P−		P−
Observer			

such as location and time of the record, which can all be verified using modern digital technology [1, 5].

Results relating to observation of surface sightings of minke whales are clearly dependent on the whales being at the surface. Therefore, factors which influence the dive time of the whales will also be apparent in terms of the surface sightings. For example, surface intervals of minke whales are known to vary throughout the year and throughout the day [17], and this seems to be correlated to the type of food the whales are foraging on and the local bathymetry of the foraging site [18]. In this study, both date and sea surface temperature had opposite effects (with a positive relationship between sightings and date, and a negative relationship with sea surface temperature). Such a finding is consistent with previous findings, indicating that surface intervals may be longer in the spring, since whales may feed on plankton blooms, but may also associate themselves with areas of upwelling (cooler water) where these blooms, or other food, may be more abundant [19].

Overall, presence-only data do provide some useful, biological information regarding the sightings of Cetaceans, and the ability to collect a greater volume of data should offset concerns over its value. However, whether data are collected by citizen scientists or trained scientists, there are limitations of the use of tourist vessels in this type of explanatory variable approach. A first step in many analyses of data using model-reduction approaches is the determination of whether sightings or distributions of a species can be explained by chance, by testing the distribution data against a Poisson distribution [20]. Given that tourist vessels automatically head to previous sightings and communicate with each other as to the location of recently seen whales, any approximation to a regular grid, or random sampling, cannot be assumed, and the fact that distribution of whale sightings could be entirely random cannot be ruled out. Furthermore, although for minke whales, much published research has been conducted from whale-watching vessels (e.g., [18, 19]), the behaviour of other cetacean species to whale-watching boats can be variable, with some species, such as humpback whales, avoiding vessels by increasing dive time [21] and others actively approaching boats [22]. In particular, whale-watching vessels appear to record more "active" behaviour, such a leaping out of the water, and also from younger individuals [23]. However, the use of presence-only data, with recordings and understanding of the implications of the behaviour (avoiding, approaching, jumping, etc.), may still allow useful data on factors such as habitat selection to be collected from whale-watching vessels.

As considered elsewhere, the use of digital technology and internet storage facilities can both increase uptake and accuracy of citizen science work [1, 5], and while collection of presence-only data has some disadvantages, some limitations cannot be improved by the use of trained personnel and require dedicated random surveys. In some cases, presence-only data even have some advantages and are worthy of consideration given the current resource cuts to science budgets and the need to greater engage the public with scientific research.

Acknowledgments

The authors would like to thank the Elding Whale Watching Company, with special thanks to G. Vignir Sigursveinsson and Rannveig Grétarsdóttir without whose support, providing a platform for all survey activities, this research would not have been possible. They are grateful to CSI for funding the data collection in the year 2010, and also to the Faxaflói Cetacean Research volunteer Mirjam Held, who helped with

An Evaluation of Ad Hoc Presence-Only Data in Explaining Patterns of Distribution: Cetacean Sightings from
Whale-Watching Vessels

137

data collection during the 2010 field season in Faxaflói Bay. They would also like to thank the anonymous reviewer for helpful suggestions to improve the paper.

References

[1] R. Stafford, A. G. Hart, L. Collins et al., "Eu-social science: the role of internet social networks in the collection of bee biodiversity data," *PloS one*, vol. 5, no. 12, p. e14381, 2010.

[2] J. Franklin, *Mapping Species Distributions: Spatial Inference and Prediction*, Cambridge University Press, Cambridge, UK, 2009.

[3] M. W. Tingley and S. R. Beissinger, "Detecting range shifts from historical species occurrences: new perspectives on old data," *Trends in Ecology and Evolution*, vol. 24, no. 11, pp. 625–633, 2009.

[4] C. Hassall and D. J. Thompson, "Accounting for recorder effort in the detection of range shifts from historical data," *Methods in Ecology and Evolution*, vol. 1, no. 4, pp. 343–350, 2010.

[5] C. L. Catlin-Groves, "Submitted to this special issue. The citizen science landscape: from volunteers to citizen sensors and beyond," *International Journal of Zoological Research*. In press.

[6] A. E. Goodenough, A. G. Hart, and R. Stafford, "Regression with empirical variable selection: description of a new method and application to ecological datasets," *PLoS One*, vol. 7, no. 3, Article ID e34338, 2012.

[7] M. J. Whittingham, P. A. Stephens, R. B. Bradbury, and R. P. Freckleton, "Why do we still use stepwise modelling in ecology and behaviour?" *Journal of Animal Ecology*, vol. 75, no. 5, pp. 1182–1189, 2006.

[8] J. Fan and J. Lv, "A selective overview of variable selection in high dimensional feature space," *Statistica Sinica*, vol. 20, no. 1, pp. 101–148, 2010.

[9] M. J. Crawley, *Statistics: An Introduction Using R*, Wiley, Chichester, UK, 2005.

[10] P. G. H. Evans and P. S. Hammond, "Monitoring cetaceans in European waters," *Mammal Review*, vol. 34, no. 1-2, pp. 131–156, 2004.

[11] D. Palka, "Effects of Beaufort sea state on the sightability of harbor porpoises in the Gulf of Maine," *Forty-Sixth Report of the International Whaling Commission*, pp. 575–582, 1996.

[12] M. Ridout and C. Demetrio, "Generalized linear models for positive count data," *Revista de Matematica e Estatstica*, vol. 10, pp. 139–148, 1992.

[13] M. J. Crawley, *The R Book*, Wiley, Chichester, UK, 2007.

[14] D. J. Pasta, "Learning when to be discrete: continuous vs. categorical predictors," Paper 248, 2009, SAS Global Forum 2009.

[15] S. Dawson, P. Wade, E. Slooten, and J. Barlow, "Design and field methods for sighting surveys of cetaceans in coastal and riverine habitats," *Mammal Review*, vol. 38, no. 1, pp. 19–49, 2008.

[16] R. L. Williams, S. Porter, A. G. Hart, and A. E. Goodenough, "Submitted to this special issue. The accuracy of behavioural data collected by visitors in a zoo environment Can visitors collect meaningful data?" *International Journal of Zoological Research*. In press.

[17] K. A. Stockin, R. S. Fairbairns, E. C. M. Parsons, and D. W. Sims, "Effects of diel and seasonal cycles on the dive duration of the minke whale (Balaenoptera acutorostrata)," *Journal of the Marine Biological Association of the United Kingdom*, vol. 81, no. 1, pp. 189–190, 2001.

[18] K. Macleod, R. Fairbairns, A. Gill et al., "Seasonal distribution of minke whales Balaenoptera acutorostrata in relation to physiography and prey off the Isle of Mull, Scotland," *Marine Ecology Progress Series*, vol. 277, pp. 263–274, 2004.

[19] P. C. Gill, M. G. Morrice, P. Brad, P. Rebecca, A. H. Levings, and C. Michael, "Blue whale habitat selection and within-season distribution in a regional upwelling system off southern Australia," *Marine Ecology Progress Series*, vol. 421, pp. 243–263, 2011.

[20] A. E. Goodenough, S. L. Elliot, and A. G. Hart, "Are nest sites actively chosen? Testing a common assumption for three non-resource limited birds," *Acta Oecologica*, vol. 35, no. 5, pp. 598–602, 2009.

[21] A. Schaffar, B. Madon, V. Garrigue, and R. Constantine, "Avoidance of whale watching boats by humpback whales in their main breeding ground in New Caledonia," Paper SC/61/WW/6, International Whaling Commission, Cambridge, UK, 2009.

[22] F. Ritter, *Interactions of Cetaceans with Whale Watching Boats—Implications for the Management of Whale Watching Tourism*, M.E.E.R.e.V., Berlin, Germany, 2003.

[23] M. Weinrich, "Are behavioral data from whalewatch boats biased?" Paper SC/61/WW/3, International Whaling Commission, Cambridge, UK, 2009.

17

Reproductive Strategy of *Labeobarbus batesii* (Boulenger, 1903) (Teleostei: Cyprinidae) in the Mbô Floodplain Rivers of Cameroon

Claudine Tekounegning Tiogué,[1] Minette Tabi Eyango Tomedi,[2] and Joseph Tchoumboué[3]

[1] *The University of Dschang, Faculty of Agronomy and Agricultural Sciences, Laboratory of Applied Ichthyology and Hydrobiology, P.O. Box 222, Dschang, Cameroon*
[2] *The University of Douala, Institute of Fisheries and Aquatic Sciences of Yabassi, P.O.Box 2701, Douala, Cameroon*
[3] *The University of Mountains, P.O. Box 208, Banganté, Cameroon*

Correspondence should be addressed to Claudine Tekounegning Tiogué; tekou_claudine@yahoo.fr

Academic Editor: Michael Thompson

Aspects of the reproductive strategy of African carp, *Labeobarbus batesii*, were investigated from May 2008 to October 2009 in the Mbô Floodplain of Cameroon. Samples were collected monthly from artisanal fishermen. The total length and total body mass of each specimen were measured to the nearest mm and 0.01 g, respectively. Sex was determined by macroscopic examination of the gonads after dissection. The sex ratio was female skewed (overall sex ratio: 1 : 1.42). Females reach sexual maturity at a larger size (213 mm) than the males (203 mm). The mean gonadosomatic index ranges from $0.32 \pm 0.17\%$ to $1.91 \pm 1.15\%$, whereas the mean K factor ranges from 0.90 ± 1.09 to 1.10 ± 0.13. These two parameters are negatively correlated. The reproduction cycle begins in mid-September and ends in July of the next year, and they are reproductively quiescent for the rest of the year. *Labeobarbus batesii* is a group-synchronous spawner with pulses of synchronised reproduction spread over a long period. The mean absolute, potential, and relative fecundities are 2898 ± 2837 oocytes, 1016 ± 963 oocytes, and 9071 ± 7184 oocytes/kg, respectively. The fecundity is higher and positively correlated with the gonad mass than with body size. Its reproductive biology suggests that *L. batesii* is suitable for pond culture.

1. Introduction

In many parts in the world, cyprinid fish species are important in aquaculture, representing 61% of world production in 2008 [1, 2]. Although Asia is the largest aquaculture producer of cyprinids in the world, with its endogenous species, the diversity of freshwater fishes in Africa (3200 species) is comparable to that in Asia (3000 species) [3], but Africa lacks significant cyprinid aquaculture.

In general, the problems for aquaculture in sub-Saharan Africa are related to poor breeding techniques and a limited number of suitable species [4]. Thus, despite the diversity of African fishes, the main species that are farmed are imported from outside of Africa. Of the nearly 500 species of African cyprinids [5], only *Labeo parvus* (Boulenger) is

used for aquaculture [6–9]. The technological support that accompanies domesticated exotic species contributes to the neglect of indigenous species that would require a long process of domestication for aquaculture [3]. Moreover, the harmful impact of the introduction of exotic species is likely to divert attention from the aquaculture potential of native species [10]. Indeed, the introduction of new species is the main cause of extinction of native freshwater fishes in Africa [11]. Endogenous fish species found in several agroecological zones of Cameroon are important candidates for aquaculture.

The African carp, *Labeobarbus batesii* (Boulenger, 1903), is a common and widespread species of the Cyprinidae family in Lower Guinea. Elsewhere, it is known from the Dja and Tibesti in Tchad [12]. In Cameroon, the Mbô Floodplain is an important scientific and socioeconomic centre, where the fishery improves the incomes and the consumption rates of animal protein of the local people [13]. *Labeobarbus batesii* is a high nutritional value fish in the Mbô Floodplain and other Cameroonian zones. Four families of fish species (Channidae, Cichlidae, Clariidae, and Cyprinidae) are fished in this ecosystem. The reproductive biology of a few species has been studied in this zone [13, 14], but apart from one small study on its growth [15], nothing is known about the reproductive biology of *Labeobarbus batesii*, even though it is economically important. Our aim, therefore, is to describe reproductive traits of the African carp *Labeobarbus batesii* to assess its suitability for aquaculture in this area.

2. Material and Methods

2.1. Physical Environment. The study was carried out from May 2008 to October 2009 in the Mbô Floodplain (MF) (NL 5°10′, LE 9°50′) in Cameroon, an area of 390 km^2 with an altitude of about 700 m. It is located between the littoral and west regions of Cameroon. The soil is volcanic, sandy and favourable to agriculture all year round. MF has a hot and humid climate characterised by two seasons; the dry season starts in mid-November and ends in mid-March. The temperature ranges from 17°C to 30°C and the relative humidity varies from 49% to 98% in dry and rainy seasons, respectively. The average rain fall is about 1860 mm. The MF Rivers descend from the Bambouto Mountains (Menoua River), which is part of the Manengouba Massif (Nkam and Black Water Rivers). Several streams, such as Metschie and Mfouri, descend from the Bana Massif (Figure 1). All are drained by the Wouri River, which flows into the Atlantic Ocean [15].

2.2. Fish Samples. Monthly samples, totalling 448 samples of *Labeobarbus batesii* (387 mature and 61 undetermined sexes), were obtained from artisanal fishermen from May 2008 to October 2009. Ten sampling sites were identified and were gathered into two zones (confluence and interconfluence). Fishes were collected by means of traditional fishing gear (bow nets, hooks, and gill nets). Collected fish were counted, rinsed and anesthetized in the solution of tricaine methanesulfonate (MS 222), prepared by dissolving 4 g of MS 222 in 5 L tap water, and then preserved in 10% formalin. The samples were transported to the laboratory for analysis. Fishes were identified according to the criteria of [12].

Fishes were measured for total length to the nearest 1.0 mm. The total body mass was measured using an electronic balance (Sartorius Competence) to the nearest 0.01 g. The morphometric characteristics recorded for each fish were used to calculate Fulton's K condition factor by the formula $K = 100W * TL^{-3}$, where W is the total body mass and TL is the total length.

The sex of fish was determined by macroscopic examination of the gonads after dissection and the sex ratio compared to 1:1 using a Chi-square test (χ^2) [16]. The length at which 50% (L_{50}) of fish matured was determined by visual estimation after plotting the percentage of mature fish against their lengths. The gonadosomatic index (GSI) was calculated

by expressing the gonad mass as a percentage of body mass [17]. Spawning and breeding periodicity was determined from the inverse trend of GSI and K condition factor and by an examination of the GSI monthly variation during 18 months: higher values of GSI showed the breeding season while the lowest values showed a sexual quiescence.

The oocytes were removed from gravid fish, weighed, and preserved in modified Gilson's fluid (nitric acid 17 mL, acetic acid 4 mL, mercuric chloride 20 g, ethanol 95%, and distilled water 900 mL) [18]. The preserved ovaries were washed several times to get rid of the preservative and the oocytes were separated from ovaries in Petri dish. The potential fecundity (PF) and absolute fecundity (AF) were estimated by counting the number of mature oocytes or all oocytes, respectively, from a known weight of subsamples collected from the ovaries. These two fecundities were calculated by multiplying the total mass of oocytes by the number of oocytes per gram [19]. The relative fecundity (RF) was obtained as the number of oocytes per unit fish mass. The relationship between the absolute fecundity and the fish length was determined by power regression technique with the following equation: AF $= aX^b$. The relationship between the absolute fecundity and the fish body mass or between the absolute fecundity and the gonad mass was determined by linear regression. The equation was AF $= a + bX$, where AF = absolute fecundity; X = fish length, fish body mass, or gonad mass; a = regression constant; b = regression coefficient.

3. Results

3.1. Relative Abundance Distribution. Fishes were only caught between July and October in the interconfluences (Table 1), but the number and the biomass were significantly higher ($P < 0.01$) than those caught at the confluences. The number of fishes caught and their biomass in the inter-confluences or in the rainy season were significantly higher ($P < 0.01$) than those in the confluences or in the dry season, respectively. Similarly, the number and biomass of fishes collected in 2008 were significantly higher ($P < 0.01$) than those in 2009.

3.1.1. Sex Ratio. The sex ratio of fish varied from month to month (Table 2). The pooled sex ratio (160 males and 228 females, mean ratio of 1:1.42) over all months differed significantly from the expected ratio ($P < 0.05$). Within months, males were dominant only in December 2008 (1:0.60). The expected sex-ratio (1:1) was observed in only two months (July 2008 and February 2009).

3.1.2. Sexual Maturity Size. Gravid fishes ($n = 37$) ranged from 230 to 505 mm in length and from 105 to 1300 g in mass. Males matured at 203 mm total length (TL) and females at 213 mm total length. In overall population, L_{50} were 200 mm total length (Figure 2).

3.1.3. Gonadosomatic Index, K Condition Factor, and Spawning Periods. The mean gonadosomatic indices (GSIs) ranged from 0.32 ± 0.17% in August to 1.91 ± 1.15% in December 2008. There was a significant ($P < 0.01$) monthly variation

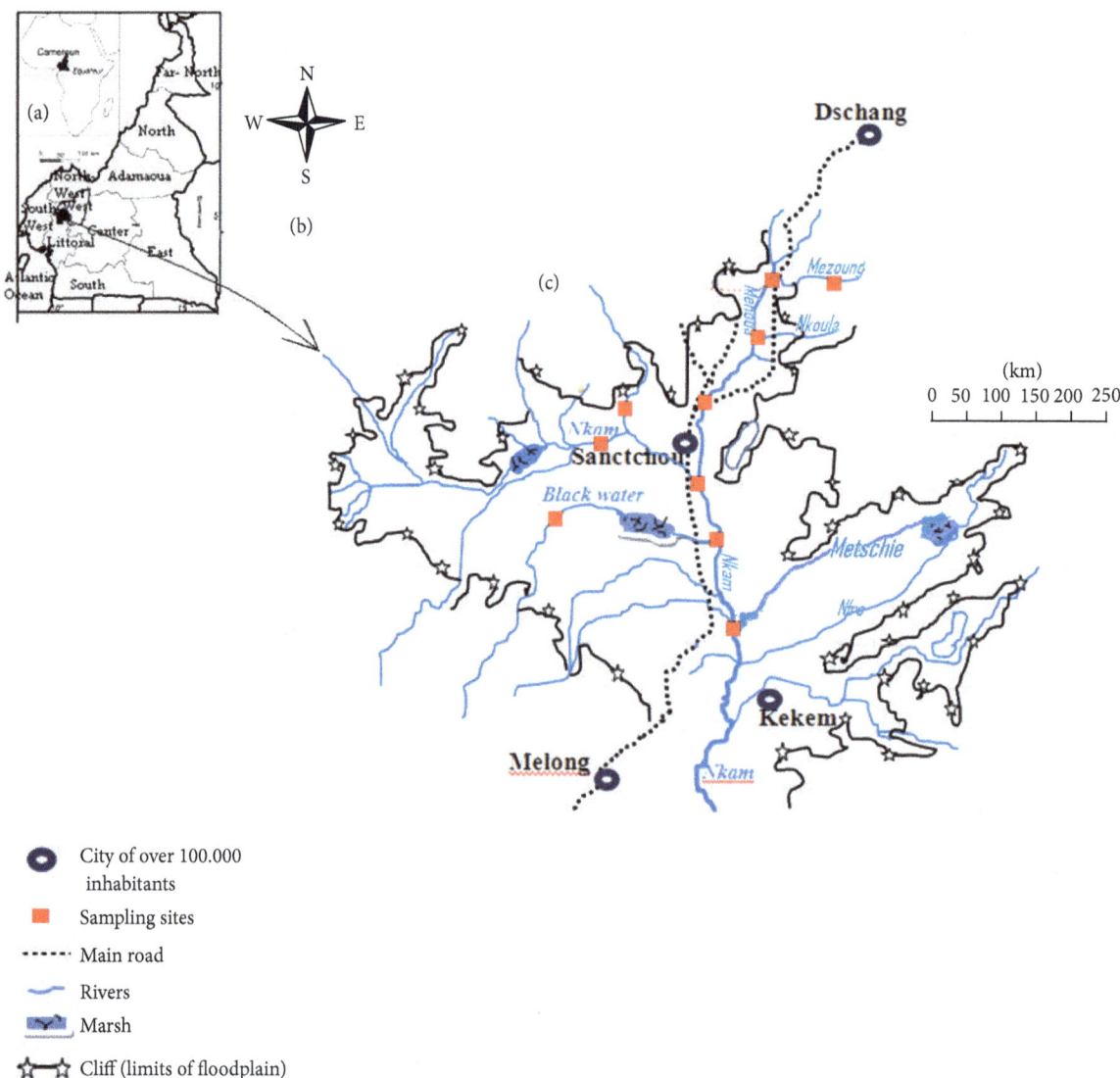

FIGURE 1: Study area: (a) Africa, (b) Cameroon, (c) Mbô Floodplain.

in the gonadosomatic index. Both years recorded the highest values between November 2008 and June 2009 and the lowest values from July to October. Months with high standard deviations occur when the fishes are reproductively active (Figure 3(a)). The monthly changes in mean GSIs revealed an increase in reproductive activity in both sexes in September (Figure 3(b)). The monthly mean GSIs of female *Labeobarbus batesii* varied from 0.36% in August 2008 to 1.94% in March 2009 and of males varied from 0.54% in October 2009 to 2.97% in February 2009 (Figure 3(a)). The mean male GSIs $(1.31 \pm 0.81\%)$ were significantly higher $(P < 0.01)$ than those of females $(0.97 \pm 0.53\%)$. Male mean GSIs, which peaked in February, June, and November, were significantly higher $(P < 0.05)$ than in other months. Female mean GSIs peaked in March and December. Female mean GSIs were significantly lower $(P < 0.01)$ in the rainy season $(0.91 \pm 0.37\%)$ than in dry season $(1.25 \pm 0.62\%)$. The mean GSIs of male in the dry season $(1.92 \pm 0.76\%)$ were significantly higher $(P < 0.01)$ than in the wet season $(1.01 \pm 0.65\%)$.

The mean female GSIs were lower $(P > 0.05)$ in 2009 $(0.91 \pm 0.50\%)$ than in 2008 $(1.05 \pm 0.58\%)$, whereas the mean male GSIs in 2008 $(1.17 \pm 0.82\%)$ and in 2009 $(1.44 \pm 0.83\%)$ were significantly different $(P < 0.05)$ (Figure 3(b)).

The mean condition factor, K, ranged from 0.90 ± 1.09 in March 2009 to 1.10 ± 0.13 in September 2008 with significant differences $(P < 0.05)$ between years and months. K was significantly negatively correlated $(r = -0.31; P < 0.05)$ with the gonadosomatic index (Figure 4).

3.1.4. Fecundity. Out of 227 females sampled, only 37 (16.30%) were gravid with ripe oocytes for fecundity estimation. Absolute and potential fecundities varied from 484 to 14034 oocytes and from 240 to 4043 oocytes respectively. Relative fecundity varied from 1492 to 37188 oocytes/kg. Mean absolute and potential fecundities were significantly $(P < 0.05)$ higher in 2008 than in 2009. These fecundities were not significantly different $(P > 0.05)$ in the dry season

Reproductive Strategy of Labeobarbus batesii (Boulenger, 1903) (Teleostei: Cyprinidae) in the Mbô Floodplain Rivers of Cameroon

141

TABLE 1: Number and mass of *Labeobarbus batesii* captured in each month and in each zone of the Mbô Floodplain.

Period		Zone				Season				Overall	
		Confluence		Inter-confluence		Rainy		Dry			
Year	Month	n	Biomass(g)	n	Biomass (g)	n	Biomass (g)	n	Biomass (g)	n	Biomass (g)
2008	M	13 (4i)	2160			13 (4i)	2160			13 (4i)	2160
	J	10 (1i)	1543			10 (1i)	1543			10 (1i)	1543
	J			35	5153	35	5153			35 (11i)	5153
	A			61 (6i)	12430	61 (6i)	12430			61 (6i)	12430
	S			85 (5i)	17577,5	85 (5i)	17577,5			85 (5i)	17577,5
	O	16	4336			16	4336			16	4336
	N	12	3267					12	3267	12	3267
	D	32	6511					32	6511	32	6511
	Overall	**83 (5i)**	**17817**[b]	**181 (11i)**	**35160,5**[a]	**220 (16i)**	**43199,5**[a]	**44**	**9778**[b]	**264 (27i)**	**52977,5**[a]
2009	J	34 (5i)	4786					34 (5i)	4786	34 (5i)	4786
	F	2	380					2	380	2	380
	M	9 (3i)	2480					9 (3i)	2480	9 (3i)	2480
	A	8	2365			8	2365			8	2365
	M	13 (1i)	3170			13 (1i)	3170			13 (1i)	3170
	J	7	1215			7	1215			7	1215
	J			5	975	5	975			5	975
	A	10	1615	7	1110	17	2725			17	2725
	S	6	1005	37 (1i)	5080	43 (1i)	6080			43 (1i)	6080
	O			46 (24i)	8085	46 (24i)	8085			46 (24i)	8085
	Overall	**82 (9i)**	**17016**[a]	**95 (25i)**	**15250**[a]	**139 (26i)**	**24620**[a]	**45 (8i)**	**7646**[b]	**184 (34i)**	**32266**[b]
Overall		172 (14i)	34833[b]	276 (36i)	50410,5[a]	359 (42i)	67819,5[a]	89 (8i)	17424[b]	448 (61i)	85243,5

[a,b]Numbers with same letters in exponent are not significantly different within years, seasons, or capture zones ($P > 0.05$), (i): undetermined sex (immature or rottenness).

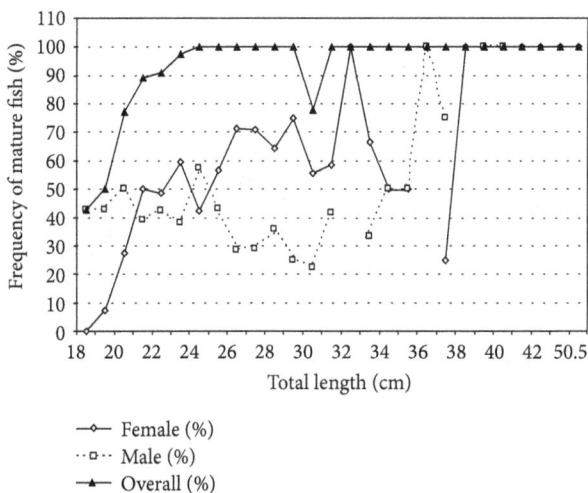

FIGURE 2: Percentage occurrence of mature *Labeobarbus batesii* at different lengths.

than the rainy season, or in the confluences than in the inter-confluences. Similarly, relative fecundity was not significantly different in 2008 than in 2009, or in the dry season and in the confluences than in the rainy season and in the interconfluences (Table 3).

The number of eggs per fish all increased with total mass, gonad mass, and absolute fecundity (Figures 5(a), 5(b), and 5(c)). The best fit relationship for total mass was a power function, whereas linear regressions were used for the other two factors. Coefficients of determination (R^2) values were 0.36, 0.43, and 0.80 respectively for total length, total mass and gonad mass relationships.

4. Discussion

4.1. Relative Abundance Distribution of Labeobarbus batesii. The lack of fishes in the inter-confluence in the dry season and the beginning of the rainy season can be explained by the fact that the water level decreases and the fishes migrate to other areas of catchments or prepare for spawning. Additionally, there is reduced activity of fishing during this period.

4.2. Sex Ratio. The overall sex ratio of 1 : 1.42 in favour of females in the fishery is similar to that observed in other species: *Labeo coubie* (1 : 1.67) and *Rasbora tawarensis* (1 : 3.39) [14, 20]. These results might signify that the reproductive

TABLE 2: Annual and monthly variation in sex ratio (male : female) of *Labeobarbus batesii*.

Year	Month	Total sample (n)	Male (n)	Female (n)	Sex ratio	χ^2
2008	May	9	4	5	1:1.25	0.020
	June	9	4	5	1:1.25	0.020
	July	24	12	12	1:1	
	Aug.	55	22	33	1:1.5	3.156
	Sept.	80	38	42	1:1.10	0.006
	Oct.	16	1	15	1:15	3.480
	Nov.	12	3	9	1:3	0.530
	Dec.	32	20	12	1:0.60	0.120
	Overall in 2008	**237**	**104**	**133**	**1:1.28**	**1.75**
2009	Jan.	29	14	15	1:1.07	0.051
	Feb.	2	1	1	1:1	
	March	6	2	4	1:2	0.180
	April	8	2	6	1:3	0.480
	May	12	3	9	1:3	0.530
	June	7	4	3	1:0.75	0.141
	July	5	0	5	0:5	4.35*
	Aug.	17	6	11	1:1.83	3.95*
	Sept.	42	18	24	1:1.33	3.042
	Oct.	22	6	16	1:2.66	5.46*
	Overall in 2009	**150**	**56**	**95**	**1:1.70**	**5.02***
	Overall in the floodplain	387	160	227	1:1.42	4.05*

n: number of fish, significant at * = 0.05, and χ^2: Chi-square.

strategy of fish species should be the polygamy or an r-selected reproductive strategy where the number of larvae is high but with less parental care [21]. The female biased sex ratio in the sample may be due to the differential fishing factors related to seasons and schooling of fishes in the feeding and spawning grounds, or to selective fishing for the large fish, rather than reflecting a real population sex ratio. Sex ratio divergence might also be explained by partial segregation of mature individuals through the preference of school formation, rendering one sex more vulnerable to capture [14]. Additionally, once fertilization of eggs is completed, males may move from spawning to feeding areas located in the shallows where they are not easily caught. In contrast to our data, males outnumber females in the cyprinid fish, *Garra rufa* (1.10 : 1) [19], and the sex ratios are equal in *Labeo senegalensis* [22]. The monthly variation in sex ratios might be an adaptation of reproductive strategies of tropical species to their hydrological environment [22].

4.3. Sexual Maturity Size.
Females reach sexual maturity at a larger size (213 mm) than males (203 mm), which presumably allows for an increase for egg production. Thus, males may grow slower than the females or males may mature earlier than the females. *Labeo senegalensis* shows similar size differences at maturity (290 mm and 257 mm) in

the Oueme Bassin in Benin [22] to *L. batesii*, whereas *Labeo parvus* is smaller at sexual maturity (155 mm and 129 mm) in Benin [23] than *L. batesii* in our study. The minimum length recorded for specimens sampled was 170 mm, which may indicate overfishing. Hence, close and less intense fishing during October 2008–June 2009 (period of abundance of gravid *L. batesii*) would help the conservation of the natural stock of fishes by allowing fishes to breed at least once in their lifetime.

4.4. Gonadosomatic Index, K Condition Factor, and Spawning Periods.
The cyclic monthly variation of the condition factor, K, is inversely correlated to GSI, indicating that *L. batesii* uses muscular or fat reserves and the viscera to fuel reproduction. Similar observations have been reported for, *Barbus callensis* and *Barbus fritschi* [24]. Multiple GSI peaks observed on February, March, June, and December characterise a fractional multiple spawning [25]. An intermittent breeding cycle has the advantages of reduced larval crowding and a decreased impact of predation and unfavourable environmental conditions on eggs and larvae [25]. Thus, according to [26], the oogenesis is the group-synchronous type, with pulses of reproduction spread over a long period. The reproductive cycle begins in mid-September 2008 and ends in July 2009, with no reproduction for the rest of the year (end-July 2009 to mid-September 2009). As multiple spawning over a protracted season has major benefits in aquaculture because it provides a consistent supply of high quality larval [25], the breeding biology of *L. batesii* makes it a suitable candidate for aquaculture. The status of the gonads of fishes caught during the nonreproductive period revealed that most of them were spent, but juveniles were abundant at that time.

4.5. Fecundity.
A lower proportion of gravid female *L. batesii* (37 out of 227) was recorded in this study than for the cyprinid *Labeo coubie* (58 out of 205) [14], but it was higher than for *Labeo parvus* (41 out of 461) [23]. The absolute fecundity was low compared to other cyprinids, *Barbus grypus* (16000–235784 oocytes) [27], *Labeo senegalensis* (12948–74832 oocytes) [22], and *Labeo parvus* (8723–124363 oocytes) [28], but was higher than for *Garra rufa* (283–3794 oocytes) [19]. The potential fecundity recorded for *L. batesii* was higher than 128 oocytes reported for the cyprinid *Liza klunzingeri* [29]. The mean relative fecundity was higher than the 1994–15920 oocytes/kg reported for *Barbus holotaenia* [29] but lower than *Garra rufa* (109430 oocytes/kg) [19] and *Labeo Parvus* (357000 ± 22000 oocytes/kg) [23].

Fecundity varied within years, months, and zones, which may be due to the differences in fish sizes and food availability [30]. Fecundity is higher in the confluences than in the inter-confluences, which suggests that the fishes migrate from inter-confluences to confluences for spawning, mainly during rainy season floods. The relationships between fecundity and fish sizes or gonad mass recorder for *L. batesii* are also observed for the fish species in a natural west African Lake by [30]. The positive relationships between fecundity and length ($R^2 = 0.36$) and fecundity and body mass ($R^2 = 0.43$) were

Reproductive Strategy of Labeobarbus batesii (Boulenger, 1903) (Teleostei: Cyprinidae) in the Mbô Floodplain
Rivers of Cameroon

143

TABLE 3: Female *Labeobarbus batesii* fecundity according to the year, season, and capture zone.

F	Year		Season		Zone		Overall (37)
	2008 (21)	2009 (16)	Dry (13)	Wet (24)	Conf. (27)	Inter. (10)	
RF	6530 ± 4061^b	12400 ± 9001^a	9770 ± 5420^a	8487 ± 5720^a	9290 ± 4001^a	7955 ± 7400^a	9071 ± 7184
AF	4062 ± 3708^a	2014 ± 1635^b	2977 ± 2623^a	2521 ± 2376^a	2900 ± 2760^a	2930 ± 2348^a	2898 ± 2837
PF	1381 ± 1211^a	739 ± 619^b	1108 ± 638^a	893 ± 328^a	954 ± 512^a	943 ± 418^a	1016 ± 963

[a,b]Numbers with same letters in exponent are not significantly different within years, seasons, or capture zones ($P > 0.05$); F: fecundity, RF: relative fecundity (number of oocytes per kilogram body mass), AF: absolute fecundity (total number of oocytes per female), PF: Potential fecundity (number of mature oocytes per female), ( ): number of females, Conf.: confluence, Inter.: Inter-confluence.

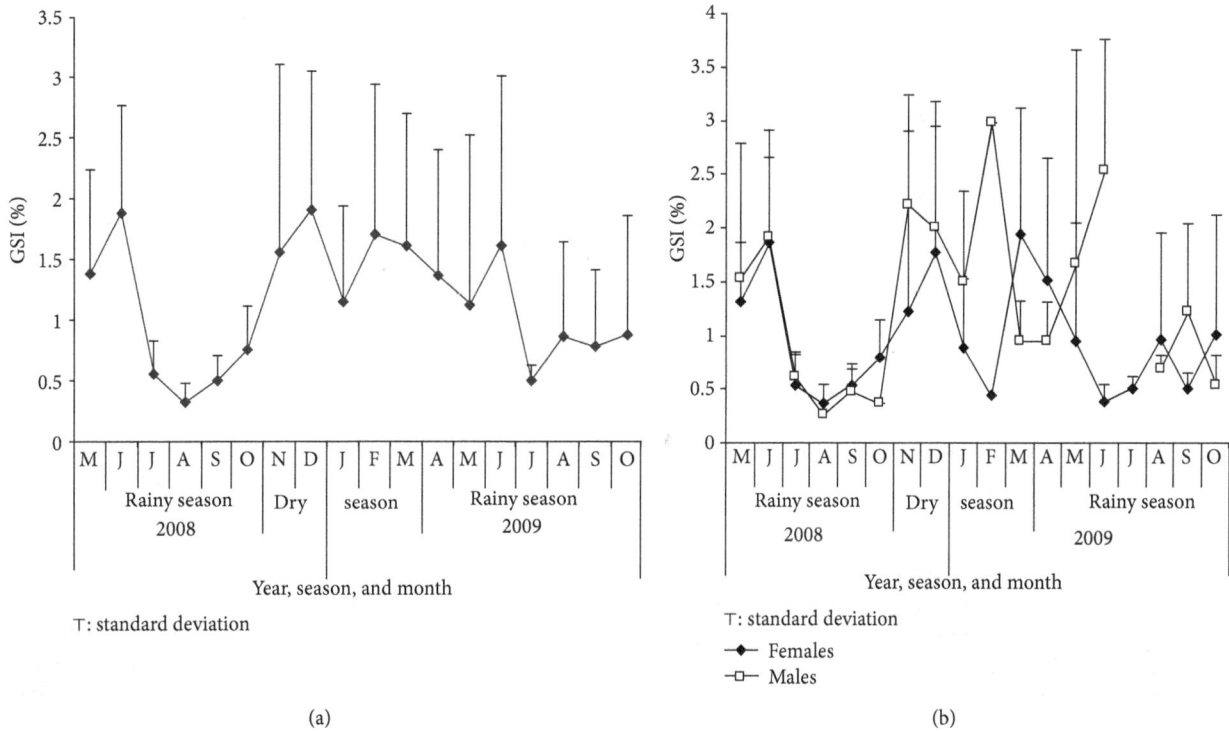

(a)

(b)

FIGURE 3: Monthly variation of mean gonadosomatic indices (GSI) for both sexes between seasons and years (a), mean gonadosomatic indices of females and males of *Labeobarbus batesii* (b).

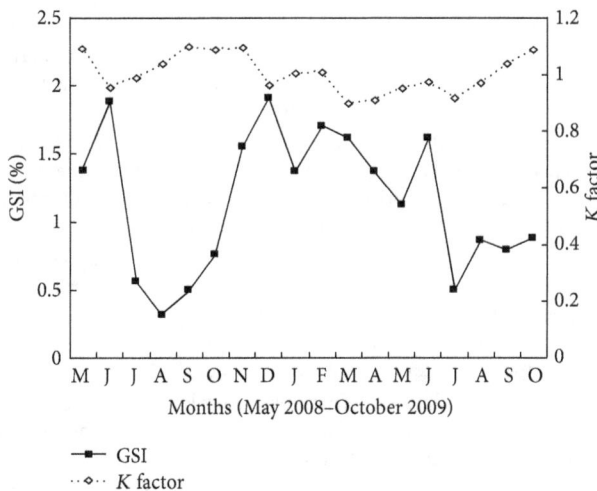

FIGURE 4: Monthly variation of mean gonadosomatic index (GSI) and condition factor K for both sexes of *Labeobarbus batesii*.

comparable to those of $R^2 = 0.48$, recorded for *Labeo parvus* [28]. However, these values were higher than those recorded for *Labeo coubie* ($R^2 = 0.18$ and 0.10, resp. [14]). The coefficient of determination for the relationship between fecundity and gonad mass ($R^2 = 0.80$) for *L. batesii* is higher than the values obtained with the fish size. Thus, it is clear that the gonad weight is better correlated with reproductive capacity than fish size. Fish species with these types of relationships have rapid growth and high fertility [31], characteristics that are important for aquaculture species. Similar results were reported for other cyprinids such as *Garra rufa* [19], *Barbus holotaenia* [32], and *Labeo parvus* ($R^2 = 0.87$) [23].

5. Conclusion

Females of *Labeobarbus batesii* are caught more frequently than males in the MF of Cameroon, and they reach sexual maturity at a larger size than the males. Multiple peaks of GSI

(a)

(b)

(c)

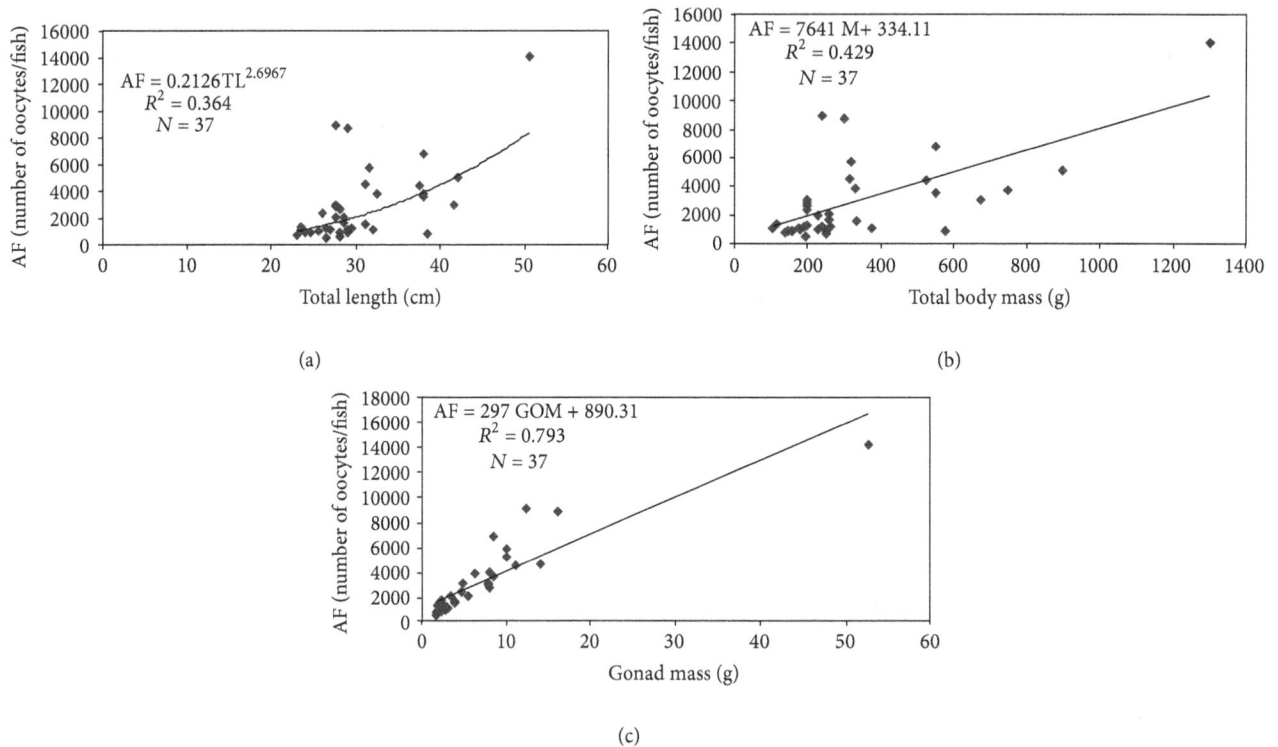

FIGURE 5: (a) Fecundity/total length, (b) total mass, and (c) gonad mass relationships of female *Labeobarbus batesii*. N = number of females, R^2 = coefficient of determination, AF = absolute fecundity, TL = total length, M = total body mass, and GOM = gonad mass.

observed characterise a fractional multiple spawning. Thus *L. batesii* has a group-synchronous oogenesis. Fecundity varies within years, months, and captures zones and is correlated with gonad mass. These features make *L. batesii* a suitable candidate for aquaculture.

Acknowledgments

The authors are grateful to the fishermen who helped in collecting the data and to the anonymous referees for their constructive review of this paper.

References

[1] FAO, "Statistics and information service of the fisheries and aquaculture department," in *FAO Yearbook*, Fisheries and Aquaculture Statistics, Rome, Italy, 2010.

[2] P. Fontaine, M. Legendre, M. Vandeputte, and A. Fostier, "Domestication de nouvelles espèces et développement durable de la pisciculture," *Cahiers Agricultures*, vol. 18, no. 2-3, pp. 119–124, 2009.

[3] J. Lazard and C. Levêque, "Introduction et transferts d'espèces de poissons d'eau douce," *Cahiers Agricultures*, vol. 18, no. 2-3, pp. 157–163, 2009.

[4] O. Mikolasek, B. Barlet, E. Chia, V. Pouomogne, and E. T. M. Tomedi, "Développement de la petite pisciculture marchande au Cameroun: la recherche-action en partenariat," *Cahiers Agricultures*, vol. 18, no. 2-3, pp. 157–163, 2009.

[5] Anonymous, *CLOFFA: Catalogue des poissons d'eau douce d'Afrique*, Edited by J. Daget, J. P. Gosse; D. F. E Thys

[6] E. Montchowui, P. Laleye, J. C. Philippart, and P. Poncin, "Reproductive behaviour in captive African Carp, *Labeo parvus* Boulenger, 1902 (Piscies: Cyprinidae)," *Journal of Fisheries International*, vol. 6, no. 1, pp. 6–12, 2011.

[7] E. Montchowui, A. C. Bonou, P. Lalèye, and J. C. Philippart, "Successful artificial reproduction of the African carp: *Labeo parvus* Boulenger, (Pisces : Cyprinidae)," *International Journal of Fisheres and Aquaculture*, vol. 3, no. 3, pp. 35–40, 1902.

[8] E. Montchowui, P. Lalèye, E. N'tcha, J. C. Philippart, and P. Poncin, "Larval rearing of African carp, *Labeo parvus* Boulenger, 1902 (Pisces : Cyprinidae), using live food and artificial diet under controlled conditions," *Aquaculture Research*, vol. 43, no. 8, pp. 1243–1250, 2012.

[9] J. F. Guégan, D. Paugy, and C. Lévêque, "L'étude des *Barbuss* dans le cadre d'un programme PICADOR: programme international sur les Cyprinidae Africains- Distribution et Origine," *Cahiers d'Ethologie*, vol. 13, no. 2, pp. 83–184, 1993.

[10] J. Lazard, "Introduction/Domestication d'espèces de poissons: Quelques éléments de réflexion. Unité propre de Recherche 'Aquaculture et gestion des ressources aquatiques,'" Département emvt. CIRAD. Avenue Agropolis, TA 30/01,34398 Montpellier Cedex 5, Montpellier, France, 2006.

[11] Anonymous, Union Internationale pour la Conservation de la Nature (UICN). Wetlands, water and the law. Using law to advance wetland conservation and wise use. Clare Shine and Cyrille de Klemm. IUCN Evironmental policy and law, Paper no. 38, 2003, http://www.fao.org/docrep/009/a0113e/A0113Eo3.

[12] M. L. J. Stiassny, G. G. Teugels, and C. D. Hopkins, "Poissons d'eaux douces et saumâtres de la basse Guinée, Ouest de

Van den Audenaerde, ISNB, ORSTOM, MRAC. Collection/Coordonateurs, 1984.

l'Afrique Centrale," in *Faune et Flore Tropicales, Paris*, vol. 1, p. 805, MRAC, Tervuren, Belgium, IRD edition, 2007.

[13] V. Pouomogne, "Capture-based aquaculture of *Clarias catfish*: case study of the Santchou fishers in western Cameroon," in *Capture-Bsed Aquaculture. Global Overview*, A. Lovatelli and P. F. Holthus, Eds., FAO Fisheries Technical Paper no. 508, pp. 93–108, FAO, Rome, Italy, 2008.

[14] U. Ikpi and B. I. Okey, "Estimation of dietary composition and fecundity of African carp, *Labeo coubie*, Cross River, Nigeria," *Journal of Applied Sciences and Environmental Management*, vol. 14, no. 4, pp. 19–24, 2010.

[15] T. C. Tiogué, M. T. E. Tomedi, D. Nguenga, and J. Tchoumboué, "Caractéristiques de morphologie générale et de croissance du Cyprinidae africain *Labeobarbus batesii* dans la plaine inondable des Mbô, Cameroun," *International Journal of Biological Sciences*, vol. 4, no. 6, pp. 1988–2000, 2010.

[16] V. R. Suresh, B. K. Biswas, G. K. Vinci, K. Mitra, and A. Mukherjee, "Biology and fishery of barred spiny eel, *Macrognathus pancalus* Hamilton," *Acta Ichthyologica et Piscatoria*, vol. 36, no. 1, pp. 31–37, 2006.

[17] M. E. Allison, F. D. Sikoki, and I. F. Vincent-Abu, "Fecundity, sex-ratio, maturity stages, size at first maturity, breeding and spawning, of *parailla pellucida* (Boulenger, 1901) in the lowe Nun River, Niger Delta, Nigeria," *Caderno de Pesquisa Srie Biologia*, vol. 20, no. 2, pp. 31–47, 2008.

[18] S. B. Ekanem, "Some reproductive aspects of *Chrysichthys nigrodigitatus* (Lacepede) from Cross River Nigeria," *Naga, the ICLARM quarterly*, vol. 23, no. 2, pp. 24–27, 2000.

[19] M. Abedi, A. H. Shiva, H. Mohammadi, and R. Malekpour, "Reproductive biology and age determination of *Garra rufa* Heckel, 1843 (Actinopterygii: Cyprinidae) in central Iran," *Turkish Journal of Zoology*, vol. 35, no. 3, pp. 317–323, 2011.

[20] Z. A. Muchlisin, M. Musman, and M. N. S. Azizah, "Spawning seasons of *Rasbora tawarensis* (Pisces: Cyprinidae) in Lake Laut Tawar, Aceh Province, Indonesia," *Reproductive Biology and Endocrinology*, vol. 8, article 49, 2010.

[21] "Comportements et stratégies de reproduction chez les poissons," http://www.e-ocean.fr/index.php?option=com_content&task=view&id=21&Itemid=44.

[22] E. Montchowui, P. Lalèyè, P. Poncin, and J. C. Philippart, "Reproductive strategy of *Labeo senegalensis* valenciennes 1842 (Teleostei: Cyprinidae) in the Ouémé basin, Benin," *African Journal of Aquatic Science*, vol. 35, no. 1, pp. 81–85, 2010.

[23] E. Montchowui, M. Ovidio, P. Laleye, J. C. Philippart, and P. Poncin, "Stratégies de reproduction et structure des populations chez *Labeo parvus* Boulenger, 1902 (Cypriniformes: Cyprinidae) dans le bassin du fleuve Ouémé au Bénin," *Annales des Sciences Agronomiques*, vol. 15, no. 2, pp. 153–171, 2011.

[24] S. Bouhbouh, *Bio-écologie de Barbus Callensis (Valencienne 1842) and Barbus fritschi (Günther 1874) au niveau du réservoir Allal El Fassi (Maroc) [Ph.D. thesis]*, Université SIDI Mohamed Ben Abdallah, Faculté des Sciences Dhar El Mehrazfes, 2002.

[25] M. Dorostghoal, R. Peyghan, F. Papan, and L. Khalili, "Macroscopic and microscopic studies of annual ovarian maturation cycle of Shirbot *Barbus grypus* in Karoon river of Iran," *Iranian Journal of Veterinary Research*, vol. 10, no. 2, pp. 172–179, 2009.

[26] V. L. De Vlaming, "Oocytes development patterns and hormonal involvements among Teleosts," in *Control Process in Fish Physiology*, J. C. Ranchin, T. J. Petcher, and R. Duggan, Eds., pp. 176–199, London Croom Helm, 1983.

[27] A. S. Oymak, N. Dogan, and E. Uysal, "Age, growth and reproduction of the Shabut *Barbus grypus* (Cyprinidae) in

Atatürk dam lake (Euphrates river), Turkey," *Cybium*, vol. 32, no. 2, pp. 145–152, 2008.

[28] E. Montchowui, P. Lalèye, J. C. Philippart, and P. Poncin, "Biologie de la reproduction de *Labeo parvus* Boulenger, 1902 (Cypriniformes: Cyprinidae) dans le bassin du fleuve de l'Ouémé au Bénin (Afrique de l'Ouest)," *Cahiers d'Ethologie*, vol. 22, no. 2, pp. 61–80, 2007.

[29] F. Abou-Seedo and S. Dadzie, "Reproductive cycle in the male and female grey mullet, *Liza klunzingeri* in the Kuwaiti waters of the Arabian Gulf," *Cybium*, vol. 28, no. 2, pp. 97–104, 2004.

[30] N. M. Inyang and H. M. G. Ezenwaji, "Size, length-weigt relationship, reproduction and trophic biology of *Chrysichthys nigridigitatus* and *Chrysichthys auratus* (Siluriformes: Bagridae) in a Natural West African Lake," *Bio-Research*, vol. 2, no. 1, pp. 47–58, 2004.

[31] K. Demska-Zakes and M. Dlugosz, "Fecundity of vendace from two lakes of Mazurian district," *Rybna*, vol. 31, pp. 37–50, 1995.

[32] S. Mutambue, "Biologie et écologie de Barbus holotaenia, boulenger, 1904, du bassin de la rivière luki (ZAÏRE)," *Bulletin Francais de la Peche et de la Protection des Milieux Aquatiques*, vol. 69, no. 340, pp. 25–41, 1996.

18

Appennino: A GIS Tool for Analyzing Wildlife Habitat Use

Marco Ferretti,[1] Marco Foi,[2] Gisella Paci,[1] Walter Tosi,[3] and Marco Bagliacca[1]

[1] *Department of Animal Production, University of Pisa, Viale delle Piagge 2, 56100 Pisa, Italy*
[2] *Department of the Earth Science, University of Milan, Via Mangiagalli 34, 20133 Milan, Italy*
[3] *Geographic Information System Office, Province of Pistoia, Corso Gramsci 110, 51100 Pistoia, Italy*

Correspondence should be addressed to Marco Ferretti, ferretti@vet.unipi.it

Academic Editor: Hynek Burda

The aim of the study was to test Appennino, a tool used to evaluate the habitats of animals through compositional analysis. This free tool calculates an animal's habitat use within the GIS platform for ArcGIS and saves and exports the results of the comparative land uses to other statistical software. Visual Basic for Application programming language was employed to prepare the ESRI ArcGIS 9.x utility. The tool was tested on a dataset of 546 pheasant positions obtained from a study carried out in Tuscany (Italy). The tool automatically gave the same results as the results obtained by calculating the surfaces in ESRI ArcGIS, exporting the data from the ArcGIS, then using a commercial spreadsheet and/or statistical software to calculate the animal's habitat use with a considerable reduction in time.

1. Introduction

Wildlife management studies identify the resources (e.g., food items or habitats) used by animals and document their availability. Resource availability is defined as the quantity accessible to the animal or populations of animals and is distinguished from abundance, which is defined as the resources in the environment [1]. Resource usage is the "quantity" taken by an animal or population of animals. Resources may be consumed, in the case of food items, or simply visited, in the case of habitats [2]. A wide variety of methods are available to study animal resource selection [3]. One such method is compositional analysis, which is often used to analyze habitat preference [4, 5]. It studies the animal's preference both in terms of "home range" and "fix" (single positions within the home range). At present, positions and surfaces calculated in GIS programs must be exported to other free or commercial software such as spreadsheets (LibreOfficeCalc, OpenOffice, Microsoft-Excel, and so on), general statistical programs (R Project for Statistical Computing, JMP, SPSS, and so on), or specific programs (Compos Analysis v.6.3-Smith ecology, Biotas-Ecological Software Solutions LCC).

The development of VHF- and GPS-radio collars to track animal movements [6–8] led to the need to store and transpose hundreds to thousands of positions (fixes) for each animal onto digital maps. Managing this dataset manually is complex and susceptible to errors by the use of simple spreadsheets; therefore, we produced a freely available tool, Appennino, that is completely operable within the GIS suite (ArcGIS) to calculate the animal's preferences by using compositional analysis. We decided to make the tool for "ArcGIS" since most game managers use this particular program for GIS management. However, it should be stressed that we do not have any direct financial relationship with ESRI, the producers of ArcGIS.

2. Materials and Methods

The tool was first coded in Visual Basic for Application (VBA) directly using the facilities provided by the ESRI ArcGIS environment [9, 10]. The tool was based on compositional analysis [3, 11, 12].

Appennino can be downloaded, free of charge, from http://biblio.unipi.it/content/servizio-bibliotecario/risorse-web or http://www.marcoferretti.altervista.org/index_file/Page419.htm or http://bagliacca.altervista.org/GIStool.html. The tool needs at least four "shapes": the land use polygon layer, the home range polygon shape of every individual animal, the land use circular random plots, and the fix layer

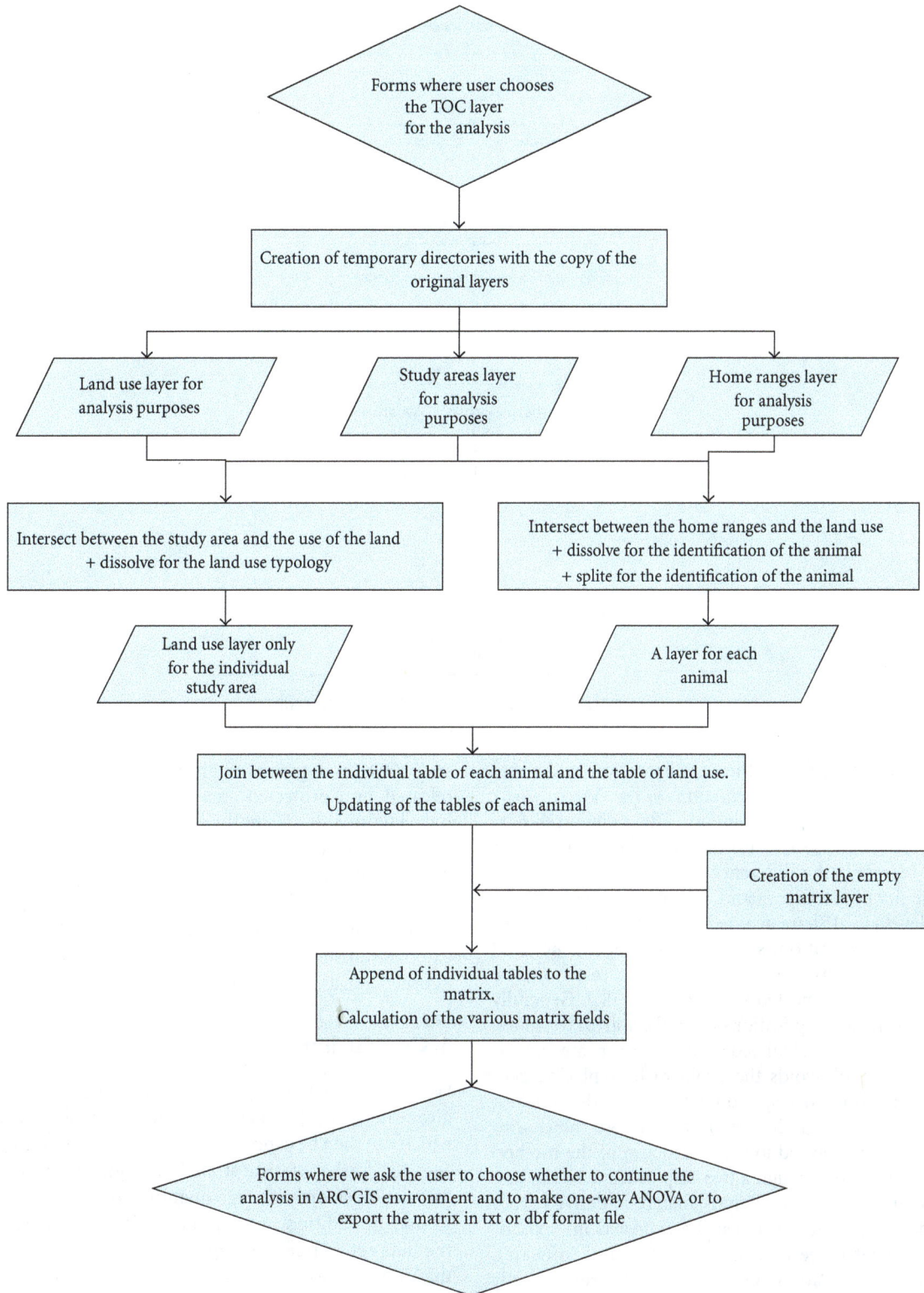

FIGURE 1: Flow chart of the first step of the tool.

of the animals. The tool works both on home range and fix analyses.

The tool works as follows. It assumes there are D types of resource units available, and that the individual animal's proportional resource usage is described by the composition $o_{u1}, o_{u2}, \ldots, o_{uD}$, where o_{ui} is the estimated proportion of the i type resource used by the individual. The proportions sum to one. Similarly, the analysis sets the available proportions for the same animals as $\pi_{a1}, \pi_{a2}, \ldots, \pi_{aD}$. The analysis cannot be applied when resources units are described by

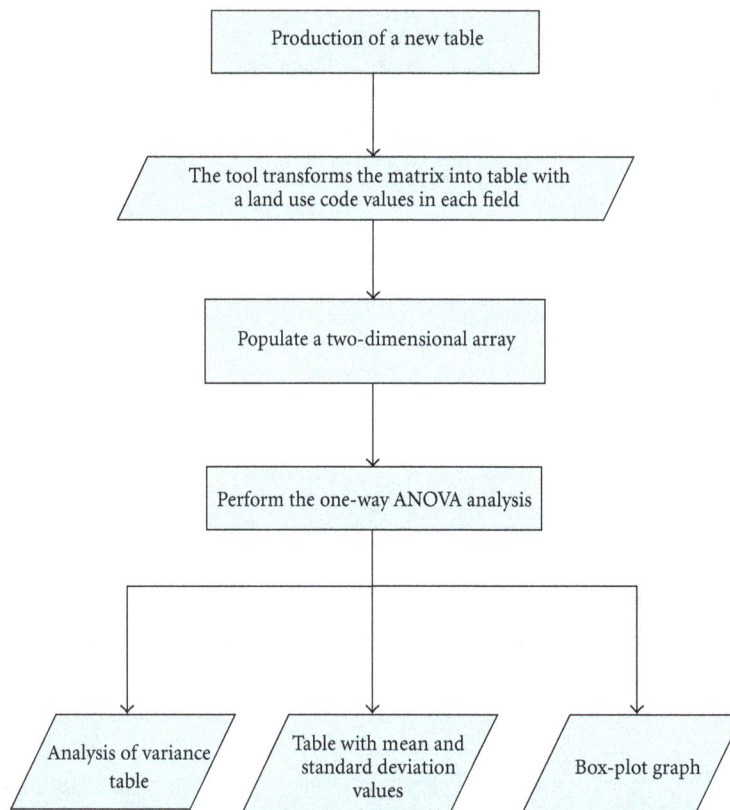

FIGURE 2: Flow chart of the second step of the tool.

continuous variables, unless they are expressed as discrete classes. The log-ratio transformation $y_i = \log_e(o_{ui}/o_{uj})$ is calculated for any component o_j. The differences $i = \log_e(o_{ui}/o_{uj}) - \log_e(\pi_{ai}/\pi_{aj})$ are then calculated for the ith animal to represent the difference between the relative use and availability of resources i and j. Problems related to the sampling level [13–15] are avoided, since the animal is used as the unit of observation, so that data independence and multivariate normality are ensured. With no selection, the mean value of d_i is expected to be zero for all i. Generally speaking, in software applications when there are zero values, land use cannot be calculated and an error message is generated. Our tool avoids the problem by replacing zero values with a value corresponding to 1% of the smallest value observed, although the substitution of zero values with arbitrary constants has led to some criticism of the method [12]. The tool works through a two-step process. In the first step, the definition of the analysis matrix is derived from primary data (this data array can be exported in ".txt" or ".dbf" file format to be used in other statistical software). In the second step, an ANOVA test can be run directly on the newly created matrix, without leaving the ArcGIS environment.

Figures 1 and 2 show how the program works.

3. Results and Discussion

Figures 4 and 5 show the final output of the tool when it was applied to data from a study carried out between 2008 and 2009 on a group of pheasants (Phasianus colchicus) released in a protected area (PA) near Florence (Tuscany, Italy, Figure 3) [6, 16] at the end of the two steps.

The Appennino tool is completely operable within the ArcGIS suite to evaluate animal preferences by compositional analysis. This enables it to be maintained within the GIS software and avoids having to export the database to any external statistical software, while producing the same results as other statistical software.

4. Conclusions

We have presented Appennino as a tool that automatically gives the complete matrix of the compositional analysis, which can then be exported in other statistical software packages for further statistical analysis. Our tool thus prevents calculation errors with high quantities of data and is also easy to use. In addition, Appennino performs basic statistics of the data set, is free of charge, and can be downloaded with the VBA source code for further improvements.

Acknowledgments

The authors would like to thank Daniel S. Soper, PhD. Assistant Professor Department of Information Systems and Decision Sciences Mihaylo College of Business and Economics California State University Fullerton and Provincia di Pistoia and Ambito Territoriale di Caccia Firenze 5 for

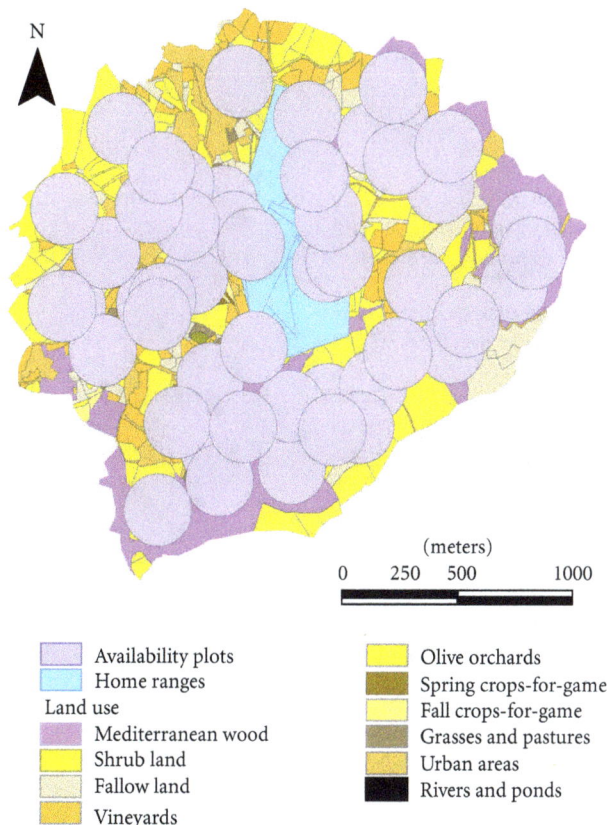

FIGURE 3: Land use polygon layer with the home range polygon shape of each individual animal and the land use circular plots generated with the Hawth tool (http://www.spatialecology.com/htools/), from the ArcGIS table of contents. Note: the land use layer, performed for clarity, is not required by the tool. With this set of data, 50 random circular plots of the average home range size of the birds in the case study (110 m) were chosen across the study area.

FIGURE 4: Analysis of variance table.

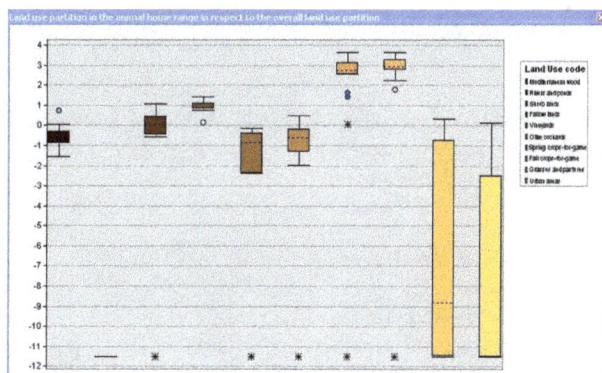

FIGURE 5: Box-plot graph of the land uses. Note: with this data set, the land uses which give maximum values of occurrence are "Spring crops-for-game" and "Fall crops-for-game." The land uses which give the least uses are the "Rivers and ponds" and "Urban areas" ($P < 0.05$).

their cooperation. The authors of the paper do not have any financial association with ESRI, the producers of the program mentioned in the paper.

References

[1] J. R. Alldredge, D. L. Thomas, and L. L. Mcdonald, "Survey and comparison of methods for study of resource selection," *Journal of Agricultural, Biological, and Environmental Statistics*, vol. 3, no. 3, pp. 237–253, 1998.

[2] B. F. J. Manly, "Comments on design and analysis of multiple-choice feeding-preference experiments," *Oecologia*, vol. 93, no. 1, pp. 149–152, 1993.

[3] B. F. J. Manly, L. L. McDonald, D. L. Thomas, T. L. McDonald, and W. P. Erickson, *Resource Selection By Animals: Statistical Design and Analysis For Field Studies*, Kluwer Academic Publishers, Boston, Mass, USA, 2nd edition, 2002.

[4] T. M. Fearer and D. F. Stauffer, "Relationship of ruffed grouse Bonasa umbellus to landscape characteristics in southwest Virginia, USA," *Wildlife Biology*, vol. 10, no. 2, pp. 81–89, 2004.

[5] M. Ferretti, G. Paci, S. Porrini, L. Galardi, and M. Bagliacca, "Habitat use and home range traits of resident and relocated hares [Lepus europaeus, Pallas)," *Italian Journal of Animal Science*, vol. 9, no. 3, pp. 278–284, 2010.

[6] M. Ferretti, F. Falcini, G. Paci, and M. Bagliacca, "Captive rearing technologies and survival of pheasants (*Phasianus colchicus* L.)," *Italian Journal of Animal Science*, vol. 11, no. e2, pp. 160–164, 2012.

[7] B. Zweifel-Schielly and W. Suter, "Performance of GPS telemetry collars for red deer Cervus elaphus in rugged Alpine terrain under controlled and free-living conditions," *Wildlife Biology*, vol. 13, no. 3, pp. 299–312, 2007.

[8] M. Pellerin, S. Said, and J. M. Gaillard, "Roe deer Capreolus capreolus home-range sizes estimated from VHF and GPS data," *Wildlife Biology*, vol. 14, no. 1, pp. 101–110, 2008.

[9] R. Burke, *Getting To Know Arcobjects: Programming ArcGIS With VBA*, ESRI, 2003.

[10] K. T. Chang, *Programming ArcObjects With VBA: A task-Oriented Approach*, CRC, 2007.

[11] N. J. Aebischer, P. A. Robertson, and R. E. Kenward, "Composition analysis of habitat use from animal radio-tracking data," *Ecology*, vol. 74, no. 5, pp. 1313–1325, 1993.

[12] G. W. Pendleton, K. Titus, E. Degayner, C. J. Flatten, and R. E. Lowell, "Compositional analysis and GIS for study of habitat selection by goshawks in Southeast Alaska," *Journal of Agricultural, Biological, & Environmental Statistics*, vol. 3, no. 3, pp. 280–295, 1998.

[13] S. H. Hurlbert, "Pseudoreplication and the design of ecological field experiments," *Ecological Monographs*, vol. 54, pp. 187–211, 1984.

[14] R. E. Kenward, "Quality versus quantity: programmed collection and analysis of radio-tracking data," in *Wildlife Telemetry: Remote Monitoring and Tracking of Animals*, I. G. Priede and S. M. Swift, Eds., pp. 231–246, Ellis Horwood, Chichester, UK, 1992.

[15] R. Kenward, *Wildlife Radiotagging. Equipment, Field Techniques and Data Analysis*, Academic Press, London, UK, 1993.

[16] M. Ferretti, F. Falcini, G. Paci, and M. Bagliacca, "Radiotracking of Pheasants (*Phasianus colchicus* L.) to test captive rearing technologies," in *Telemetry*, O. Krejcar, Ed., vol. 5, pp. 403–422, Animal Telemetry Tech, 2011.

19

Mouthpart Morphology of Three Sympatric Native and Nonnative Gammaridean Species: *Gammarus pulex*, *G. fossarum*, and *Echinogammarus berilloni* (Crustacea: Amphipoda)

Gerd Mayer, Andreas Maas, and Dieter Waloszek

Workgroup Biosystematic Documentation, University of Ulm, Helmholtzstrasse 20, 89081 Ulm, Germany

Correspondence should be addressed to Gerd Mayer, gerd.mayer@uni-ulm.de

Academic Editor: Thomas Iliffe

In the last 20 years several nonnative amphipod species have immigrated inland waters of Germany and adjacent central European countries. Some of them have been very successful and could establish stabile populations. In some places, they have even replaced native or earlier established species. The gammarid *Echinogammarus berilloni* originates from the Atlantic region of France and the north-western part of Spain and coexists in some central European waters with the native *Gammarus pulex* and *G. fossarum*. Here, we describe and compare the mouthparts and other structures involved in food acquisition of these three sympatric gammaridean species. Our hypothesis was that differences in the mode of feeding of the three species could be the reason for their coexistence and that these differences would be expressed in differences in mouthpart morphology. The results of our SEM study demonstrate that there are indeed interspecific differences in details of the morphology of the feeding structures. This is especially true for the setation of antennae, maxillulae, gnathopods, and third uropods, which can be interpreted as adaptations to special modes of feeding. Generally, all three species are omnivorous, but specializations in details point to the possibility to use some food resources in a special effective way.

1. Introduction

The gammaridean fauna of middle European inland waters has dramatically changed in the last two decades; particularly, Ponto-Caspian gammarideans arrived in middle European rivers, canals, and lakes [1–11]. Immigration has happened and is still ongoing via three main corridors: (i) along the Danube, Main-Donau Canal and Main into the Rhine system; (ii) via the Pipet-Bug connection from the east, and from the north and along the Baltic coast via ships [12]; (iii) also from the Mediterranean region, freshwater gammarideans have enlarged their range of distribution towards western and middle Europe [5, 13–16]. Some of these nonnative gammaridean species could establish stabile populations and occur in high densities, and several species have a severe impact on the ecology of the invaded regions by reducing and even eliminating native and earlier established gammaridean species (therefore, the immigrants are called invasive).

One well-examined example of such invasive species is *Dikerogammarus villosus* (Sowinsky, 1894) [17]. This species is now the dominant gammaridean species in many rivers, canals, and larger lakes all over central Europe, affecting also other members of the macrozoobenthos [1, 6, 18–20]. But not all nonnative species are invasive, and invasive species are, vice versa, not always able to eliminate native species in every habitat. Under certain conditions, coexistence of native and nonnative species may be possible. In some places, even the very successful invasive species *D. villosus* occurs together with other species in the same waters. In Lake Constance, Germany, for example, no less than four sympatric species can be found: the invasive *D. villosus*, the earlier established *Gammarus roeselii* Gervais, 1835 [21], the native *Gammarus lacustris* Sars, 1863 [22], and the recently discovered *Crangonyx pseudogracilis* Bousfield, 1958 [23]. In this lake, like in other waters with a diverse amphipod fauna, coexistence of these closely related species may be possible because it is rich in its ecological structure, that is, by offering different microhabitats, which can be inhabited by different species according to their substrate preferences [10, 24–34]. But also different food preferences might be a factor, which enables individuals of these species to cooccur in the same habitat. If sympatric species are specialized to feed on different types of food, we hypothesize that such specializations should be expressed in the morphology of their mouthparts.

Therefore, we investigated the mouthparts and other structures involved in food acquisition of the three gammarideans *Gammarus fossarum* Koch in Panzer, 1836 [35], *G. pulex* (Linnaeus, 1758) [36], and *Echinogammarus berilloni* (Catta, 1878) [37] using scanning electron microscopy (SEM). These three species, the two native *Gammarus* species and *E. berilloni*, originating from the Atlantic region of France and Spain, occur sympatrically in some middle European rivers, such as the river Meuse in France and Belgium [9], the Viroin, Belgium [38], and in a karstic stream system in the Paderborn Plateau, Westphalia, Germany [39]. Moreover, *G. pulex* and *E. berilloni* are sympatric in a Rhine tributary near Iffezheim, Germany [30], in smaller rivers in Brittany, France [10], and in the Loire and its tributaries, Region Centre, France [40]. Sympatric occurrences of *G. pulex* and *G. fossarum* have been reported from some waters in Germany, for example, Fulda-Elder Basin [41], Schlitz [42], and from a forest brook, Rimbach [43]. However, in most waters, the populations of these two species live separately, with *G. fossarum* inhabiting springs and the upper reaches of small streams and rivers and tolerating high current and lower temperatures, whereas *G. pulex* prefers sections of brooks and smaller rivers with lower currents [44, 45].

Because these three species are closely related, we do not expect major differences in the morphology of their mouthparts, but more likely modifications in detail should be found, for example, size and shape of particular structures and, in particular, differences in limb setation as signs of specialization to a specific kind of food. Such morphological modifications of the mouthparts have already been demonstrated for *G. roeselii* and *D. villosus* [46]. We therefore aimed at investigating the mouthpart morphologies of the three mentioned species. We expect specific differences that enable the species to live sympatrically on an, at least, slightly different food source that would explain their coexistence.

2. Material and Methods

Specimens of *Gammarus pulex* were obtained from a spillway of a gravel pit filled with groundwater, draining into a side canal of the river Danube near Ulm (N 48° 18′ 40.5″, E 9° 52′ 9.9″) in 04/2007. Specimens of *G. fossarum* were collected from the river Nau near Langenau (N 48° 30′ 3.4″, E 10° 8′ 18.7″) in 07/2008. Specimens of *Echinogammarus berilloni* were obtained from the collection of Gerhard Maier, Senden, collected in 05/2004 from a Rhine tributary near Iffezheim (N 48° 50′ 15″, E 8° 7′ 11.9″). For identification of species, the taxonomic key of Eggers and Martens [47, 48], the original taxonomic descriptions, and redescriptions were used [14, 35–37, 49–51]. Macrophotographs of specimens

stored in 70% ethanol were taken with a Canon Macro Photo Lens EF-S 60 mm mounted on a Canon EOS 450D digital camera. Specimens were illuminated with cold-light lamps. For reducing reflections, both the lamps and the lens were equipped with polarizing filters.

For SEM studies, approximately 30 adult males of each species were anesthetized with carbon dioxide by adding a small amount of sparkling mineral water and fixed and stored in 70% ethanol. After dissection, debris was removed from the specimens by using an ultrasonic cleaner. The specimens were dehydrated in an alcohol series, critical-point dried, and sputter-coated with a mixture of gold and palladium. SEM work was performed with a Zeiss DSM 962 scanning electron microscope of the Central Unit Electron Microscopy at the University of Ulm. Digital images obtained from the SEM were trimmed in Adobe Photoshop, and plates were arranged using Adobe Illustrator.

3. Results

Besides the mouthparts (mandibles, maxillulae, maxillae, maxillipeds; also considered are the paragnaths), we also describe several more structures, which are involved in food acquisition and are of potential significance. These are the antennulae, the antennae, the third pair of uropods, and the first and second pairs of gnathopods. We start the description with the latter appendages. All specimens illustrated and described herein are adult males. Description is complete for the first species, while we applied an abbreviated style thereafter to focus on differences and cut descriptions short.

3.1. Gammarus pulex (Linnaeus, 1758) [36] (Figure 1(a)). The antennulae (Figure 2(a)) are about half as long as the body of the animal. They consist of a three-jointed peduncle and a 23-jointed flexible flagellum. In addition, a short five-part accessory flagellum is present at the junction of peduncle and flagellum. The proximal portion of the peduncle is the longest; the distal portion is slightly more than half as long as the second. Setation of both peduncle and flagellum is short and sparse.

The antennae (Figure 2(d)) are shorter than the antennulae. They consist of a two-part proximal section, the protopod with coxa and basipod, and a distal endopod. The endopod is made up of three long tubular portions and a flagellum consisting of 16 annuli. The annuli of the flagellum are anteroposteriorly flattened and broadened in mediolateral dimension. The posterior surface of the proximal 12 flagellar annuli is armed with a transverse row of about 12 long simple setae, together forming a kind of flag-like brush. Moreover, each flagellar annulus 2 to 11 bears one calceolus on their medioposterior margin (Figure 2(g)). The distal 4 annuli are only weakly setated. The (excretory) gland cone on peduncle segment 2 is rather long, nearly reaching the distal end of peduncle segment 3.

In contrast with the styliform first and second pairs of uropods, those of the third pair (Figures 3(a) and 3(b)) are foliaceous and articulate more flexibly. The one-part endopod is shorter than the bipartite exopod with its small second distal part. In *G. pulex*, the endopod reaches about

Mouthpart Morphology of Three Sympatric Native and Nonnative Gammaridean Species: Gammarus pulex, G. fossarum, and Echinogammarus berilloni (Crustacea: Amphipoda)

153

FIGURE 1: Photographs of males of (a) *Gammarus pulex*; (b) *Gammarus fossarum*; (c) *Echinogammarus berilloni*.

3/4 of the length of the proximal part of the exopod (Figure 3(a)). Most of the setae on the median and lateral margin of endopod and exopod are plumose (with setulae in two opposing rows along the shaft of the seta) (Figure 3(b)). The two slender lobes of the deeply notched telson overhang the peduncles of the third uropods. Their distal ends are armed with two spines and about 5 long simple setae each (Figure 3(a)).

The first two pairs of pereiopods (=2nd and 3rd pairs of thoracopods) are modified to subchelate gnathopods. Mainly because the ischium has the shape of an elbow, the gnathopods are flexed anteriorly, covering the mouthparts ventrally with their distal four podomeres. The propodus of the first gnathopods (Figure 4(a)) is piriform. The curved dactylus is slightly more than half as long as the propodus. The propodus of the second gnathopods (Figures 4(b) and 4(c)) is less piriform, nearly rectangular. The dactylus is arranged almost transverse. Setation of the second gnathopods is much denser and the setae are much longer than in the first gnathopods. Carpus and propodus bear many groups of long and distally curved setae on their margins. These setae are directed medioventrally in the natural position of the gnathopods.

The maxillipeds (=first pair of thoracopods) (Figures 5(a)–5(d)) are bent anteriorly in their natural position, covering most of the other mouthparts and the labrum (Figure 5(a)). Their coxae are fused medially, so that they act as one unit for feeding and handling food. The basipods stem from a triangular socket, built by the fused coxae. Each basipod is medially drawn out into a distally directed spatulate endite (Figure 5(b)). The five-partite

endopods are well developed. The first portion, the ischium, of each side also bears a distally directed spoon-shaped endite. The remaining four portions of each endopod build two opposing "palps." Their distal portions, the dactyli, taper into medially directed claw-like spines.

Basipod, ischium, and merus of the maxilliped bear one group of six to seven simple setae on their posterior sides each. Several groups of medioposteriorly directed simple setae are sited on the posteromedian margins of carpus and propodus (Figure 5(a)). On their anterior sides, the median margins of the endites of the basipods are bent anteriorly, together forming a keel-like elevation. Here several long pappose setae (with setulae randomly arranged along the shaft) are sited, which are anteriorly directed (Figure 5(b)). The distal end of each basipodal endite is, on their anterior sides, armed with a row of distally directed short pappose setae (Figure 5(c)). In addition, four distally directed tooth-like cuspidate setae insert on the medial section of the distal margin of each endite. The endite of each ischium bears a row of mediodistally directed, flattened, and hook-shaped cuspidate setae on the posterior side of its medial margin (Figure 5(d)). This row is accompanied by two rows of short, tape-like setae on each endite.

The small coxal elements of the maxillae (Figure 8) arise from a common sternal elevation (Figures 8(a) and 8(b)). The coxae are predominantly membranous and do not give rise to any enditic extensions. The basipod is medially drawn out into a distally pointing spatulate endite, the so-called "inner plate." Additionally, the so-called "outer plate," possibly representing the endopod, stems from the outer side of each basipod. The median margin of the inner plate

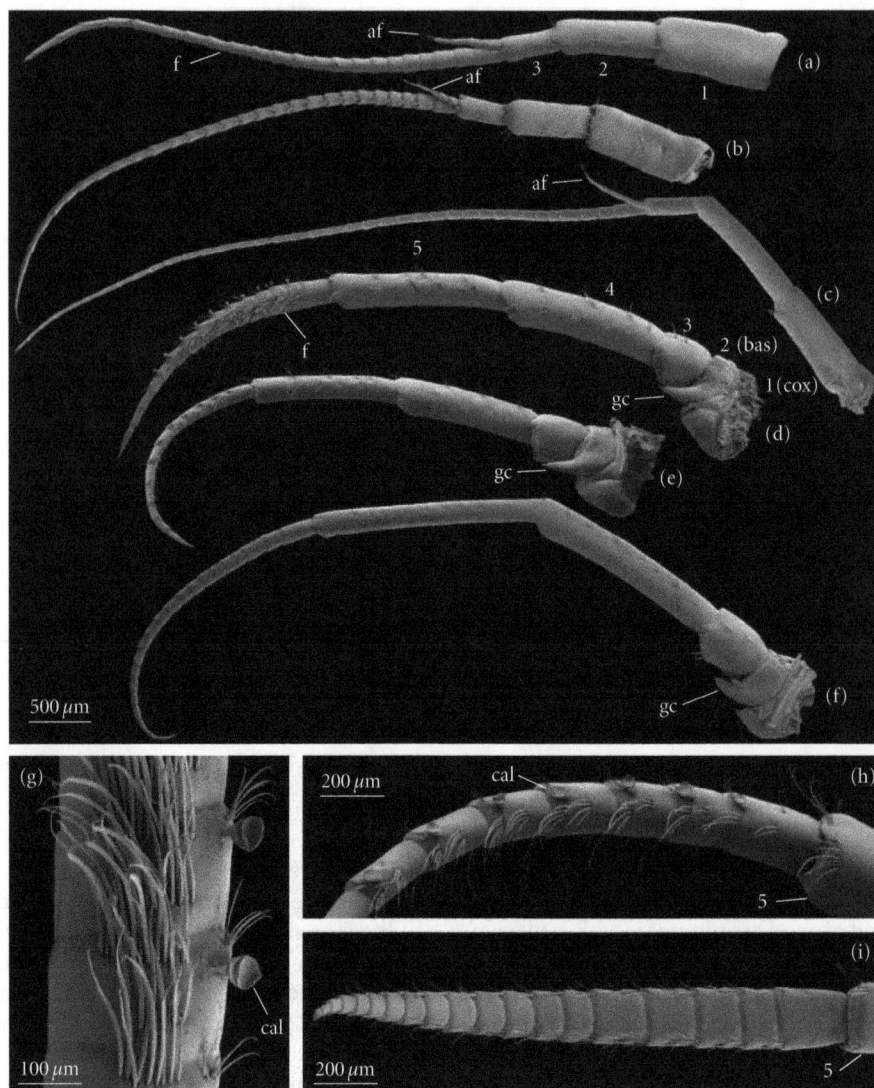

FIGURE 2: : SEM images of antennulae and antennae: (a–c) right antennula in median view of (a) *Gammarus pulex*, (b) *Gammarus fossarum*, and (c) *Echinogammarus berilloni*; (d–f) right antenna in median view of (d) *G. Pulex*, (e) *G. Fossarum*, and (f) *E. berilloni*; (g) proximal part of right antenna of *G. pulex* in posterior view; (h) proximal part of flagellum of right antenna of *G. fossarum* in median view; (i) flagellum of left antenna of *E. berilloni* in posterior view. af: accessory flagellum; bas: basipod; cal: calceolus; cox: coxa; f: flagellum; gc: gland-cone; 1, 2, 3, 4, 5: number of podomeres of peduncle.

carries three rows of setae. The anterior row of setae curves from the median margin onto the anterior surface of the inner plate towards the distal margin of the inner plate (Figure 8(b)). The shafts of these setae are on their lateral and posterior sides equipped with long thin setulae (Figure 8(d)). The medio-distally pointing setae of the posterior row are short, straight, and bear only scale-like setulae on one side of the distal third of their shafts. The median row comprises pappose setae. The median and posterior rows of setae follow the margin along the distal end of the inner plate. As a result of that, these setae change their shape. Those of the posterior row become more and more similar from proximal to distal to those of the other plate, those of the median row are only partly equipped with scale-like setulae. The outer plates are movable in the mediolateral plane and partly cover the

smaller inner plates posteriorly (Figure 8(a)). On their distal margin, each outer plate is armed with two rows of setae (Figure 8(c)). The setae of the anterior row are flattened and bear no setulae. The shafts of the setae of the posterior row are, on their distal third, flattened and equipped with closely arranged, triangular lobes.

The maxillulae (Figure 11) consist of a coxa, inner plate, outer plate, and palp (Figures 11(a) and 11(b)). The spherical coxa inserts in an ample membrane on the cephalothorax with plenty of muscle fibres (Figure 11(b)). The distally directed coxal endite, the so-called "inner plate," has the shape of an isosceles triangle in anterior and posterior perspective (Figure 11(c)). It stems from the coxa with a small and short connexion. It bears a row of medio-distally directed long pappose setae on its longer median margin. In

Mouthpart Morphology of Three Sympatric Native and Nonnative Gammaridean Species: Gammarus pulex, G. fossarum, and Echinogammarus berilloni (Crustacea: Amphipoda)

155

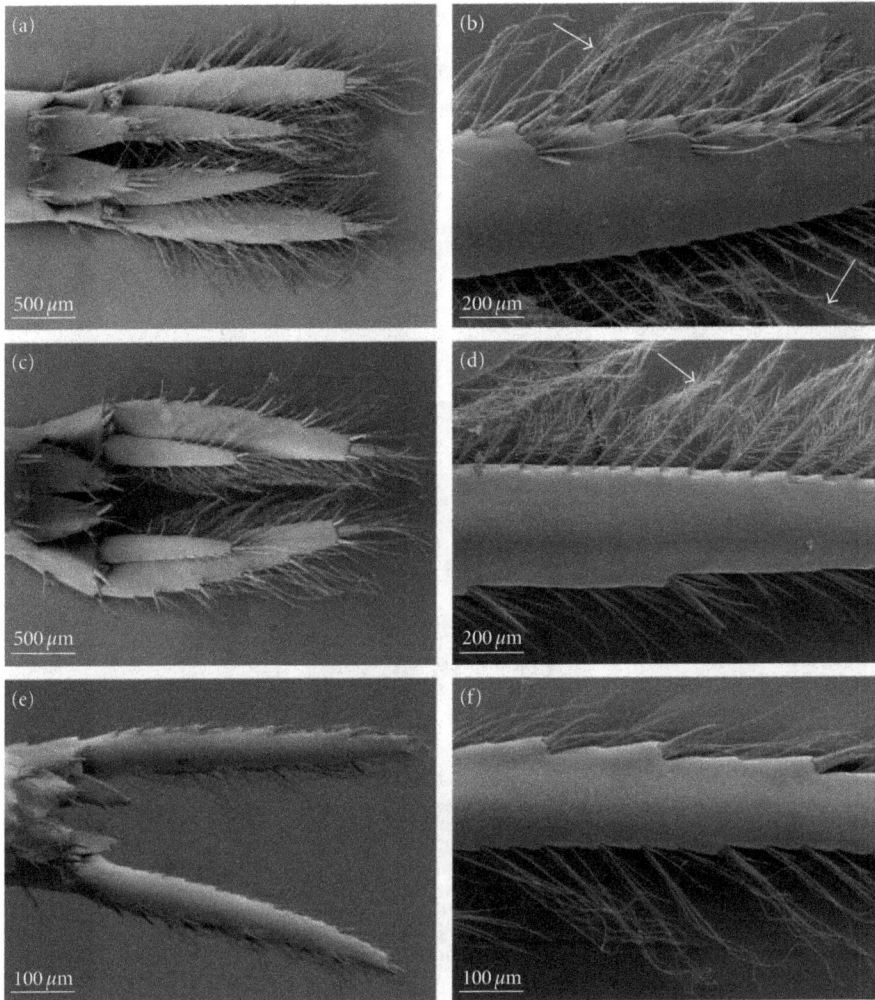

FIGURE 3: SEM images of telson and third uropods: (a) telson and third pair of uropods of *G. pulex in situ* in dorsal view; (b) exopod of left third uropod of *G. pulex* in anterior view; (c) telson and third pair of uropods of *G. fossarum in situ* in dorsal view; (d) exopod of left third uropod of *G. fossarum* in posterior view; (e) telson and third pair of uropods of *E. berilloni in situ* in dorsal view; (f) exopod of left third uropod of *E. berilloni* in anterior view. Arrows indicate presence or absence of plumose setae.

their natural position, the coxal endites of the two maxillulae look like two opposing hand brushes, building a dense net. The basipods are anteriorly directed in their natural position. The articulation with the coxa allows movement in medio-lateral plane. The basipod carries a broad spatulate endite medio-distally. The compound of basipod and its endite is traditionally termed "outer plate." The latter is anteroposteriorly flattened with a concave anterior side and a convex posterior side. The distal margin of the basipodal endite is armed with 10 very robust, medio-distally directed cuspidate setae which are arranged in two rows (Figures 11(e) and 11(f)). Each of these setae, for their part, is armed with one row of up to 12 medio-posteriorly directed secondary spines. Along the distal margin of the outer plate, the number of secondary spines of the setae decreases from median to lateral of the row of setae. Therefore, the median setae look like coarsely elaborated combs with up to 12 prongs, whereas those of the lateral parts of the rows are equipped with three finger-like secondary spines (Figure 11(f)). The sockets of these robust cuspidate setae are

still membranous and therefore elastic. The maxillular palp, the endopod, consists of two portions (Figure 11(b)). The proximal portion is small and cylindrical, whereas the distal portion is nearly four times as long, flattened, and medially bent. The spatulate distal portions of the endopods of the left and right maxillulae are asymmetric. That of the right one is broader and its distal margin is armed with a row of six stout, triangular, tooth-like, cuspidate setae. They are accompanied by two conical setae on the anterior end of the row (Figure 11(e)). In contrast, on the left endopod, there is a row of eight robust, conical setae on the distal margin flanked by five simple setae posterodistally (Figure 11(f)).

The paragnaths (Figures 5(e) and 5(f)), traditionally often termed "lower lip," are a pair of flap-like, medially fused extensions of the sternum of the mandibular segment. The paragnaths are the posterior limitation of the mouth area and they build, together with the labrum, the space in which the mandibles operate. There is a deep cut between the two flaps. In this region, the flaps are armed with a dense field of short, thin, and scaled setae. On the anterior

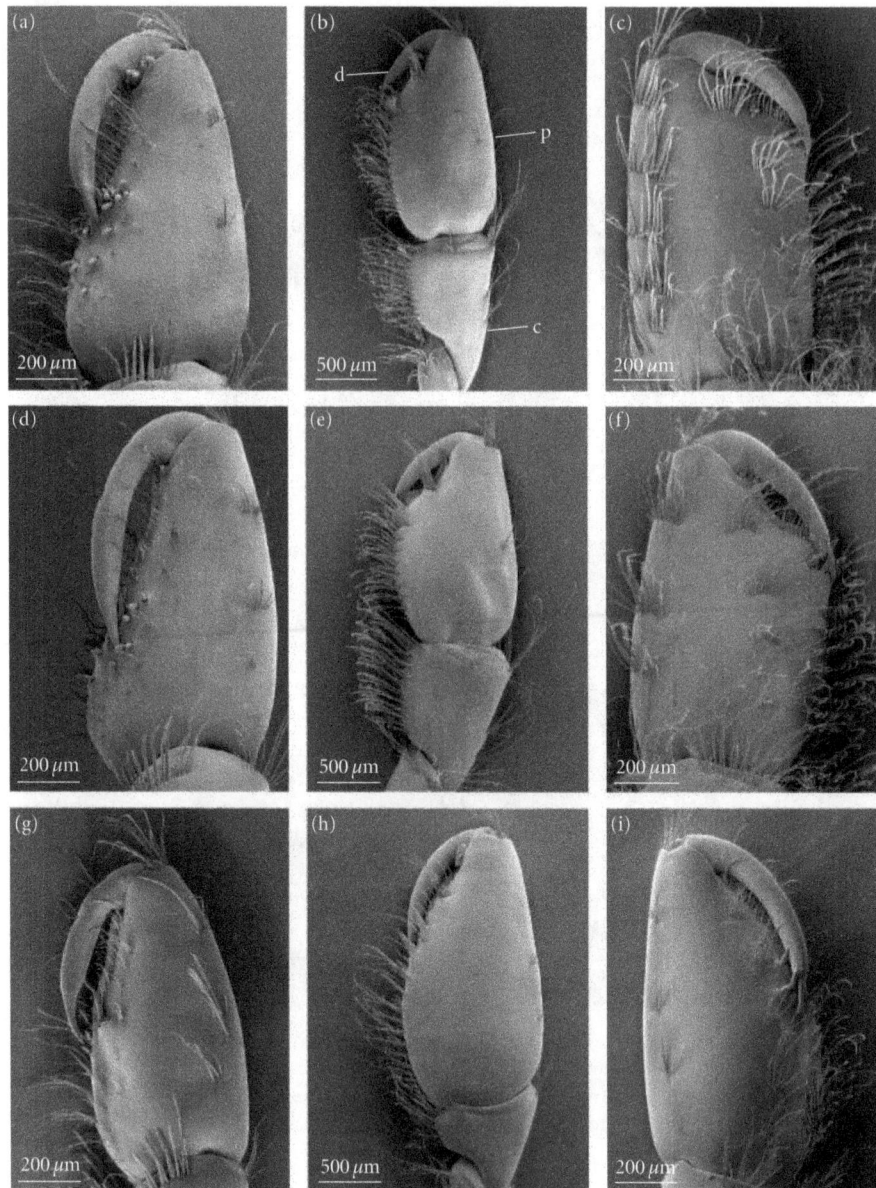

FIGURE 4: SEM images of distal part of gnathopods: (a) right first gnathopod of *G. pulex* in anterior view; (b) left second gnathopod of *G. pulex* in posterior view; (c) left second gnathopod of *G. pulex* in anterior view; (d) right first gnathopod of *G. fossarum* in anterior view; (e) left second gnathopod of *G. fossarum* in posterior view; (f) left second gnathopod of *G. fossarum* in anterior view; (g) right first gnathopod of *E. berilloni* in anterior view; (h) left second gnathopod of *E. berilloni* in posterior view; (i) left second gnathopod of *E. berilloni* in anterior view. Abbreviations other than in previous figures: c: carpus; d: dactylus; p: propodus.

side (Figure 5(f)), near the entrance to the oesophagus, there are two depressions, in which the molars of the mandibles fit, when these are adducted. On their posterior side (Figure 5(e)), the paragnaths are drawn out into a proximolaterally directed cusp on each side. The flaps of the paragnaths are asymmetric, with the left being slightly wider than the right in proximodistal extension. This corresponds with the asymmetry of the mandibles. In all three species investigated here, the paragnaths are very similar. In *G. pulex*, the proximo-laterally directed cusps are slender, elongated, and pointed. The distal and proximolateral margins of the paragnaths are curved.

The mandibles (Figure 14) comprise a prominent proximal portion, the coxa and a distal portion, the palp consisting of the basipod, and a two-segmented endopod (for overview see, Figure 16(a)). In their natural position, only the distal part of the coxa surmounts the labrum (Figures 14(a) and 14(b)). The coxa is medially drawn out into a proximodistally extending protrusion. This gnathal edge is divided into a distal pars incisiva (or incisor process), a lacinia mobilis, a spine row, and a proximal pars molaris (or molar process) (Figures 14(c) and 14(d)). There is a distinct asymmetry of left and right mandible and their components. On the right mandible, the whole coxal body as well as the gnathal

Mouthpart Morphology of Three Sympatric Native and Nonnative Gammaridean Species: Gammarus pulex, G. fossarum, and Echinogammarus berilloni (Crustacea: Amphipoda)

157

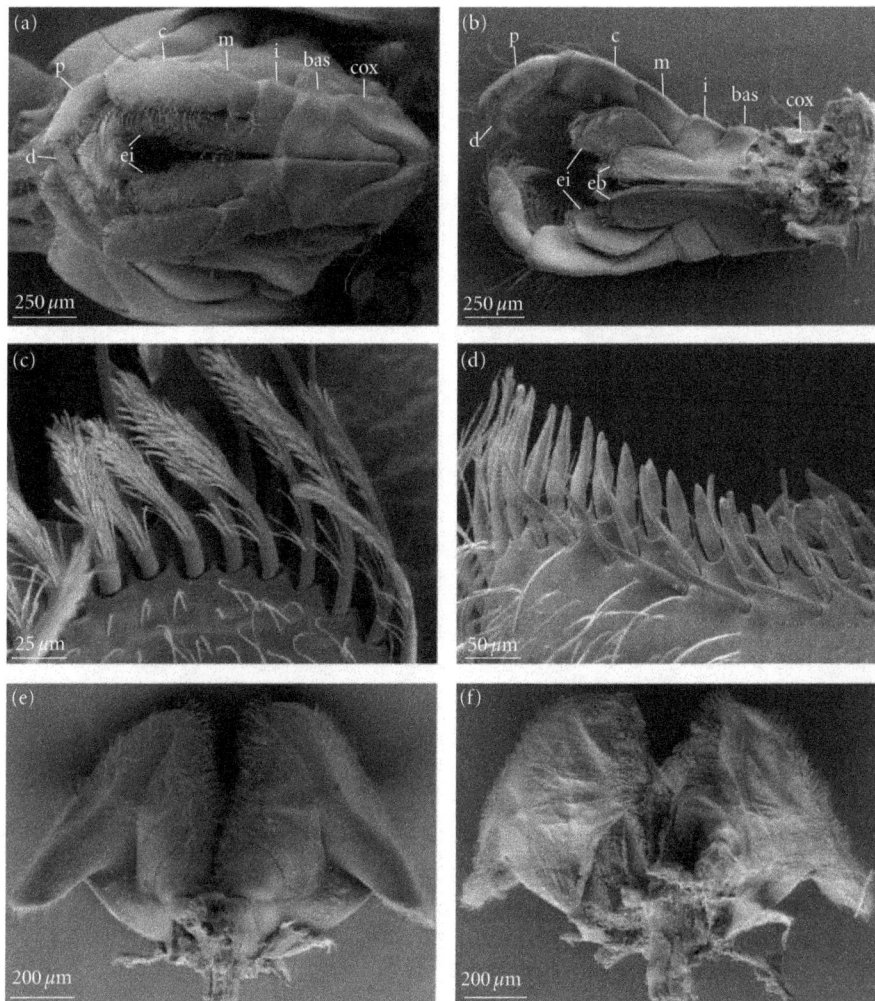

FIGURE 5: SEM images of maxillipeds and paragnaths of *Gammarus pulex*: (a) maxillipeds in posterior view; (b) maxillipeds in anterior view; (c) distal setation of endite of left basipod in anterior view (cf. eb in (b)); (d) median setation of endite of right ischium in posterior view (cf. ei in (a)); (e) paragnaths in posterior view; (f) paragnaths in anterior view. Abbreviations other than in previous figures: eb: endite of basipod; ei: endite of ischium; i: ischium; m: merus.

edge is smaller in proximo-distal extension than on the left mandible. Also the angles, in which the molar surfaces are oriented, are different, being nearly rectangular on the left, but about 60° on the right mandible (Figure 14(b)). The left incisor is stout and five toothed. The well-developed stout and broad left lacinia mobilis is blade shaped and four toothed. The base of the lacinia mobilis is slightly protruded against the incisor. This so-called articular condylus reaches into a cavity on the base of the incisor (Figure 14(b)). On the posterior side of the lacinia, the articular condylus is well developed, whereas on the anterior side, it cannot be detected, when the lacinia mobilis is in its upright position parallel to the incisor (Figures 14(c) and 14(d)). The right incisor is stout and four toothed. The right lacinia mobilis is smaller than the left, and its articular condylus is poorly developed and visible only in posterior aspect. The right lacinia mobilis is distally notched, therefore divided into

two distal parts which are arranged parallel to the incisor (Figure 14(d)). The part adjacent to the incisor is slightly bent inwards; its distal edge consists of 3–5 small spines flanked by two somewhat longer lateral spines. The other part is longer and nearly straight; its edge consists of 4–6 spines which are homogeneous in size. The setae of the spine rows are directed mediodorsally *in situ* (Figure 14(b)). Each spine row directly starts at the molar with a group of short pappose setae, together building a tuft of fine hair-like setae. Towards the incisor, the setae of the spine row successively change from pappose with setulae on the entire shaft to spinelike with only few setulae on the distal end of their shafts. These setulae are located on the proximal side of the shaft facing the molar and are distally directed. All setae of the spine row are, at least their proximal part, flattened and therefore band shaped. This shape only allows deflexion of the setae in the plane between incisor and molar

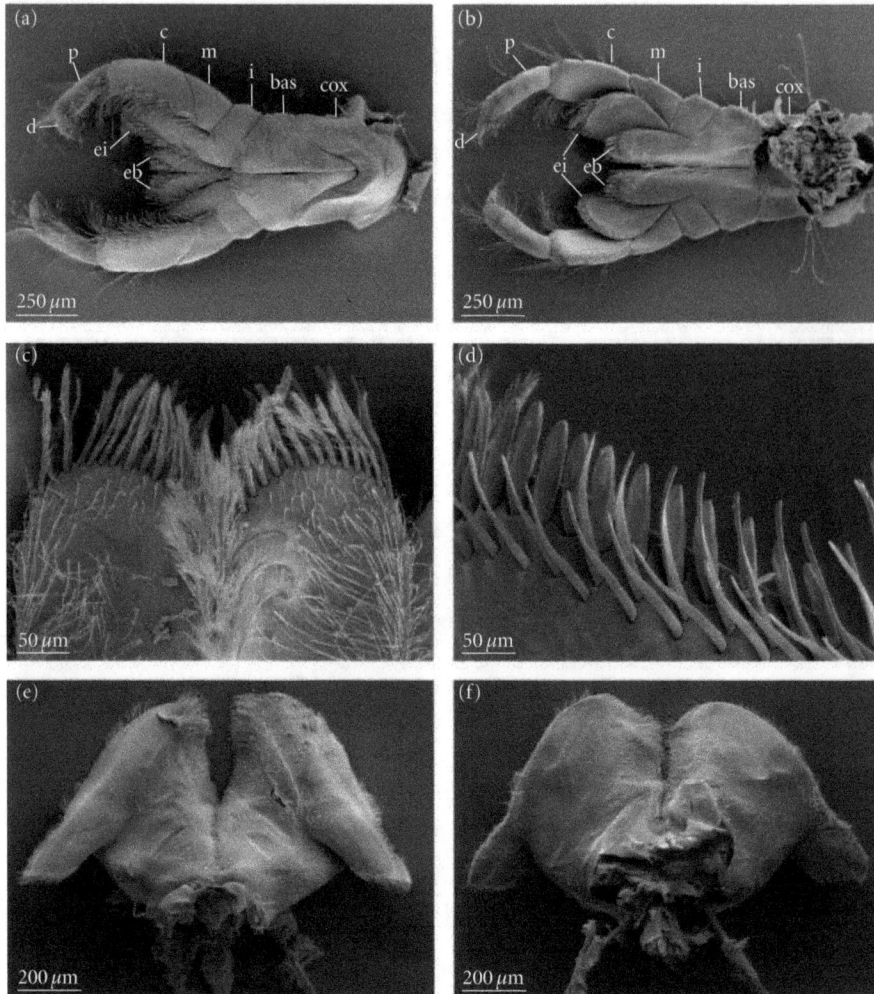

FIGURE 6: SEM images of maxillipeds and paragnaths of *Gammarus fossarum*: (a) maxillipeds in posterior view; (b) maxillipeds in anterior view; (c) distal setation of endite of left basipod in anterior view (cf. eb in (b)); (d) median setation of endite of right ischium in posterior view (cf. ei in (a)); (e) paragnaths in posterior view; (f) paragnaths in anterior view.

(Figures 14(c) and 14(d)). The succession of setae is more distinct on the spine row of the right mandible. Here, some of the setae are stiletto shaped with very broad bases.

The anterior side of each molar bears a single gnathobasic seta pointing anteromedially into the oesophagus in the live animal (Figures 14(c) and 14(d)). The left molar area is kite shaped in median view, whereas the right is more ellipsoidal. The molar surface is slightly concave with parallel edges, which are arranged vertically to the gnathal edge (Figure 14(e)). These edges are probably built by laterally conjoined feathered setae with free setulae only on one side. Therefore, an alternating sequence of hard compact cuticular mass and flexible separate setulae together builds the rasp-like structure of the surface of the molars (Figure 14(f)). On the surface of the right molar, these free setulae are directed proximally, those of the left molar distally. The short first portion of the mandibular palp, the basipod, is cylindrical and bears no setae. The mediolaterally compressed proximal portion of the endopod is about three times as long as the basipod. It is armed with a row of simple setae on the lateral side of its posterior margin. These setae are becoming longer

towards the distal end of this row. The distal portion of the palp is also medio-laterally compressed and has about two-thirds of the length of the second portion. Its tip is armed with a group of simple setae, the longest being nearly as long as the distal portion. The posterior margin of the distal portion bears a regular row of setae. These setae are adorned with medio-distally directed setulae on their distal halves. On the lateral and median surface of the distal portion of the palp, there is one, in some of the investigated specimens two, groups of up to five simple setae and a dense field of short hair-like setae near its posterodistal margin of the lateral surface (compare Figures 16(g) and 16(h)).

3.2. Gammarus fossarum Koch in Panzer, 1836 (Figure 1(b)).
Antennulae (Figure 2(b)): almost half as long as body; second portion of peduncle about twice as long as distal portion, proximal portion nearly as long as second and third portion together; flagellum 29-partite; accessory flagellum consisting of 4 annuli; setation of flagellum and peduncle poorly developed.

Mouthpart Morphology of Three Sympatric Native and Nonnative Gammaridean Species: Gammarus pulex, G. fossarum, and Echinogammarus berilloni (Crustacea: Amphipoda)

159

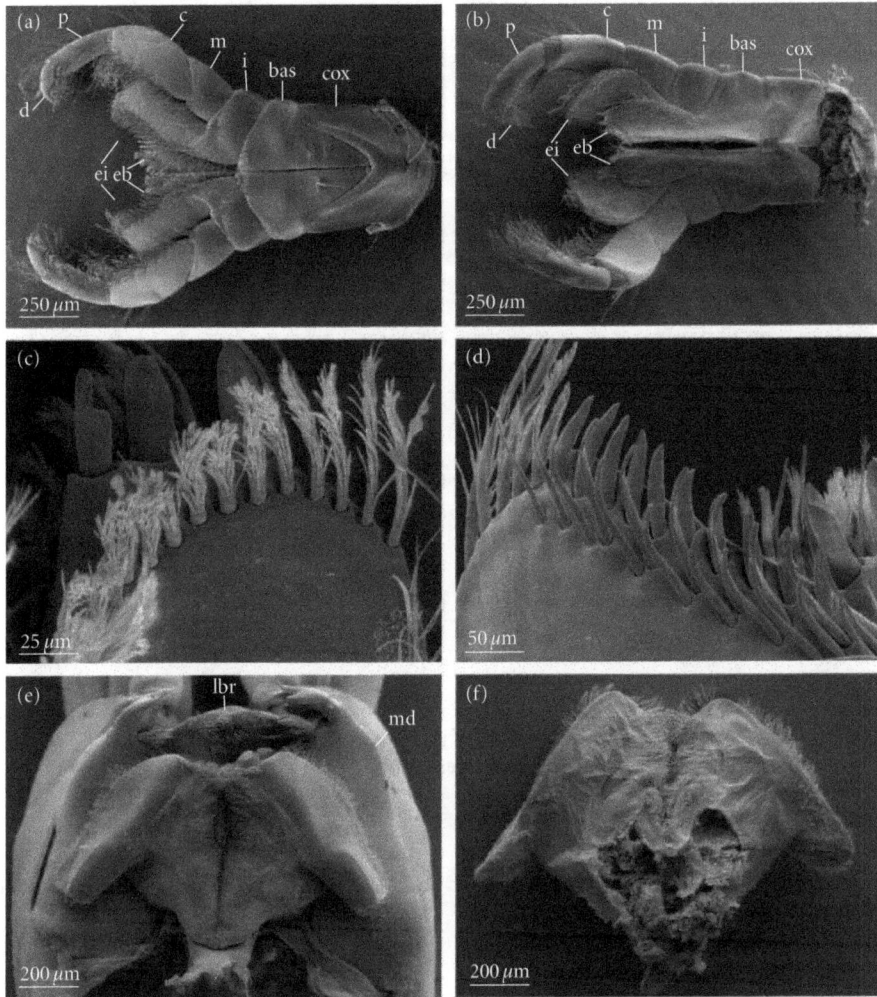

FIGURE 7: SEM images of maxillipeds and paragnaths of *Echinogammarus berilloni*: (a) maxillipeds in posterior view *in situ*; (b) maxillipeds in anterior view; (c) distal setation of endite of left basipod in anterior view (cf. eb in (b)); (d) median setation of endite of right ischium in posterior view (cf. ei in (a)); (e) paragnaths in posterior view demonstrating also the position against labrum and mandibles *in situ*; (f) paragnaths in anterior view. Abbreviations other than in previous figures: lbr: labrum; md: mandible.

Antennae (Figure 2(e)): shorter than antennulae; gland-cone pointed, nearly reaching the distal end of peduncle segment 3; flagellum slender, consisting of 13 tubular annuli; each annulus with two transversely arranged groups of 2–5 simple setae on median and lateral side, respectively; one calceolus present on median side of each flagellar annulus 1–7 (Figure 2(h)).

Third uropods: endopod about half as long as exopod (Figure 3(c)); most of the setae on the median margin of both endopod and exopod plumose; lateral margin of exopod bears simple setae; here, if any, only scattered plumose setae (Figure 3(d)); lobes of telson only little longer than peduncles of third uropods (Figure 3(c)).

Gnathopods (Figures 4(d)–4(f)): propodus of first gnathopod (Figure 4(d)) very similar to that of *G. pulex*; dactylus not as strongly curved as in *G. pulex*, more than half as long as propodus; shape and setation of second gnathopod (Figures 4(e) and 4(f)) closely resemble those of *G. pulex*.

Maxillipeds (Figures 6(a)–6(d)): in shape and setation very similar to those in *G. pulex*.

Maxillae (Figure 9): very similar to those of *G. pulex*, differences exist in details of setation; setae of posterior row of inner plate with short scale-like setulae on two opposing sides (Figure 9(d)); some setae on distal margin of inner plates with broad rounded scale-like setulae all around their shafts.

Maxillulae (Figure 12): shape of inner and outer plate (Figures 12(a)–12(c)) very similar to those of *G. pulex*, but distinct differences in setation of distal margin of outer plates (Figures 12(a) and 12(e)); two to three setae on lateral end of the two rows medially bent, overhanging the other comb-like setae posteriorly (Figure 12(a)); these two setae distally flattened and broadened, like chisels with three humps on distal edge (Figure 12(e)). In some of the specimens investigated, clear signs of abrasion on these two and adjacent

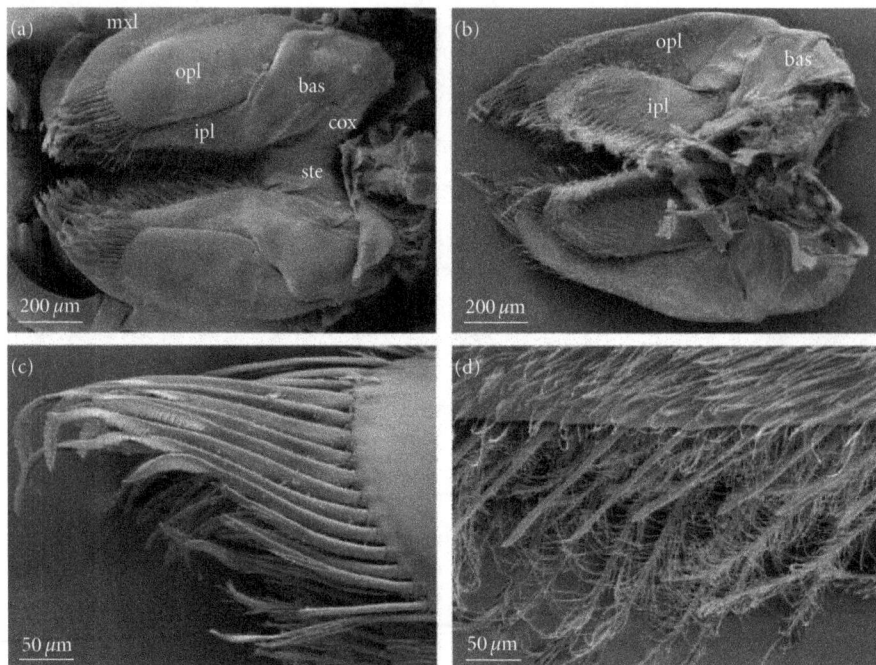

FIGURE 8: SEM images of maxillae of *Gammarus pulex*: (a) maxillae *in situ* in posterior view; (b) maxillae in anterior view; (c) distal setation of outer plate of left maxilla in posterior view; (d) median setation of inner plate of left maxilla in posterior view. Abbreviations other than in previous figures: ipl: inner plate; mxl: maxillula; opl: outer plate; ste: sternite.

setae were found. The setae are abraded to more than half of their length (Figure 12(f)).

Paragnaths (Figures 6(e) and 6(f)): very similar to those in *G. pulex*; proximo-laterally cusps not as slender and pointed as in *G. pulex*; distal margins medially edged.

Mandibles (Figure 15): incisors and left lacinia mobilis slender in posterior aspect but broad in anteroposterior dimension; left incisor five-toothed; right incisor four-toothed; left lacinia mobilis four-toothed, with well-developed articular condylus; right lacinia mobilis distally notched, spines on the two edges longer and more pointed than in *G. pulex*; articular condylus on right lacinia mobilis short but stout (Figure 15(b)); spine row of right mandible (Figure 15(d)) with one or two stiletto-shaped setae without any setulae, other setae of spine rows like in *G. pulex*; molar surfaces rasp-like, with regularly arranged edges (Figures 15(e) and 15(f)); lateral surface of distal portion of mandibular palp with dense field of hair-like setae which extends as long as adjacent row of setae on posterior margin.

3.3. Echinogammarus berilloni (Catta, 1878) [37] (Figure 1(c)).
Antennulae (Figure 2(c)): more than half as long as the body of the animal; proximal and second portion of peduncle about twice as long as distal portion; flagellum 38-partite, annuli of flagellum slightly flattened anteroposteriorly; accessory flagellum consisting of 5 annuli; setation of flagellum and peduncle poorly developed.

Antennae (Figure 2(f)): peduncle segments 4 and 5 slender and elongate; gland-cone short; flagellum (Figure 2(i)) 20-partite, flagellum distinctly anteroposteriorly flattened

and therefore broadened; setae on flagellum and peduncle short; calceoli absent.

Third uropods (Figures 3(e) and 3(f)): endopod very short compared to exopod (Figure 3(e)); exopod with groups of simple setae on median and lateral margin, those on median margin longer than laterally located ones (Figure 3(f)); distal portion of exopod short, about as long as terminal spines of proximal portion; lobes of telson compressed, little shorter than peduncle of third uropods (Figure 3(e)).

Gnathopods (Figures 4(g)–4(i)): propodus of first gnathopods (Figure 4(g)) slender; dactylus only slightly bent, with nearly straight median part; propodus of second gnathopods (Figures 4(h) and 4(i)) piriform; setae on carpus and propodus less numerous and shorter than on those of *G. pulex* and *G. fossarum*.

Maxillipeds (Figures 7(a)–7(d)): in shape and setation very similar to those of *G. pulex*.

Maxillae (Figure 10): very similar to those of *G. pulex* and *G. fossarum*; differences in details of setation; setae on lateral section of distal margin of inner plate correspond to those of outer plate with shafts flattened and equipped with closely arranged, triangular lobes on their distal third (Figure 10(c)); setae of median section of distal margin of inner plate with broad, rounded, scale-like setulae all around their shafts.

Maxillulae (Figure 13): inner plates (Figure 13(c)) more oblong than triangular; median margin of outer plates (Figure 13(a)) straight; basipodal endites not as strongly bent medially as in *G. pulex*, 10-11 setae on distal margin of outer plate (Figure 13(d)), lateral 4 of these setae with markedly

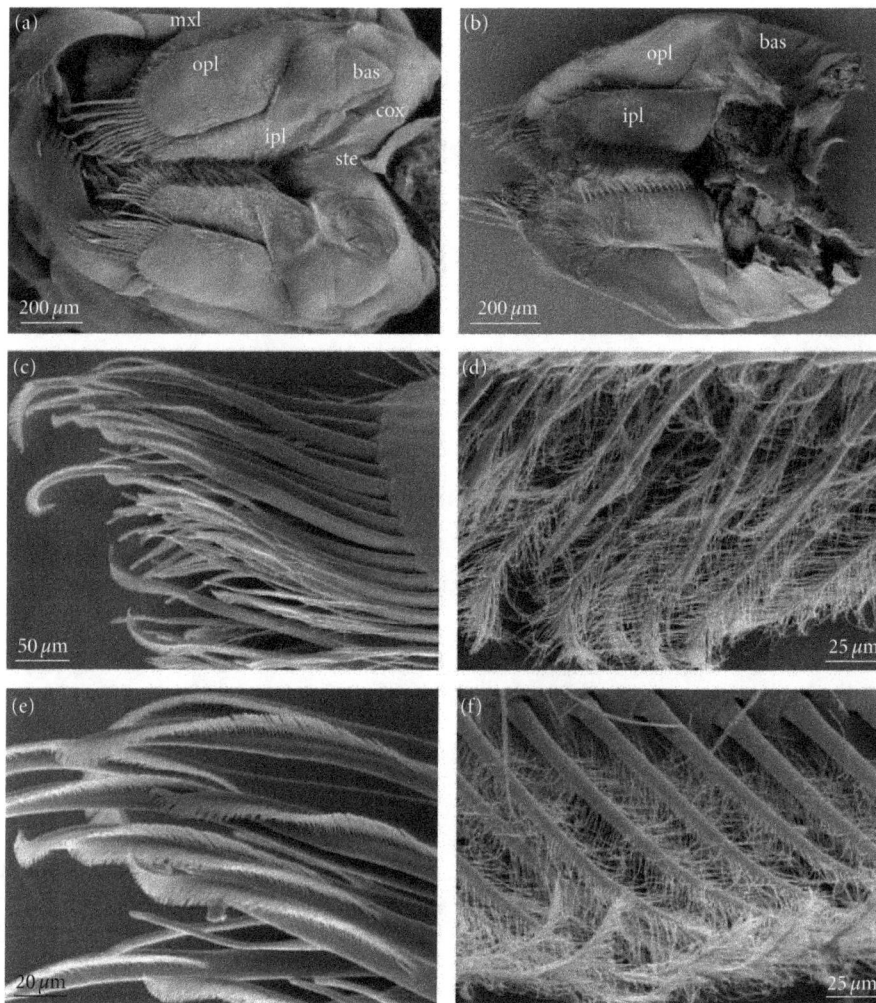

FIGURE 9: SEM images of maxillae of *Gammarus fossarum*: (a) maxillae *in situ* in posterior view; (b) maxillae in anterior view; (c) distal setation of outer plate of left maxilla in posterior view; (d) median setation of inner plate of left maxilla in posterior view; (e) distal setation of outer plate of left maxilla in posterior view; (f) median setation of inner plate of left maxilla in anterior view.

thickened shafts, distal end flattened and broadened, distal edge blunt with 2-3 humps; shafts of these 4 setae without secondary spines laterally; in some specimen, these setae display signs of abrasion (Figure 13(f)); setae of median part of distal row of setae on outer plate with up to 15 medio-posteriorly directed, pointed secondary spines (Figures 13(d) and 13(f)); the latter arranged nearly rectangular to shaft of setae and are longer than those in *G. pulex*; distal margin of left palp (Figure 13(e)) with 8 robust conical setae, flanked by a posterior row of 5 setae; these distally angled, with posterodistally directed setulae; distal margin of right palp (Figure 13(d)) with a row of 4-5 very robust triangular cuspidate setae; additionally on anterolateral end of row there is one conical seta and one seta which is distally provided with setulae.

Paragnaths (Figures 7(e) and 7(f)): more angular in shape; distal margins with hump; proximo-distally pointing cusps broader than in the two *Gammarus* species.

Mandibles (Figure 16): incisors and left lacinia mobilis very slender in posterior aspect; left incisor five-toothed; left lacinia mobilis four-toothed, with well-developed articular condylus (Figures 16(c) and 16(d)); right incisor four-toothed; right lacinia mobilis distally notched, both distal parts straight with pointed spines of different length; articular condylus on right lacinia mobilis well developed, its cavity on the base of the incisor small but well elaborated (Figures 16(b) and 16(e)); setae of spine rows (Figures 16(b) and 16(c)) comparable to those of *G. pulex*; molar surfaces (Figure 16(f)) rasp-like, with regularly arranged edges; distal portion of mandibular palp (Figures 16(g) and 16(h)) with 3 groups of up to 4 simple setae, distal half of posterior margin with a row of serrate setae, tip armed with group of long serrate setae; field of dense hair-like setae on lateral side of distal portion about half as long as row of setae on posterior margin; setae on posterior margin of second portion of palp also serrated.

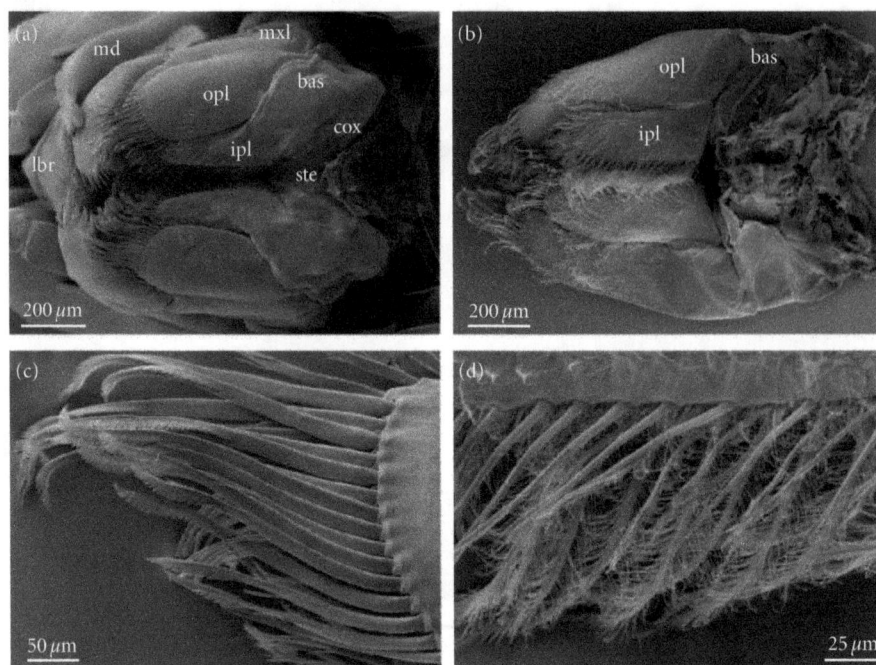

FIGURE 10: SEM images of maxillae of *Echinogammarus berilloni*: (a) maxillae *in situ* in posterior view; (b) maxillae in anterior view; (c) distal setation of outer plate of left maxilla in posterior view; (d) median setation of inner plate of left maxilla in posterior view.

The main differences between the three species in morphology of mouthparts and other structures involved in food acquisition are listed in Table 1.

4. Discussion

Specialization on a certain kind of food such as carrion, sponges, algae, or periphyton is mainly known from marine amphipods. The mouthparts of these nutrition specialists exhibit morphological adaptation for collecting, handling, processing, and ingestion of this specific kind of food [52–60].

In context with the dramatic change of the nonmarine gammaridean fauna in central Europe, several investigations on the ecology of native and invasive species were performed. The results of these field and laboratory experiments changed our view on the feeding habit of non-marine gammarideans. It was demonstrated that these animals, formerly presumed to feed mainly on dead plant material and therefore assigned to the functional feeding group of shredders [43, 61, 62], are, in fact, able to use a much wider variety of food [63–72].

For the very successful invasive species *Dikerogammarus villosus*, the following feeding-related activities were identified: detritus feeding, coprophagy, grazing, particle feeding, predation on free-swimming animals, benthic animals, and fish eggs and feeding on byssus threads of zebra mussels (*Dreissena polymorpha* Pallas, 1771) [73]. However, morphological investigations demonstrated that also the mouthparts of several non-marine gammarideans possess morphological adaptations, which enable the animal to use a certain food resource in an especially effective manner [46, 74, 75].

Until now, detailed descriptions of mouthparts and other structures involved in food acquisition of non-marine gammarideans are scarce. In taxonomic descriptions, mouthparts have often been neglected, possibly because preparation is necessary. The structures are also often very small, and differences can only be found in details. This is likewise true for the descriptions of the three species investigated herein. The original taxonomic description of *Gammarus pulex* by Linnaeus [36] is so short that Sars [76] questioned whether Linnaeus had actually described specimens of *G. pulex*. Pinkster [49] redescribed *G. pulex*, but did not describe the mouthparts except for the mandibular palp. In their redescription of *G. pulex*, Karaman and Pinkster [51] described all mouthparts, but also this description lacks enough details to detect functional relevant differences between the mouthparts of the various species. The latter is also true for the work of Agrawal [77], who described and illustrated the feeding appendages and the digestive system of *G. pulex*. Lastly, the original descriptions as well as the redescriptions of *Gammarus fossarum* [35, 50, 51] and of *Echinogammarus berilloni* [14, 37] give comparably little information on details of the morphology of the mouthparts.

The mouthparts of specimens of the three species investigated herein are very similar, but nevertheless various fine-graded differences could be found between these species.

Gammarus pulex is the most widespread freshwater amphipod in mainland Europe as well as the most widespread and abundant freshwater amphipod in Britain [78, 79] and is regarded as one of the most important invertebrate species in chalk streams in terms of biomass and food for fish [80]. It predominantly occurs in middle

Mouthpart Morphology of Three Sympatric Native and Nonnative Gammaridean Species: Gammarus pulex, G. fossarum, and Echinogammarus berilloni (Crustacea: Amphipoda)

163

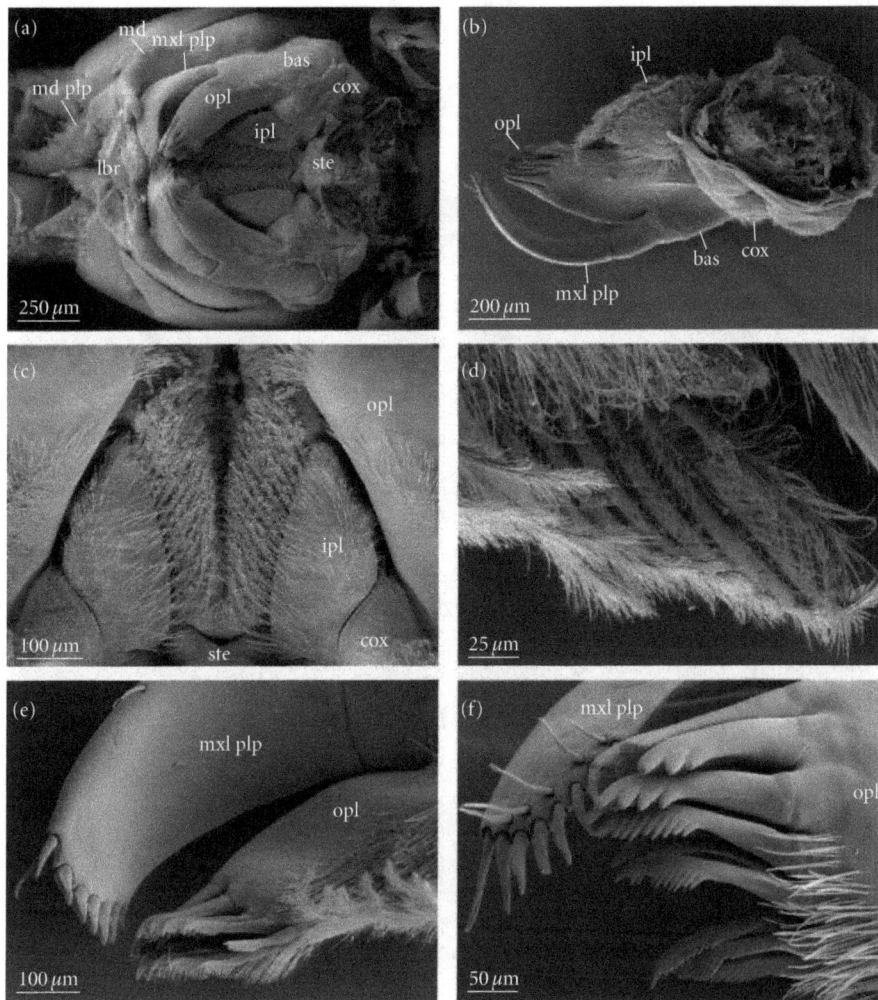

FIGURE 11: SEM images of maxillulae of *Gammarus pulex*: (a) maxillulae *in situ* in posterior view; (b) left maxillula in anterior view; (c) inner plates of maxillulae *in situ* in posterior view; (d) median setation of inner plate of left maxillula in medio-posterior view; (e) outer plate and palp of right maxillula in anteromedian view; (f) outer plate and palp of left maxillula in posterior view. Abbreviations other than in previous figures: md plp: palp of mandible; mxl plp: palp of maxillula.

TABLE 1: Main differences in morphology of mouthparts and other structures involved in food acquisition.

Structure	Gammarus pulex	Gammarus fossarum	Echinogammarus berilloni
Antennal flagellum	Annuli medio-laterally broadened; each with a row of 12 posteriorly directed setae together building a flag-like brush	Each annulus with a group of 2–5 simple setae on median and lateral side	Antero-posteriorly flattened and therefore broadened; setation sparse; setae short
Cuspidate setae on distal margin of basipodal endites of maxillula	Lateral setae with three finger-like secondary spines	2-3 lateral setae distally flattened and broadened like chisels with three humps on distal edge	4 lateral setae with thickened shafts, distally flattened; distal margin blunt with 2-3 humps
Setation of carpus and propodus of 2nd gnathopod	Long and closely arranged setae with curled distal ends	Long and closely arranged setae with curled distal ends	Less numerous and shorter setae than in *G. pulex* and *G. fossarum*
Third uropod	Endopod 3/4 as long as exopod; plumose setae on medial and lateral margins of endopod and exopod	Endopod half as long as exopod, plumose setae on median margin of endopod and exopod; simple setae on lateral margin of exopod	Endopod very short; only simple setae on lateral and median margin of exopod

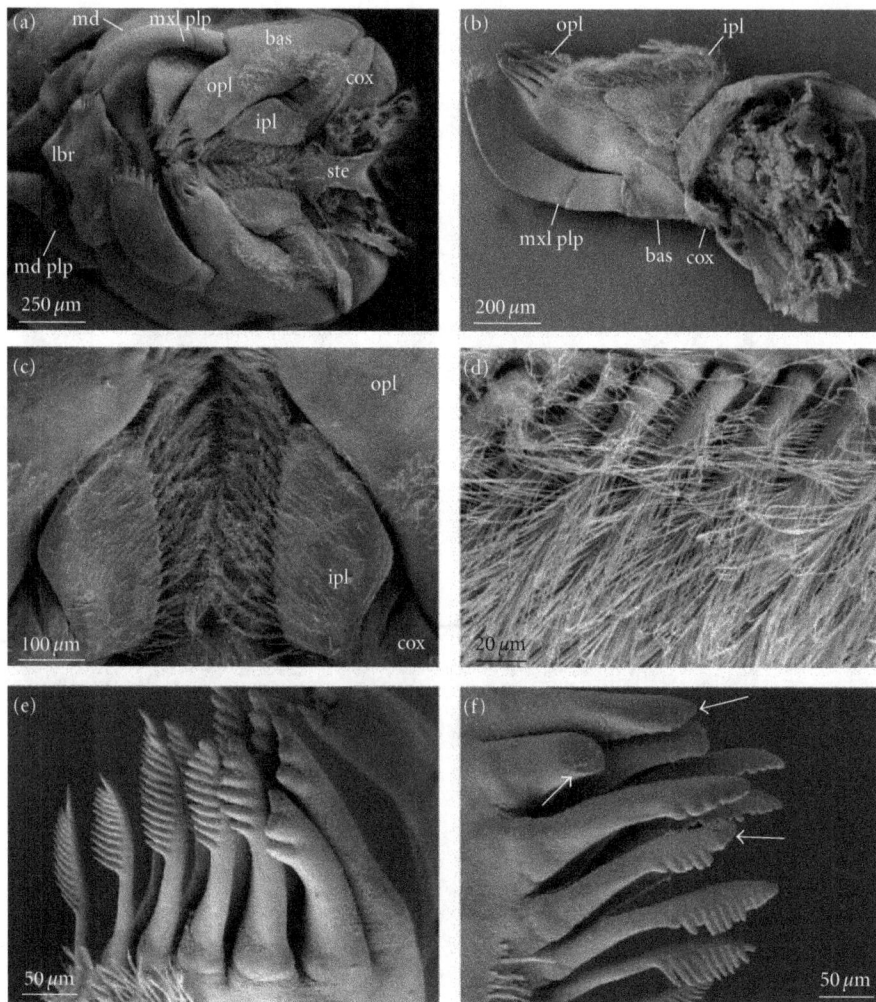

FIGURE 12: SEM images of maxillulae of *Gammarus fossarum*: (a) maxillulae *in situ* in posterior view; (b) left maxillula in anterior view; (c) inner plates of maxillulae *in situ* in posterior view; (d) median setation of inner plate of left maxillula in posterior view; (e) distal setation of outer plate of left maxillula in posterior view; (f) distal setation of outer plate of right maxillula in posterior view. Arrows indicate signs of abrasion.

and lower reaches of streams and rivers, lowland lakes, ponds, and brooks [51]. In some publications, the limit of distribution is stated to be at about 450 m altitude [41, 81, 82], but when competing species are absent, it can be found in all sections of the waters [51]. Dusaugey [83] and Goedmakers [84] have found the species at 1340 and 1200 m, respectively. In Ireland, *G. pulex* is an invasive species where it replaces the native *G. duebeni* Liljeborg, 1852 in rocky parts of lakes, rivers, and along canals [78] and has a great impact on native macroinvertebrate community composition [85]. Microdistribution of *G. pulex* seems to be size assortative, with large animals being associated with large substrate particles like pieces of wood or accumulated fallen leaves and macrophytes [10, 77, 86, 87]. In running waters, *G. pulex* prefers sections with lower velocity [88, 89]. According to Piscart et al. [24], its preferred substratum is vegetation and leaf litter.

Agrawal [77] investigated gut contents and concluded that *G. pulex* mainly feeds on algal filaments and other vegetable matter. Also laboratory experiments indicate that plant material is an important food source, because specimens of *G. pulex* also shredded leaf material in the presence of animal prey [67]. Graca et al. [90, 91] reported that *G. pulex* feeds preferentially on conditioned rather than on unconditioned leaf material, although no significant effect of conditioning on growth was observed. Also Welton and Clarke [92] observed feeding on both decaying leaves and fresh green leaves. Investigations on the enzymes of the midgut glands yielded that *G. pulex* produces both cellulose and phenol oxidase by itself and is therefore adapted to digest plant material [93]. However, laboratory experiments show that *G. pulex* is also an effective predator on *Asellus aquaticus* Linnaeus, 1758 [94], Copepoda [95], mayfly nymphs [67], larvae of Chironomidae, Simuliidae, Ephemeroptera [68], Enchytraeidae [96], and on *Tubifex* sp. [68]. Moreover, experiments on intraguild predation demonstrated that *G. pulex* also preys on other gammarideans such as *G. tigrinus* Sexton 1939, *G. duebeni* Liljeborg, 1852, and *Crangonyx*

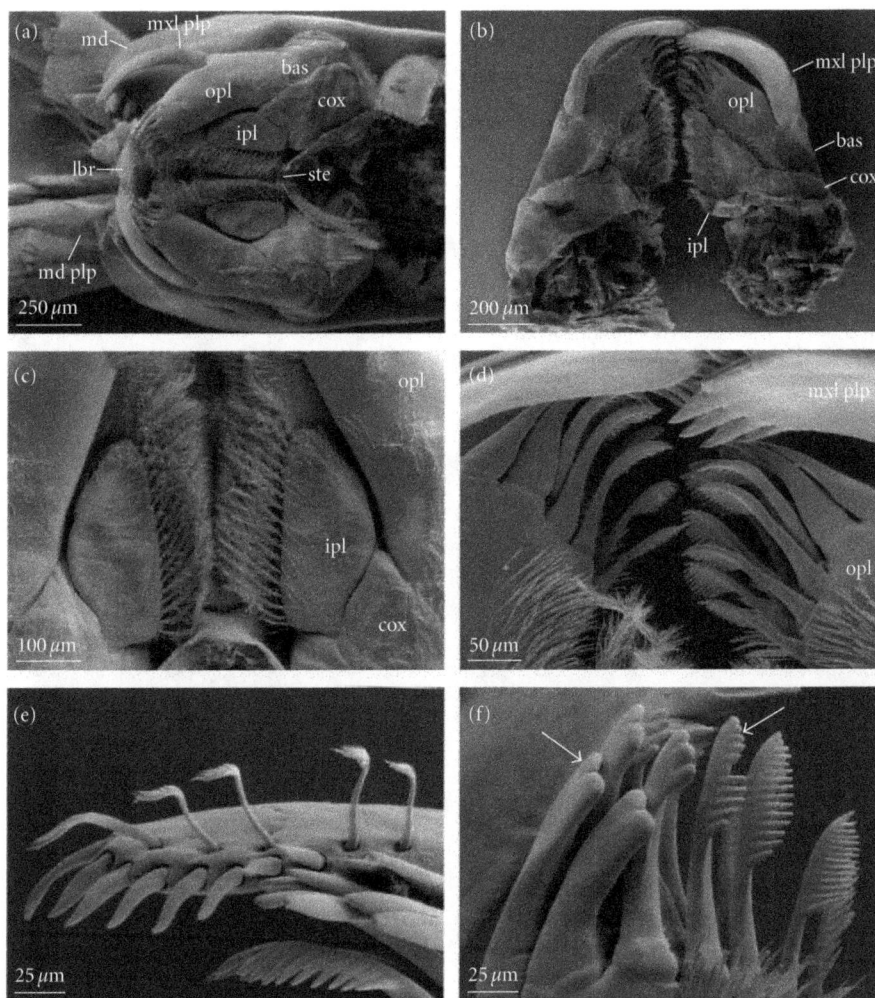

FIGURE 13: SEM images of maxillulae of *Echinogammarus berilloni*: (a) maxillulae *in situ* in posterior view; (b) maxillulae in anterior view; (c) inner plates of maxillulae *in situ* in posterior view; (d) distal setation of outer plates and palps of maxillulae in anterior view; (e) distal part of palp of left maxillula in medio-posterior view; (f) distal setation of outer plate of right maxillula in posterior view. Arrows indicate signs of abrasion.

pseudogracilis Bousfield, 1958 [27, 97]. Also analyses of stable C and N isotopes highlighted that *G. pulex* is able to feed on a broad spectrum of food sources [24].

The results of our investigation on the morphology of the mouthparts and other structures involved in food acquisition correspond with findings of the investigations mentioned above.

The antennae of *Gammarus pulex* with their rows of long and posteriorly directed setae on each medio-laterally broadened annulus of the flagellum forming a flag-like brush (Figures 2(d) and 2(g)) are well suited to collect fine particular detritus and to sieve particles out of the respiratory water current. In addition, these setae can help to create a water current for capturing free-swimming organisms when the antennae are bent towards the ventral side of the cephalothorax in a sudden movement. Such a mode of catching free-swimming organisms has been described by Platvoet et al. [73] for the pontogammarid *D. villosus*. The

antennal flagellae of this species have similar setation [74]. The endopods of the foliaceous third uropods of *G. pulex* are relatively long (Figure 3(a)). Both endopod and exopod bear plumose setae on their median and lateral margins building a broad fan (Figure 3(b)). Therefore, the third uropods are also well suited to sieve particles out of the respiratory water current and it seems possible that they are used for guiding faeces anterior toward the gnathopods (coprophagy). The closely arranged long setae of the second gnathopods with their curled distal ends (Figures 4(b) and 4(c)) together are again a structure which is suited for sieving particles out of the respiratory water current. They can also be used for cleaning the antennae and for sweeping periphyton from the substratum. Furthermore, the two opposing gnathopods with their ventromedially directed setae build a space in which food particles and organisms can be held captive and guided to the mouthparts. The maxillipeds (Figures 5(a)–5(d)) with their medially directed setae and the two claw-like

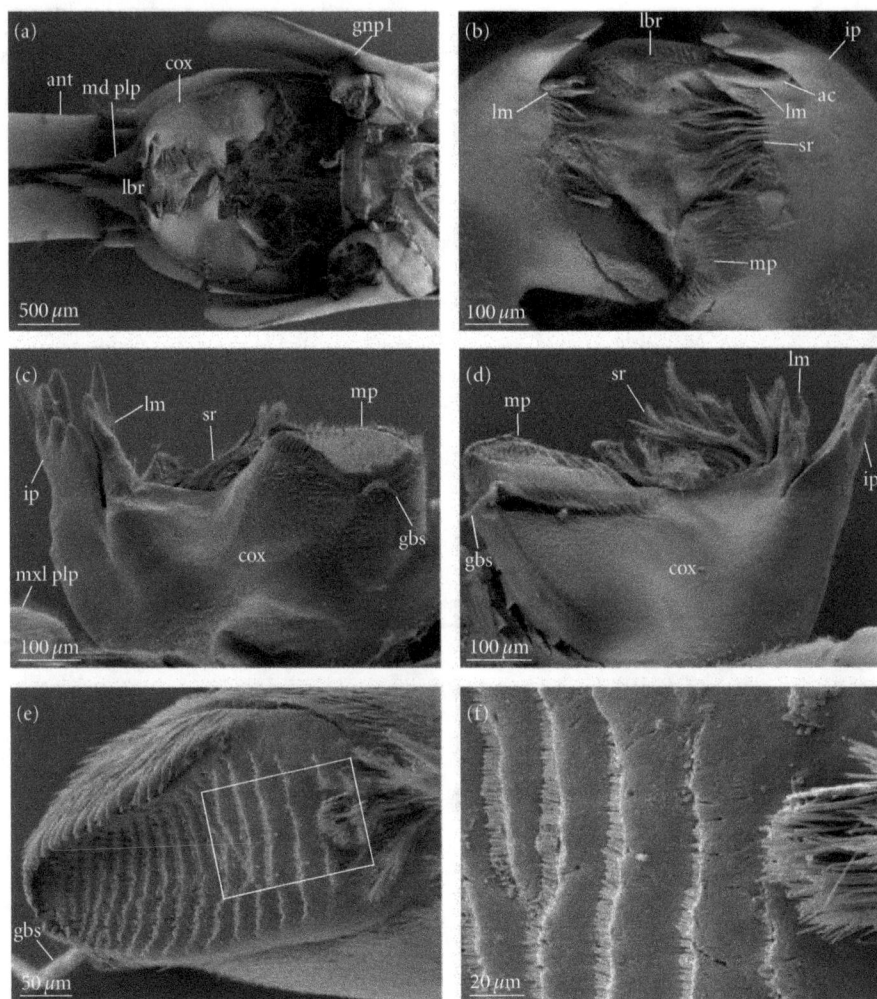

FIGURE 14: SEM images of mandibles of *Gammarus pulex*: (a) ventral view of head with exposed mandibles; maxillulae, maxillae, maxillipeds, and gnathopods removed; (b) coxa of mandibles *in situ* in posterior view; (c) coxa of left mandible in anterior view; (d) coxa of right mandible in anterior view; (e) molar of right mandible in median view; rectangle indicates close-up image in (f); (f) closeup of molar surface of right mandible in median view (cf. (e)). Abbreviations other than in previous figures: ant: antenna; gbs: gnathobasic seta; gnp1: first gnathopod; ip: incisor process; lm: lacinia mobilis; mp: molar process; sr: setal row.

setae on the distal end of the endites seem to be useful for combing out particles from the setae of the gnathopods and antennae and for transferring them to the mandibles.

The basipodal endites of the maxillae (Figures 8(a)–8(d)) with their medially directed plumose setae build a dense net for preventing food from being washed away from the mouth region and for concentrating food. The coxal endites of the maxillulae with their medially directed pappose setae (Figure 11(c)) may have the same function. The comb-like cuspidate setae on the distal margin of the basipodal endites of the maxillulae (Figures 11(e) and 11(f)) seem to be well suited for combing out particles from the setae of the antennae and gnathopods. The use of these setae for detaching periphyton from the substrate might be possible, but there are no specializations for this purpose as it could be shown for *G. roeselii* [46], and on those specimens we investigated, we did not find distinct signs of abrasion. Left and right incisors and the left lacinia of

the mandibles (Figures 14(a)–14(d)) of *G. pulex* are broad and well developed and therefore seem to be well suited for cutting off pieces of bigger food items. The rasp-like surface of the molars (Figures 14(e) and 14(f)) seems to be well suited for grinding hard plant material, although these parallel edges are not as well developed as in *G. roeselii* [46].

Gammarus fossarum is widely distributed in central Europe, the Balkan Peninsula, and Asia minor [45, 51]. It inhabits springs, brooks, and upper reaches of smaller rivers with low content of nutrients and low conductivity [44, 81, 98, 99]. It tolerates and prefers low temperature and high currents [45, 82, 100–104]. In gut contents of *G. fossarum*, Helesic et al. [105] found filamentous algae and cyanobacteria, parts of leaves, moss, plankton organisms, and macroinvertebrates living in periphyton. If no other food is available, *G. fossarum* also feeds on fresh leaves [43]. However, in laboratory experiments, growth was best with decaying leaves of lime and elm, whereas growth and survival

Mouthpart Morphology of Three Sympatric Native and Nonnative Gammaridean Species: Gammarus pulex, G. fossarum, and Echinogammarus berilloni (Crustacea: Amphipoda)

167

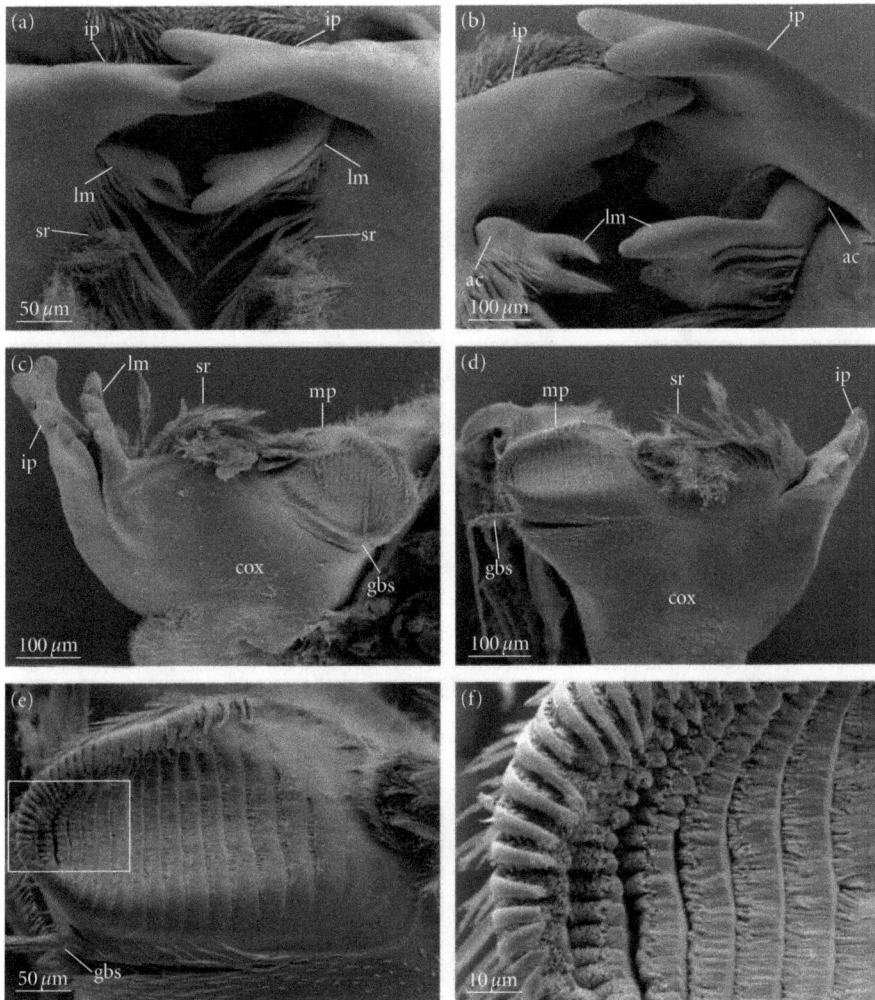

FIGURE 15: SEM images of mandibles of *Gammarus fossarum*: (a) incisor processes, lacinia mobiles, and setal rows of mandibles *in situ* in posterior view; (b) incisor processes and lacinia mobiles of mandibles *in situ* in posterodorsal view; (c) coxa of left mandible in anterior view; (d) coxa of right mandible in anterior view; (e) molar of right mandible in medioanterior view, rectangle indicates close-up image in (f); (f) closeup of molar surface of right mandible in median view (cf. (e)). Abbreviations other than in previous figures: ac, articular condylus.

rate was significantly lower with fresh macrophytes and algae [106]. Pieper [107] observed that adult specimens mainly feed on dead leaves. In their natural habitats, specimens of *G. fossarum* were most abundant in accumulations of leave litter and other dead plant material [43, 108, 109]. Gut content analyses performed by Felten et al. [72] showed that the investigated individuals of *G. fossarum* fed on detritus, diatoms, filamentous algae, leaf litter, woody debris, and animal matter.

The setation of the slender antennal flagellum (Figures 2(e) and 2(h)) of *G. fossarum* is sparse and therefore not as effective for sieving particles out of the respiratory water current and for catching free-swimming organisms as in *G. pulex*. The endopods of the third uropods (Figure 3(c)) are not as long as those in *G. pulex*, and there are plumose setae only on the median margin of the exopods (Figure 3(d)). Therefore, the third uropods of *G. fossarum* seem to be not as well suited for sieving particles out of the respiratory water current as those of *G. pulex*.

Coprophagy in the same manner as in *G. pulex* seems also to be possible in *G. fossarum*. The long and closely arranged distally curved setae of the second gnathopods (Figures 4(e) and 4(f)) closely resemble those in *G. pulex*. Therefore, the second gnathopods of *G. fossarum* are also suited for cleaning the antennae and for sweeping periphyton from the substratum. Maxillipeds (Figures 6(a)–6(d)) and maxillae (Figures 9(a)–9(f)) of *G. fossarum* are very similar to those in *G. pulex*, so we expect no functional differences. In contrast, clear differences can be observed in the setation of the maxillulae (Figures 12(a)–12(f)). The lateral two or three setae of the row of setae on the distal margin of the basipodal endopods are distally flattened like chisels overhanging the other comb-like setae posteriorly (Figure 12(e)). With these setae, the maxillulae seem to be well suited for scraping off periphyton from the substratum. Signs of abrasion on these setae support this conclusion (Figure 12(f)). In contrast, Felten et al. [72] calculated from their results of gut content analyses that the investigated *G. fossarum* population

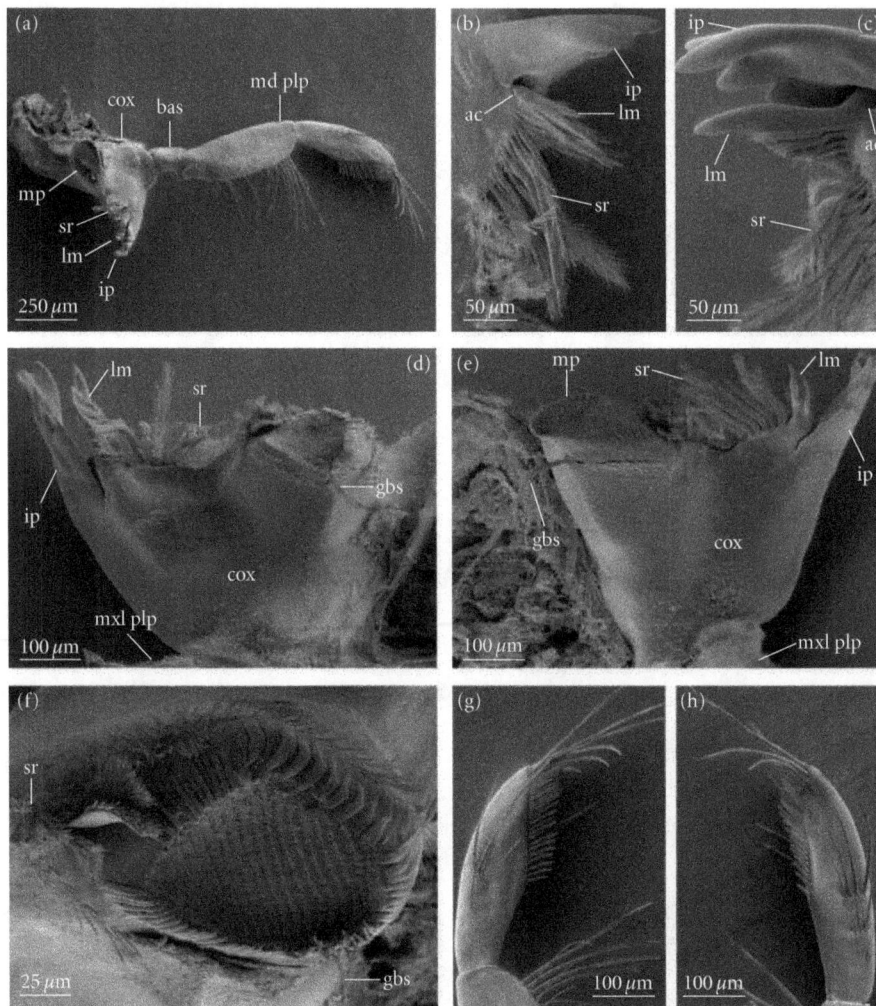

Figure 16: SEM images of mandibles of *Echinogammarus berilloni*: (a) left mandible in median view; (b) incisor process, lacinia mobilis, and setal row of right mandible in posterior view; (c) incisor process, lacinia mobilis, and setal row of left mandible in posterior view; (d) coxa of left mandible in anterior view; (e) coxa of right mandible in anterior view; (f) molar of left mandible in median view; (g) distal portion of left mandibular palp in median view; (h) distal portion of left mandibular palp in lateral view.

could be considered as 48% collector, 43% shredder, 8.3% predator, and only 1.3% scraper. The reason for the marginal importance of scraping in the investigated population might be the abundant supply of leave litter and bryophytes in this particular stream. In waters with less availability of these food resources, as is often the case in headwaters, scraping might play a more important role in the nutrition of *G. fossarum*. Furthermore, a point to be taken into account is that *G. fossarum* can live in the hyporheic interstitial where biofilm represents a very important portion of food available. This might be important especially in winter when water level and temperature is low. The mandibles (Figures 15(a)–15(f)) of *G. fossarum* are similar to those of *G. pulex* and seem also to be well suited for cutting off pieces of bigger food items and to grind hard plant material.

Catta [37] described *Echinogammarus berilloni* from a spring of a brook on the Mandarin mount, Atlantic Pyrenean mountains. This species originates from the Atlantic region of France and the north-western part of Spain [14, 16] and

extended its area of distribution to the middle and lower reaches of larger streams, channels, and rivers of the north of France [10], the Channel Islands [110], Luxembourg, Belgium, the southern part of the Netherlands [13], and local populations exist in Germany [5, 14]. Kley and Maier [30] found specimens associated with near-shore submersed macrophytes in a tributary of the river Rhine. According to Piscart et al. [10, 24], *E. berilloni* shows strong preference for vegetation and leaf litter as substratum when existing in single-species populations, but there is a shift to a more diverse use of substrates including pebbles when it coexists with *G. pulex*. *E. berilloni* is salt tolerant and eurythermous and can withstand a high amount of organic pollution [2, 39]. Since the invasion of *Dikerogammarus villosus*, there has been a dramatic reduction of the relative abundance of *E. berilloni* [6].

In *Echinogammarus berilloni*, the antennae with their sparse and short setation (Figures 2(f) and 2(i)) are not suited for sieving particles out of the respiratory water current and for catching free-swimming organisms. However, with their anteroventrally flattened and mediolaterally broadened flagellum, the antennae might be dedicated to collect food particles and move them into the reaching area of the gnathopods. Also the morphology of the third uropods indicates that sieving does not play a great role in the nutrition of *E. berilloni*, because their endopods are very short (Figure 3(e)), and there are only simple setae on the median and lateral margins of the exopods (Figure 3(f)). Also the setae on the second gnathopods (Figures 4(h) and 4(i)) are sparser and shorter, and therefore they seem to be less suitable for sieving and for sweeping periphyton from the substratum. Such constraints in the ability to use these food resources correlates with findings of Piscart et al.: "... their isotopic signatures highlighted a broader spectrum of food sources and a broader diversity of carbon sources assimilated by *G. pulex* than by *E. berilloni*" [24].

Maxillipeds (Figures 7(a)–7(d)) and maxillae (Figures 10(a)–10(d)) of *E. berilloni* are very similar to those of *G. pulex* and *G. fossarum* so we expect no functional differences. Again, there are differences in the morphology of the setae on the distal margin of the basipodal endites of the maxillulae. In *E. berilloni*, the lateral four of these setae are stout and do not bear lateral secondary spines, but end in a broadened distal margin with two or three humps (Figures 13(e) and 13(f)). Also here we found signs of abrasion, so it is likely that these setae are used to scrape off periphyton from the substratum. This adaptation of setae on the basipodal endite of the maxillulae with no lateral secondary spines and a broadened distal margin is comparable with the situation in *G. roeselii* [46]. However, in the latter, the distal margin of these setae is sharper, and therefore the adaptation to scraping seems to be even more effective as in *E. berilloni*. The mandibles (Figure 16(a)–16(f)) of *E. berilloni* are similar to those of *G. pulex* and *G. fossarum*. Therefore, they likewise seem to be well suited for cutting off pieces of bigger food items and to grind hard plant material.

5. Conclusions

The morphology of the mouthparts and structures involved in food acquisition indicate that *Gammarus pulex* is able to feed on a wide variety of food sources including sieving suspended organic particles out of the respiratory water current, coprophagy, collecting detritus, catching free-swimming organisms, removing periphyton from the substratum, biting off pieces from bigger food items, and grinding hard plant material. Therefore, this species can be characterized as omnivorous. Compared to *G. pulex*, in *G. fossarum* the structures investigated here indicate that these two species have very similar feeding habits. However, the setation of the maxillulae of *G. fossarum* and the severe abrasion of these structures indicate that feeding on periphyton removed from hard substratum plays a greater role in the nutrition of the latter. This coincides with the finding that *G. fossarum* typically inhabits the upper reaches of brooks and rivers, where periphyton on hard substratum is the most relevant food resource. The results of our work suggest that the ability to sieve particles out of the respiratory

water current and to catch free-swimming organisms is limited in *Echinogammarus berilloni*. This species seems to be adapted to collect food items with the antennae and to remove periphyton from the substratum with the maxillulae. Therefore, the ability to use different food sources seems to be restricted in *E. berilloni* compared to *G. pulex*.

Arndt et al. [60] hypothesised that "mouthpart morphology differs little between related amphipod species, but greater changes are encountered in the morphology of accessory feeding appendages as a consequence of trophic specialization." This is generally also true for the species investigated here, but we also found differences in the morphology of the maxillulae. These findings correspond with the situation in *Gammarus roeselii* and *Dikerogammarus villosus* where we, in an earlier study, also found differences in the morphology of the antennae, gnathopods, maxillipeds, maxillulae, and mandibles [46, 74].

Abreviations

1, 2, 3, 4, 5:   Number of podomeres of antennulae and
                        antennae
ac: Articular condylus
af: Accessory flagellum of antennula
ant: Antenna
bas: Basipod
cal: Calceolus
cox: Coxa
eb: Endite of basipod
ei: Endite of ischium
f: Flagellum of antennula and antenna
gc: Gland conus
gnp1:   First gnathopod
ip: Incisor process
ipl: Inner plate
lbr: Labrum
lm: Lacinia mobilis
md: Mandible
md plp: Mandibular palp
mp: Molar process
mxl: Maxillula
mxl plp: Maxillular palp
mxp: Maxilliped
opl: Outer plate
sr: Setal row
ste: Sternite.

Acknowledgments

The authors are grateful to Gerhard Maier, Axel Kley, and Werner Kinzler for providing us with specimens of *E. berilloni*. Many thanks are to the team of the Central Facility for Electron Microscopy, University of Ulm for their friendly support. Carolin and Joachim T. Haug and Verena Kutschera are thanked for inspiring ideas and discussions. They also thank an anonymous reviewer for valuable critical comments and suggestions on an earlier version of the paper. The study material is stored at the University of Ulm. This work was part of the project Wa-754/16-1, German Research Foundation (DFG).

References

[1] A. Bij de Vaate and A. G. Klink, "*Dikerogammarus villosus* Sowinsky (Crustacea: Gammaridae) a new immigrant in the Dutch part of the Lower Rhine," *Lauterbornia*, vol. 20, pp. 51–54, 1995.

[2] T. Tittizer, "Vorkommen und Ausbreitung aquatischer Neozoen (Makrozoobenthos) in den Bundeswasserstraßen," in *Neophyten, Neozoen—Gefahr für Die Heimische Natur?* S. Schmidt-Fischer, Ed., vol. 22 of *Beiträge der Akademie für Natur- und Umweltschutz Baden-Württemberg*, 1996.

[3] T. Tittizer, F. Schöll, M. Banning, A. Haybach, and M. Schleuter, "Aquatische neozoen im makrozoobenthos der binnenwasserstraßen deutschlands," *Lauterbornia*, vol. 39, pp. 1–72, 2000.

[4] A. Bij de Vaate, K. Jazdzewski, H. A. M. Ketelaars, S. Gollasch, and G. Van der Velde, "Geographical patterns in range extension of Ponto-Caspian macroinvertebrate species in Europe," *Canadian Journal of Fisheries and Aquatic Sciences*, vol. 59, no. 7, pp. 1159–1174, 2002.

[5] K. Wouters, "On the distribution of alien non-marine and estuarine macro-crustaceans in Belgium," *Bulletin de l'Institut Royal des Sciences Naturelles de Belgique, Biologie*, vol. 72, pp. 119–129, 2002.

[6] L. Bollache, S. Devin, R. Wattier et al., "Rapid range extension of the Ponto-Caspian amphipod *Dikerogammarus villosus* in France: potential consequences," *Archiv fur Hydrobiologie*, vol. 160, no. 1, pp. 57–66, 2004.

[7] K. Jazdzewski, A. Konopacka, and M. Grabowski, "Recent drastic changes in the gammarid fauna (Crustacea, Amphipoda) of the Vistula River deltaic system in Poland caused by alien invaders," *Diversity and Distributions*, vol. 10, no. 2, pp. 81–87, 2004.

[8] H. A. M. Ketelaars, "Range extension of Ponto-Caspian aquatic invertebrates in continental Europe," in *Aquatic Invasions in the Black, Caspian and Mediterranean Seas*, H. J. Dumont, Ed., pp. 209–236, Kluwer, Dordrecht, The Netherlands, , 2004.

[9] G. Josens, A. B. De Vaate, P. Usseglio-Polatera et al., "Native and exotic Amphipoda and other Peracarida in the River Meuse: new assemblages emerge from a fast changing fauna," *Hydrobiologia*, vol. 542, no. 1, pp. 203–220, 2005.

[10] C. Piscart, A. Manach, G. H. Copp, and P. Marmonier, "Distribution and microhabitats of native and non-native gammarids (Amphipoda, Crustacea) in Brittany, with particular reference to the endangered endemic sub-species *Gammarus* duebeni celticus," *Journal of Biogeography*, vol. 34, no. 3, pp. 524–533, 2007.

[11] M. Messiaen, K. Lock, W. Gabriels et al., "Alien macrocrustaceans in freshwater ecosystems in the eastern part of Flanders (Belgium)," *Belgian Journal of Zoology*, vol. 140, no. 1, pp. 30–39, 2010.

[12] M. Grabowski, K. Jazdzewski, and A. Konopacka, "Alien crustacea in Polish waters—amphipoda," *Aquatic Invasions*, vol. 2, no. 1, pp. 25–38, 2007.

[13] S. Pinkster, "The Echinogammarus berilloni-group, a number of predominantly iberian amphipod species (Crustacea)," *Bijdragen tot de Dierkunde*, vol. 43, pp. 1–38, 1973.

[14] S. Pinkster, "A revision of the genus *Echinogammarus* Stebbing, 1899 with some notes on related genera (Crustacea, Amphipoda)," *Memoire del Museo Civico di Storia Naturale 2*, vol. 10, pp. 9–185, 1993.

[15] R. Kinzelbach, "Neozoans in European waters - Exemplifying the worldwide process of invasion and species mixing," *Experientia*, vol. 51, no. 5, pp. 526–538, 1995.

[16] G. Van der Velde, S. Rajagopal, B. Kelleher, I .B. Musko, and A. Bij de Vaate, "Ecological impact of crustacean invaders: general considerations and examples from the Rhine River," in *The Biodiversity Crisis and Crustacea. Proceedings of the 4th International Crustacean Congress, Amsterdam, 1998*, J. C. Von Vaupel Klein and F. R. Schram, Eds., pp. 3–33, Balkema, Rotterdam, The Netherlands, 2000.

[17] V. K. Sowinsky, "Rakoobraznyia Azovskago Moria," *Zapiski Kievskago Obshchestva Estestvoispytatelei*, vol. 13, pp. 289–405, 1894.

[18] U. Mürle, A. Becker, and P. Rey, "*Dikerogammarus villosus* (Amphipoda) new in Lake Constance," *Lauterbornia*, vol. 49, pp. 77–79, 2004.

[19] B. Lods-Crozet and O. Reymond, "Bathymetric expansion of an invasive gammarid (*Dikerogammarus villosus*, Crustacea, Amphipoda) in Lake Léman," *Journal of Limnology*, vol. 65, no. 2, pp. 141–144, 2006.

[20] S. Casellato, G. La Piana, L. Latella, and S. Ruffo, "*Dikerogammarus villosus* (Sowinsky, 1894) (Crustacea, Amphipoda, Gammaridae) for the first time in Italy," *Italian Journal of Zoology*, vol. 73, no. 1, pp. 97–104, 2006.

[21] M. Gervais, "Note sur deux especes de Crevettes qui vivent aux environs de Paris," *Annales des Sciences Naturelles*, vol. 2, no. 4, pp. 127–128, 1835.

[22] G. O. Sars, "Beretning om en I Sommeren 1862 foretagen zoologisk Reise I Christianias og Trondhjems Stifter," *Nyt Magazin for Naturvidenskaberne*, vol. 12, pp. 193–340, 1863.

[23] E. L. Bousfield, "Freshwater amphipod crustaceans of glaciated North America," *Canadian Field Naturalist*, vol. 72, pp. 55–113, 1958.

[24] C. Piscart, J. M. Roussel, J. T. A. Dick, G. Grosbois, and P. Marmonier, "Effects of coexistence on habitat use and trophic ecology of interacting native and invasive amphipods," *Freshwater Biology*, vol. 56, no. 2, pp. 325–334, 2011.

[25] S. Kolding, "Habitat selection and life cycle characteristics of five species of the amphipod genus *Gammarus* in the Baltic," *Oikos*, vol. 37, pp. 173–178, 1981.

[26] A. Skadsheim, "Coexistence and reproductive adaptations of amphipods: the role of environmental heterogeneity," *Oikos*, vol. 43, no. 1, pp. 94–103, 1984.

[27] J. T. A. Dick, "Post-invasion amphipod communities of Lough Neagh, Northern Ireland: influences of habitat selection and mutual predation," *Journal of Animal Ecology*, vol. 65, no. 6, pp. 756–767, 1996.

[28] J. T. A. Dick and D. Platvoet, "Intraguild predation and species exclusions in amphipods: the interaction of behaviour, physiology and environment," *Freshwater Biology*, vol. 36, no. 2, pp. 375–383, 1996.

[29] C. MacNeil, J. T. A. Dick, R. W. Elwood, and W. I. Montgomery, "Coexistence among native and introduced freshwater amphipods (Crustacea); habitat utilization patterns in littoral habitats," *Archiv fur Hydrobiologie*, vol. 151, no. 4, pp. 591–607, 2001.

[30] A. Kley and G. Maier, "An example of niche partitioning between *Dikerogammarus villosus* and other invasive and native gammarids: a field study," *Journal of Limnology*, vol. 64, no. 1, pp. 85–88, 2005.

Mouthpart Morphology of Three Sympatric Native and Nonnative Gammaridean Species: Gammarus pulex, G. fossarum, and Echinogammarus berilloni (Crustacea: Amphipoda)

171

[31] C. MacNeil and D. Platvoet, "The predatory impact of the freshwater invader *Dikerogammarus villosus* on native *Gammarus pulex* (Crustacea: Amphipoda); influences of differential microdistribution and food resources," *Journal of Zoology*, vol. 267, no. 1, pp. 31–38, 2005.

[32] C. Fiser, R. Keber, V. Kerezi et al., "Coexistence of species of two amphipod genera: *Niphargus timavi* (Niphargidae) and *Gammarus fossarum* (Gammaridae)," *Journal of Natural History*, vol. 41, pp. 2641–2651, 2008.

[33] J. Hesselschwerdt, J. Meeker, and K. M. Wantzen, "Gammarids in Lake Constance: habitat segregation between the invasive *Dikerogammarus villosus* and the indigenous *Gammarus roeselii*," *Fundamental and Applied Limnology*, vol. 173, no. 3, pp. 177–186, 2008.

[34] A. Kley, W. Kinzler, Y. Schank, G. Mayer, D. Waloszek, and G. Maier, "Influence of substrate preference and complexity on co-existence of two non-native gammarideans (Crustacea: Amphipoda)," *Aquatic Ecology*, vol. 43, no. 4, pp. 1047–1059, 2009.

[35] C. L. Koch, *Deutschlands Crustaceen, Myriapoden und Arachniden. Ein Beitrag zur Deutschen Fauna*, vol. 5, Herrich-Schäfer, Regensburg, Germany, 1st edition, 1836.

[36] C. Linnaeus, *Systema Naturae*, vol. 1, Salvius, Stockholm, Sweden, 10 edition, 1758.

[37] J. D. Catta, "Note sur le *Gammarus berilloni* (n. sp.)," *Bulletin del la Societé de Borda*, vol. 1, pp. 68–73, 1878.

[38] O. Schmit and G. Josens, "Preliminary study of the scars borne by Gammaridae (Amphipoda, Crustacea)," *Belgian Journal of Zoology*, vol. 134, no. 2, pp. 75–78, 2004.

[39] A. Meyer, N. Kaschek, and E. I. Meyer, "The effect of low flow and stream drying on the distribution and relative abundance of the alien amphipod, *Echinogammarus berilloni* (Catta, 1878) in a karstic stream system (Westphalia, Germany)," *Crustaceana*, vol. 77, no. 8, pp. 909–922, 2004.

[40] M. Chovet and J. Y. Lecureuil, "Distribution of epigean Gammaridae (Crustacea, Amphipoda) in the Loire River and in the streams of the Region Centre (France)," *Annales de Limnologie*, vol. 30, no. 1, pp. 11–23, 1994.

[41] M. P. D. Meijering, "Lack of oxygen and low pH as limiting factors for *Gammarus* in Hessian brooks and rivers," *Hydrobiologia*, vol. 223, pp. 159–169, 1991.

[42] W. Teichmann, "Lebensabläufe und Zeitpläne von Gammariden unter ökologischen Bedingungen," *Archiv für Hydrobiologie*, vol. 64, no. 2, supplement, pp. 240–306, 1982.

[43] J.-W. Haeckel, M. P. D Meijering, and H. Rusetzki, "Gammarus fossarum Koch als Fallaubzersetzer in Waldbächen," *Freshwater Biology*, vol. 3, pp. 241–249, 1973.

[44] F. Foeckler, "Das Vorkommen von Gammariden im Donauraum zwischen Geisling und Straubing," *Archiv für Hydrobiologie*, vol. 84, no. 2–4, supplement, pp. 169–180, 1992.

[45] H. Nesemann, M. Pöckl, and K. J. Wittmann, "Distribution of epigean Malacostraca in the middle and upper Danube (Hungary, Austria, Germany)," *Miscellanea Zoologica Hungarica*, vol. 10, pp. 49–68, 1995.

[46] G. Mayer, G. Maier, A. Maas, and D. Waloszek, "Mouthpart morphology of *Gammarus roeselii* compared to a successful invader, *Dikerogammarus villosus* (Amphipoda)," *Journal of Crustacean Biology*, vol. 29, no. 2, pp. 161–174, 2009.

[47] T. O. Eggers and A. Martens, "Bestimmungsschlüssel der Süßwasser-Amphipoda (Crustacea) Deutschlands," *Lauterbornia*, vol. 42, pp. 1–68, 2001.

[48] T.O. Eggers and A. Martens, "Ergänzungen und Korrekturen zum Bestimmungsschlüssel der Süßwasser-Amphipoda

(Crustacea) Deutschlands," *Lauterbornia*, vol. 50, pp. 1–13, 2004.

[49] S. Pinkster, "Redescription of *Gammarus pulex* (Linnaeus, 1758) based on neotype material (Amphipoda)," *Crustaceana*, vol. 18, pp. 177–186, 1969.

[50] A. Goedmakers, "*Gammarus fossarum* Koch, 1835: redescription based on neotype material and notes to its local variation (Crustacea, Amphipoda)," *Bijdragen tot de Dierkunde*, vol. 42, no. 2, pp. 124–138, 1972.

[51] G. S. Karaman and S. Pinkster, "Freshwater *Gammarus* species from Europe, North Africa, and adjacent regions of Asia (Crustacea: Amphipoda) Part I. *Gammarus pulex*-group and related species," *Bijdragen tot de Dierkunde*, vol. 47, no. 1, pp. 1–97, 1977.

[52] E. Dahl, "Deep-sea carrion feeding amphipods: evolutionary patterns in niche adaptation," *Oikos*, vol. 33, pp. 167–175, 1979.

[53] B. Sainte-Marie, "Morphological adaptations for carrion feeding in four species of littoral or circalittoral lysianassid amphipods," *Canadian Journal of Zoology*, vol. 62, no. 9, pp. 1668–1674, 1984.

[54] M. A. McGrouther, "Comparison of feeding mechanisms in two intertidal gammarideans, *Hyale rupicola* (Haswell) and *Paracalliope australis* (Haswell) (Crustacea: Amphipoda)," *Australian Journal of Marine and Freshwater Research*, vol. 34, no. 5, pp. 717–726, 1983.

[55] P. G. Moore and P. S. Rainbow, "Feeding biology of the mesopelagic gammaridean amphipod *Parandania boecki* (Stebbing, 1888) (Crustacea: Amphipoda: Stegocephalidae) from the Atlantic Ocean," *Ophelia*, vol. 30, no. 1, pp. 1–19, 1989.

[56] C. O. Coleman, "On the nutrition of two Antarctic acanthonotozomatidae (Crustacea: Amphipoda) - Gut contents and functional morphology of mouthparts," *Polar Biology*, vol. 9, no. 5, pp. 287–294, 1989.

[57] C. O. Coleman, "*Gnathiphimedia mandibularis* K.H. Barnard 1930, an Antarctic amphipod (Acanthonotozomatidae, Crustacea) feeding on Bryozoa," *Antarctic Science*, vol. 1, pp. 343–344, 1989.

[58] C. O. Coleman, "*Bathypanoploea schellenbergi* Holman & Watling, 1983, an Antarctic amphipod (Cruatacea) feeding on Holothuroidea," *Ophelia*, vol. 31, no. 3, pp. 197–205, 1990.

[59] D. H. Steele and V. J. Steele, "Biting mechanism of the amphipod Anonyx (Crustacea: Amphipoda: Lysianassoidea)," *Journal of Natural History*, vol. 27, pp. 851–860, 1993.

[60] C. E. Arndt, J. Berge, and A. Brandt, "Mouthpart-atlas of Arctic sympagic amphipods - Trophic niche separation based on mouthpart morphology and feeding ecology," *Journal of Crustacean Biology*, vol. 25, no. 3, pp. 401–412, 2005.

[61] M. Kostalos and R. L. Seymour, "Role of microbial enriched detritus in the nutrition of *Gammarus minus* (Amphipoda)," *Oikos*, vol. 27, pp. 512–516, 1976.

[62] K.W. Cummins and M.J. Klug, "Feeding ecology of stream invertebrates," *Annual Review of Ecology and Systematics*, vol. 10, pp. 147–172, 1979.

[63] J. T. A. Dick, I. Montgomery, and R. W. Elwood, "Replacement of the indigenous amphipod *Gammarus duebeni celticus* by the introduced G. pulex: differential cannibalism and mutual predation," *Journal of Animal Ecology*, vol. 62, no. 1, pp. 79–88, 1993.

[64] X. Gayte and D. Fontvieille, "Autochthonous vs. allochthonous organic matter ingested by a macroinvertebrate in headwater streams: *Gammarus* sp. as a biological probe," *Archiv fur Hydrobiologie*, vol. 140, no. 1, pp. 23–36, 1997.

[65] C. MacNeil, J. T. A. Dick, and R. W. Elwood, "The trophic ecology of freshwater *Gammarus* spp. (Crustacea: Amphipoda): Problems and perspectives concerning the functional feeding group concept," *Biological Reviews of the Cambridge Philosophical Society*, vol. 72, no. 3, pp. 349–364, 1997.

[66] J. T. A. Dick and D. Platvoet, "Invading predatory crustacean *Dikerogammarus villosus* eliminates both native and exotic species," *Proceedings of the Royal Society B*, vol. 267, no. 1447, pp. 977–983, 2000.

[67] D. W. Kelly, J. T. A. Dick, and W. I. Montgomery, "The functional role of *Gammarus* (Crustacea, Amphipoda): shredders, predators, or both?" *Hydrobiologia*, vol. 485, pp. 199–203, 2002.

[68] H. Krisp and G. Maier, "Consumption of macroinvertebrates by invasive and native gammarids: a comparison," *Journal of Limnology*, vol. 64, no. 1, pp. 55–59, 2005.

[69] D. Platvoet, J. T. A. Dick, N. Konijnendijk, and G. Van Der Velde, "Feeding on micro-algae in the invasive Ponto-Caspian amphipod *Dikerogammarus villosus* (Sowinsky, 1894)," *Aquatic Ecology*, vol. 40, no. 2, pp. 237–245, 2006.

[70] C. Maazouzi, G. Masson, M. S. Izquierdo, and J. C. Pihan, "Fatty acid composition of the amphipod *Dikerogammarus villosus*: feeding strategies and trophic links," *Comparative Biochemistry and Physiology A*, vol. 147, no. 4, pp. 868–875, 2007.

[71] W. Kinzler, A. Kley, G. Mayer, D. Waloszek, and G. Maier, "Mutual predation between and cannibalism within several freshwater gammarids: *Dikerogammarus villosus* versus one native and three invasives," *Aquatic Ecology*, vol. 43, no. 2, pp. 457–464, 2009.

[72] V. Felten, G. Tixier, F. Guérold, V. De Crespin De Billy, and O. Dangles, "Quantification of diet variability in a stream amphipod: implications for ecosystem functioning," *Fundamental and Applied Limnology*, vol. 170, no. 4, pp. 303–313, 2008.

[73] D. Platvoet, G. Van der Velde, J. T. A. Dick, and S. Li, "Flexible omnivory in *Dikerogammarus villosus* (Sowinsky, 1894) (Amphipoda)—Amphipod Pilot Species Project (AMPIS) report 5," *Crustaceana*, vol. 82, no. 6, pp. 703–720, 2009.

[74] G. Mayer, G. Maier, A. Maas, and D. Waloszek, "Mouthparts of the Ponto-Caspian invader *Dikerogammarus villosus* (Amphipoda: Pontogammaridae)," *Journal of Crustacean Biology*, vol. 28, no. 1, pp. 1–15, 2008.

[75] I. V. Mekhanikova, "Morphology of mandible and lateralia in six endemic amphipods (Amphipoda, Gammaridea) from Lake Baikal, in relation to feeding," *Crustaceana*, vol. 83, no. 7, pp. 865–887, 2010.

[76] S. O. Sars, *An Account of the Crustacea of Norway, I Amphipoda, Part I Description*, Cammermeyer's Forlag, Christiania, Norway, 1895.

[77] V. P. Agrawal, "Feeding appendages and the digestive system of *Gammarus pulex*," *Acta Zoologica*, vol. 46, no. 1-2, pp. 67–81, 1965.

[78] M. J. Costello, "Biogeography of alien amphipods occurring in Ireland, and interactions with native species," *Crustaceana*, vol. 65, no. 3, pp. 287–299, 1993.

[79] S. Pinkster, "On members of the *Gammarus* pulex-group (Crustacea - Amphipoda) from Western Europe," *Bijdragen tot de Dierkunde*, vol. 42, no. 2, pp. 164–191, 1972.

[80] J. S. Welton, "Life-history and reproduction of the amphipod *Gammarus pulex* in a Dorset chalk stream," *Freshwater Biology*, vol. 9, pp. 263–275, 1979.

[81] A. Schellenberg, "Der *Gammarus* des deutschen Süßwassers," *Zoologischer Anzeiger*, vol. 108, no. 9-10, pp. 209–217, 1934.

[82] W. Janetzky, "Distribution of the genus *Gammarus* (Amphipoda: Gammaridae) in the River Hunte and its tributaries (Lower Saxony, northern Germany)," *Hydrobiologia*, vol. 294, no. 1, pp. 23–34, 1994.

[83] J. Dusaugey, "Les gammares du Dauphine et leur repartition," *Travaux du Laboratoire d'Hydrobiologie et de Pisciculture de Grenoble*, pp. 9–18, 1955.

[84] A. Goedmakers, "Les Gammaridea (Crustaces, Amphipodes) du Massif Central," *Bulletin Zoologisch Museum Universiteit van Amsterdam*, vol. 3, no. 23, pp. 211–219, 1974.

[85] D. W. Kelly, J. T. A. Dick, W. I. Montgomery, and C. MacNeil, "Differences in composition of macroinvertebrate communities with invasive and native *Gammarus* spp. (Crustacea: Amphipoda)," *Freshwater Biology*, vol. 48, no. 2, pp. 306–315, 2003.

[86] J. Adams, J. Gee, P. Greenwood, S. McKelvey, and R. Perry, "Factors affecting the microdistribution of *Gammarus pulex* (Amphipoda): an experimental study," *Freshwater Biology*, vol. 17, no. 2, pp. 307–316, 1987.

[87] M. A. S. Graca, L. Maltby, and P. Calow, "Comparative ecology of *Gammarus pulex* (L.) and *Asellus aquaticus* (L.) I: population dynamics and microdistribution," *Hydrobiologia*, vol. 281, no. 3, pp. 155–162, 1994.

[88] J. Dahl and L. Greenberg, "Effects of habitat structure on habitat use by *Gammarus pulex* in artificial streams," *Freshwater Biology*, vol. 36, no. 3, pp. 487–495, 1996.

[89] V. Felten, S. Dolédec, and B. Statzner, "Coexistence of an invasive and a native gammarid across an experimental flow gradient: flow-refuge use, -mortality, and leaf-litter decay," *Fundamental and Applied Limnology*, vol. 172, no. 1, pp. 37–48, 2008.

[90] M. A. S. Graça, L. Maltby, and P. Calow, "Importance of fungi in the diet of *Gammarus* pulex and Asellus aquaticus I: feeding strategies," *Oecologia*, vol. 93, no. 1, pp. 139–144, 1993.

[91] M. A. S. Graca, L. Maltby, and P. Calow, "Importance of fungi in the diet of *Gammarus pulex* and *Asellus aquaticus*. II.Effects on growth, reproduction and physiology," *Oecologia*, vol. 96, no. 3, pp. 304–309, 1993.

[92] J. S. Welton and R. T. Clarke, "Laboratory studies on the reproduction and growth of the amphipod *Gammarus pulex* (L.)," *Animal Ecology*, vol. 49, pp. 581–592, 1980.

[93] M. Zimmer and S. Bartholmé, "Bacterial endosymbionts in *Asellus aquaticus* (Isopoda) and *Gammarus pulex* (Amphipoda) and their contribution to digestion," *Limnology and Oceanography*, vol. 48, no. 6, pp. 2208–2213, 2003.

[94] L. Bollache, J. T. A. Dick, K. D. Farnsworth, and W. I. Montgomery, "Comparison of the functional responses of invasive and native amphipods," *Biology Letters*, vol. 4, no. 2, pp. 166–169, 2008.

[95] R. Margalef, "Sobre el regimen alimentico de los animales en agua dulce. 2a communicacion," *Revista Espaniola de Fisiologia*, vol. 4, pp. 207–213, 1948.

[96] W. Wolterstorff, "Der Bachflohkrebs, *Gammarus pulex* L. im Aquarium," *Blätter für Aquarien und Terrarienkunde*, vol. 28, pp. 85–87, 1917.

[97] J. T. A. Dick, W. I. Montgomery, and R. W. Elwood, "Intraguild predation may explain an amphipod replacement: evidence from laboratory populations," *Journal of Zoology*, vol. 249, no. 4, pp. 463–468, 1999.

Mouthpart Morphology of Three Sympatric Native and Nonnative Gammaridean Species: Gammarus pulex, G. fossarum, and Echinogammarus berilloni (Crustacea: Amphipoda)

173

[98] R. Kinzelbach and W. Claus, "Die Verbreitung von *Gammarus fossarum* Koch, 1835, *G. pulex* (Linnaeus, 1758) und *G. roeseli* Gervais, 1835, in den linken Nebenflüssen des Rheins zwischen Wieslauter und Nahe," *Crustaceana*, vol. 4, supplement, pp. 164–172, 1977.

[99] M. Grabowski, K. Bacela, A. Konopacka, and K. Jazdzewski, "Salinity-related distribution of alien amphipods in rivers provides refugia for native species," *Biological Invasions*, vol. 11, no. 9, pp. 2107–2117, 2009.

[100] E. Schwedhelm, "Thermopräferenz von *Gammarus fossarum* Koch, 1835 und *Gammarus roeselii* Gervais, 1835 (Crustacea, Amphipoda) in Abhängigkeit von der Jahreszeit," *Zoologischer Anzeiger*, vol. 208, no. 5-6, pp. 367–374, 1982.

[101] M. Pockl and U. H. Humpesch, "Intra- and inter-specific variations in egg survival and brood development time for Austrian populations of *Gammarus fossarum* and *G. roeseli* (Crustacea: Amphipoda)," *Freshwater Biology*, vol. 23, no. 3, pp. 441–455, 1990.

[102] M. Pockl, "Reproductive potential and lifetime potential fecundity of the freshwater amphipods *Gammarus fossarum* and *G. roeseli* in Austrian streams and rivers," *Freshwater Biology*, vol. 30, no. 1, pp. 73–91, 1993.

[103] S. Wijnhoven, M. C. Van Riel, and G. Van der Velde, "Exotic and indigenous freshwater gammarid species: physiological tolerance to water temperature in relation to ionic content of the water," *Aquatic Ecology*, vol. 37, no. 2, pp. 151–158, 2003.

[104] J. Issartel, D. Renault, Y. Voituron, A. Bouchereau, P. Vernon, and F. Hervant, "Metabolic responses to cold in subterranean crustaceans," *Journal of Experimental Biology*, vol. 208, no. 15, pp. 2923–2929, 2005.

[105] J. Helesic, F. Kubicek, and S. Zahradkova, "The impact of regulated flow and altered temperature regime on river bed macroinvertebrates," in *Advances in River Bottom Ecology*, G. Bretschko and J. Helesic, Eds., pp. 225–243, Backhuys, Leiden, The Netherlands, 1998.

[106] M. Pockl, "Laboratory studies on growth, feeding, moulting and mortality in the freshwater amphipods *Gammarus fossarum* and *G. roeseli*," *Archiv fur Hydrobiologie*, vol. 134, no. 2, pp. 223–253, 1995.

[107] H.-G. Pieper, "Ökophysiologische und produktionsbiologische Untersuchungen an Jugendstadien von *Gammarus fossarum* Koch 1835," *Archiv für Hydrobiologie*, vol. 54, no. 3, supplelment, pp. 257–327, 1978.

[108] O. Dangles, "Aggregation of shredder invertebrates associated with benthic detrital pools in seven headwater forest streams," *Verhandlungen der Internationalen Vereinigung für Limnologie*, vol. 28, pp. 1–4, 2002.

[109] S. D. Tiegs, F. D. Peter, C. T. Robinson, U. Uehlinger, and M. O. Gessner, "Leaf decomposition and invertebrate colonization responses to manipulated litter quantity in streams," *Journal of the North American Benthological Society*, vol. 27, no. 2, pp. 321–331, 2008.

[110] T. R. R. Stebbing, "Amphipoda I. Gammaridea," in *Das Tierreich. Eine Zusammenstellung und Kennzeichnung der Rezenten Tierformen*, F. E. Schulze, Ed., vol. 21, Friedländer, Berlin, Germany, 1906.

20

Visual and Chemical Prey Cues as Complementary Predator Attractants in a Tropical Stream Fish Assemblage

Chris K. Elvidge and Grant E. Brown

Department of Biology, Concordia University, 7141 Sherbrooke St. West, Montreal, QC, Canada H4B 1R6

Correspondence should be addressed to Chris K. Elvidge, chris.k.elvidge@gmail.com

Academic Editor: Marie Herberstein

To date, little attention has been devoted to possible complementary effects of multiple forms of public information similar information on the foraging behaviour of predators. In order to examine how predators may incorporate multiple information sources, we conducted a series of predator attraction trials in the Lower Aripo River, Trinidad. Four combinations of visual (present or absent) and chemical cues (present or absent) from each of two prey species were presented. The occurrences of three locally abundant predatory species present within a 1 m radius of cue introduction sites were recorded. The relative attractiveness of cue type to each predator was directly related to their primary foraging modes, with visual ambush predators demonstrating an attraction to visual cues, benthivores to chemical cues, and active social foragers demonstrating complementary responses to paired cues. Predator species-pair counts were greatest in response to cues from the more abundant prey species, indicating that individuals may adopt riskier foraging strategies when presented with more familiar prey cues. These differences in predator attraction patterns demonstrate complementary effects of multiple sensory cues on the short-term habitat use and foraging behaviour of predators under fully natural conditions.

1. Introduction

The behavioural strategies adopted by participants in predator-prey interactions are often mediated by publicly available cues [1] conveying information with some degree of immediate contextual relevance to the receiver. Public, or non-species-specific, cues may convey qualitatively different information to and elicit quantitatively different behavioural responses from different receivers [2]. The relative importance of different types of public cues in predator-prey interactions may be mediated by interactions between receiver taxon and environmental constraints; for example, visual cues are typically limited by photoperiod [3]. In aquatic environments, visual and chemical cues have been identified as the primary sources of information eliciting short-term behavioural processes for both vertebrate (e.g., fishes, [4]) and invertebrate (e.g., crustaceans, [5, 6]) species. Although acoustic cues have been demonstrated to elicit behavioural responses in freshwater fish receivers under laboratory conditions [7], the reliability of acoustic information may be limited under conditions of relatively high background noise, as in lotic systems.

Many groups of freshwater fishes produce chemical cues in the epidermis which are released into the water following mechanical damage, as would occur during a predation event [8]. Upon detection by conspecific receivers, these cues have been shown to elicit a suite of antipredator or alarm responses [9] in centrarchid [10], salmonid [11], cyprinid [12], cyprinodontiform [2], esocid [13], and poeciliid [14] species. Due to their manner of release, these chemical cues cannot be manipulated by a predator and likely serve as reliable indicators of increased risk to receivers subject to similar predation pressures [15], which is not always the case with potentially misleading visual cues [3]. Damage-released chemical cues have been shown to elicit different responses from conspecific receivers differing in ontogenetic stage from the cue sender [16], with similarly sized receivers demonstrating alarm or antipredator responses and larger receivers demonstrating behaviours consistent with foraging responses under laboratory conditions. Similar effects have been observed in heterospecific receivers belonging to the same prey guild and subject to similar predation risks as chemical cue senders [17, 18]. Conversely, heterospecific receivers of larger size than the sender have demonstrated foraging responses following exposure to damage-released chemical cues under both laboratory [18] and field [19] conditions.

Due to the potentially ultimate costs incurred by failing to respond to ambient cues indicating elevated predation risk, the responses of prey to public cues have thus far received considerably more attention from researchers than have responses by predators [20]. In freshwater fishes, laboratory experiments have documented attraction responses to chemical cues from heterospecific prey [21], while predators under natural conditions have demonstrated preferences for areas labelled with damage-released chemical cues over longer time scales (hours, [19]). Recently, Lonnstedt et al. [22] demonstrated an attraction response in a predatory coral reef fish, the dottyback *Pseudochromis fuscus* to heterospecific damage-released chemical cues under fully natural conditions over short (minutes) timescales. In addition, under both natural and laboratory conditions, *P. fuscus* demonstrated a strong preference for chemical cues extracted from heterospecific donors belonging to the ideal prey size class [22] for gape-limited predators [17]. Similarly, Elvidge et al. [2] demonstrated significant positive linear relationships between foraging behaviours and receiver size in an opportunistic predator, Hart's rivulus *Rivulus hartii*, in response to chemical cues from Trinidadian guppies *Poecilia reticulata*. Together, these results indicate that damage-released chemical cues provide predators with information about the availability and quality of potential prey.

In order to examine the effects of different combinations of visual and chemical cues indicating the availability of prey on the behaviour of predators in fresh water, the present study focused on short-term changes in local abundances of three predatory species differing in foraging modes to the cues of two cooccurring prey species. In general, we predict that multiple complementary cues indicating the presence of familiar prey species will result in greater local abundances of predators, with the relative contribution of each type of cue (visual or chemical) mediated by the typical foraging mode of the predator.

2. Materials and Methods

2.1. Study Species and Area. Predator attraction trials were conducted at $N = 16$ sites in a series of eight pools (two sites per pool) along a 1 km stretch of the Lower Aripo River in the Caroni drainage, Northern Range Mountains, Trinidad and Tobago, W. I. (10°39′ N, 61°13′ W) 04–12 May 2009. These pools have been described in an earlier study involving free-swimming Trinidadian guppies *P. reticulata* [23]. The Lower Aripo is a species-rich, high-predation environment [24] with abundant *P. reticulata* and incidental *R. hartii* populations. These two prey species are nearly ubiquitous in streams in northern Trinidad but do not always cooccur [25]. Although *R. hartii* may grow to as much as three times the

length of *P. reticulata* (*R. hartii* maximum standard length, L_S, 100 mm, *P. reticulata* common L_S 28 mm; [26]) and *P. reticulata* can account for up to 10% of the diet of large *R. hartii*, in the presence of piscivores similarly sized *P. reticulata* and *R. hartii* are likely subject to similar predation pressures [17].

Of the three predatory species examined, the pike cichlid *Crenicichla alta* is a solitary, visually foraging ambush predator and obligate piscivore which is considered the main predator of *P. reticulata* in Trinidad whenever they co-occur [25]. The blue acara cichlid *Aequidens pulcher* is a solitary forager which typically feeds on invertebrates and benthos, displaying only opportunistic piscivory [27]. The two-spot sardine *Astyanax bimaculatus*, by contrast, is a highly social and active forager [23, 26], whose predominantly algae- and insect-based diet undergoes an ontogenetic switch to include opportunistic piscivory when individuals exceed ~50 mm total length [28]. These three species likely account for the majority of predation pressure on *P. reticulata* and any incidental *R. hartii* within the study area [29]. Based on these differences in social behaviour and foraging mode, we predict that *A. bimaculatus* incorporates information received through both visual and chemical cues into their foraging decisions, while *A. pulcher* responds more strongly to chemical than to visual cues and *C. alta* responds primarily to visual cues.

2.2. Prey Cues. Damage-released chemical cues were extracted from female *P. reticulata* (L_S 27.6 ± 2.7 mm (mean ± SD), $N = 18$) and *R. hartii* (L_S 42 ± 7.9 mm (mean ± SD), $N = 8$) donors collected from the Naranja River tributary (10°41′ N; 61°14′ W) approximately 6 km upstream from the observation sites in the Lower Aripo River using a beach seine net (length 2.5 m, height 1 m, mesh size 3 mm). Donors were collected from the Naranja River because intensive sampling via seine net did not find any *R. hartii* present at the study sites in the Lower Aripo during the course of the present experiment. Chemical cues from *P. reticulata* donors from the Naranja River have previously been demonstrated to elicit qualitatively similar responses to those of Lower Aripo donors [23] in conspecific receivers from either population.

Chemical cue donors were euthanized via cervical dislocation, measured (L_S), immediately decapitated behind the opercula, and had their tails removed at the caudal peduncle. Visceral tissues were manually extruded, and the remaining carcasses were mechanically homogenized in dechlorinated tap water, diluted to a final concentration of 0.1 cm^2 skin mL^{-1}, and filtered through polyester floss. This concentration of skin extract has previously been shown to elicit both antipredator and foraging behavioural responses in tropical stream fish under laboratory [30] and field [23] conditions. The chemical cues from each prey species, as well as a stream water control treatment, were packaged in 60 mL aliquots and frozen at −20°C until use.

Several female *P. reticulata* (L_S 27.9 ± 2.9 mm (mean ± SD), $N = 8$) and juvenile *R. hartii* (L_S 25.4 ± 4.8 mm (mean ± SD), $N = 8$) were retained from the pools of wild-caught chemical cue donors to serve as visual prey

TABLE 1: Effects of different combinations of cues from two prey species indicating potential foraging opportunities on the local abundance of three predatory species in a nested ANOVA with observation site nested within prey species*.

Predator	Prey cue	Treatment effects			Nested effects			Variance components		
		F	df	P	Nested factor	F	P	Prey cue	Prey species	Site
C. alta	Visual	19.41	2,92	<0.0001	Prey species	1.02	0.3145	73.1%	5.2%	21.7%
	Chemical	0.09	2,92	0.91	Site	0.05	0.8234			
A. pulcher	Visual	2.12	2,92	0.126	Prey species	0.92	0.3394	65.4%	17.4%	17.2%
	Chemical	6.36	2,92	0.0026	Site	23.58	<0.0001			
A. bimaculatus	Visual	6.08	2,92	0.0033	Prey species	0.45	0.5036	52.7%	47.3%	0%
	Chemical	19.21	2,92	<0.0001	Site	6.71	0.0108			

*The interactions between prey cue types were nonsignificant in all tests so the analyses were limited to main and nested effects only.

cues. Subjects were transported to the observation site and placed singly into clear plastic bottles (250 mL) that had been perforated to allow water exchange. The bottles were attached to 1 m lengths of wooden dowelling (1 cm diameter) by transparent fishing wire and held stationary in the water column approximately 5 cm off the substrate. The bottles were also presented empty to serve as a control treatment to the visual prey cues and provide estimates of ambient predator abundance in the presence of observers. Each site (N = 16) was presented with the four combinations of cues for each prey species (N = 8 trials per site) for a total of N = 128 observations.

2.3. Experimental Protocol and Analysis. Chemical stimuli consisting of 60 mL of either *P. reticulata* or *R. hartii* chemical cues (CC) or a stream water control (SW) were delivered by syringe through 2 m lengths of airline tubing anchored by a rock (5 cm diameter) placed at the site of an observation. The bottles containing the visual stimuli or visual control were introduced into the water column directly above the stimulus injection sites. The apparatus was left in place for 1-minute prior to an observation to allow nearby fish to acclimate to its presence. Subsequent observations were conducted in an upstream direction to minimize the likelihood of attracting additional predators from the downstream dispersion of the chemical prey cues.

Following the 1 minute acclimation period, a stopwatch was activated to begin a 5-minute observation, throughout which the chemical stimuli or chemical controls were delivered through the airline tubing at a rate of 10 mL min^{-1} and the numbers of individuals of the three predatory species within a 1 m radius of the cue presentation site were recorded every 15 seconds. These predator counts were then averaged by species over the 5-minute observation periods. A similar protocol has previously been used to examine the predator inspection behaviour of free-swimming *P. reticulata* under field conditions [23]. The mean counts for each predatory species were subsequently examined as univariate responses in two-way nested ANOVAs with the chemical and visual prey cues as main effects, and replicated observation site nested within prey species. All analyses were conducted as linear mixed-effects *lme* models using the *nlme* statistical package [31] for R (version 2.12.1; [32]). The models were then decomposed to determine the relative influence of

model components on the variance in response using the *varcomp* command in the *ape* library [33]. Secondarily, the dataset was split into two parts by prey species and analyzed as univariate two-way ANOVAs to enable direct comparisons of the attractiveness of visual and chemical prey cues to each predator species. Additionally, in order to examine the possibility that the presence of the top predator, *C. alta*, at an observation site may have inhibited the attraction of the other predatory species to the area, predator species-pair counts were square-root transformed and compared using Pearson's correlation analyses.

3. Results

3.1. Predator Attraction. In no predator versus prey species combination was there a significant interaction between chemical and visual prey cues on predator species counts ($P > 0.05$), so further analyses examined main effects of prey cue types only. Results of two-way nested ANOVAs on the attraction of each predatory species to the combinations of prey cues with observation site as a factor nested within prey species are presented in Table 1. Despite the likelihood of a high degree of spatial heterogeneity in the distribution of predatory species in the Lower Aripo, observation site as a nested factor accounted for ≤21.7% of the variability in mean predator counts (Table 1). The least spatial variability in predator counts was demonstrated by the highly active *A. bimaculatus*, with the less motile and solitary *A. pulcher* and *C. alta* demonstrating greater heterogeneity in their distributions throughout the study sites.

The response patterns of each predator to the chemical and visual cue combinations appear to be similar for both *P. reticulata* and *R. hartii* cues (Figure 1). Prey species accounts for 5–47% of the variance in mean predator counts within study areas with *C. alta* demonstrating the least difference in response between prey species and *A. bimaculatus* demonstrating the greatest difference (Table 1). Overall, there appears to be a nonstatistically significant preference for the *P. reticulata* cues as suggested by the greater mean counts of predators within an observation radius relative to the *R. hartii* cues (Figure 1). Prey cue treatments (main effects) accounted for 52.7% of the variability in mean counts of *A. bimaculatus* and 73.1% of the variability in *C. alta* counts. As with the effect of prey species, the portions of variance in *A.*

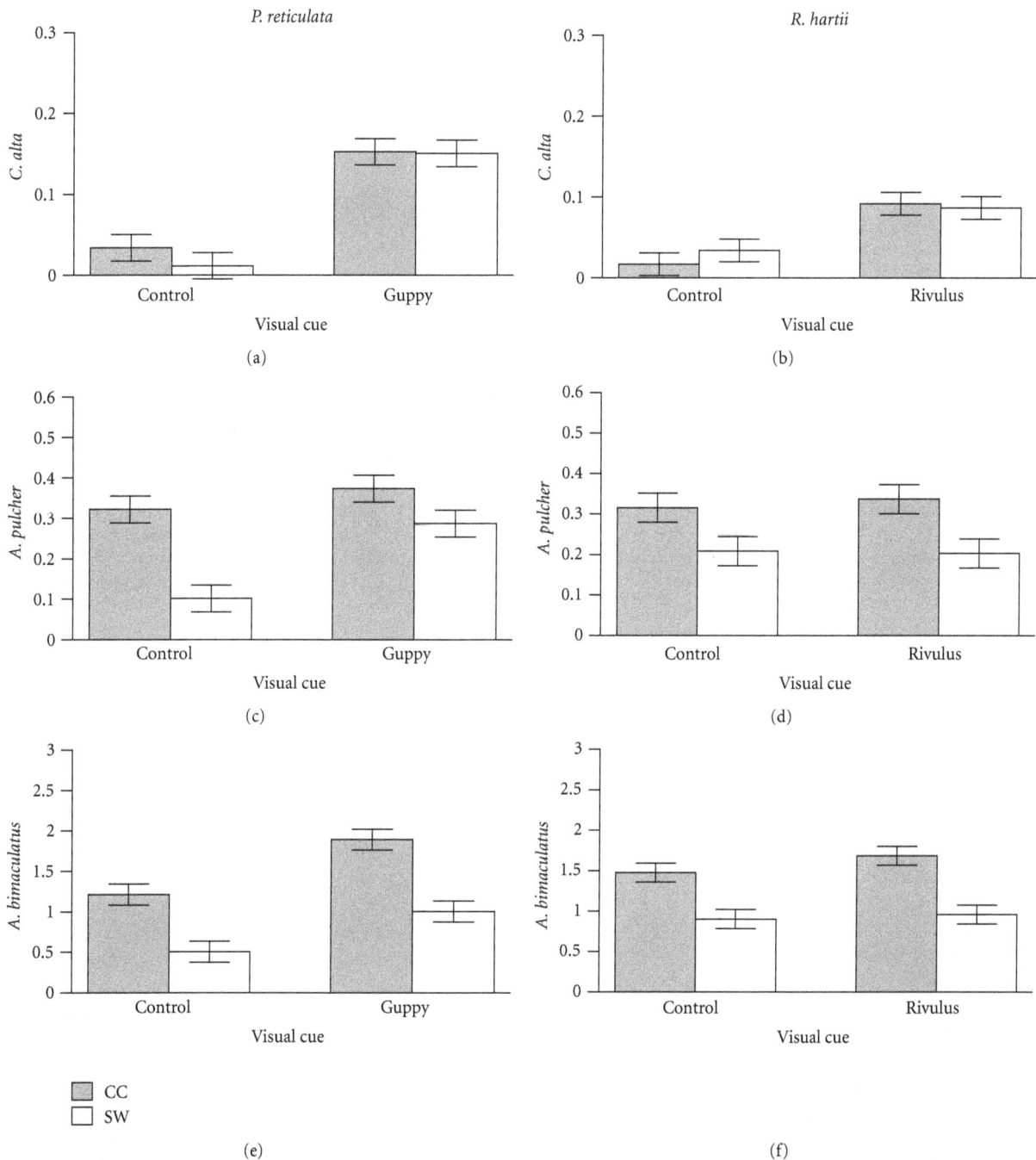

FIGURE 1: Mean ($\pm$SE) number of predators present within a 1.5 m radius of prey cue presentation sites in the Lower Aripo over 5 minutes. *Crenicichla alta* attraction to (a) guppy *Poecilia reticulata* and (b) rivulus *Rivulus hartii* cues. *Aequidens pulcher* attraction to (c) guppy and (d) rivulus cues. *Astyanax bimaculatus* attraction to (e) guppy and (f) rivulus cues. Visual cues (horizontal axes) were paired with either conspecific chemical cues (shaded bars) or a stream water control (open bars). $N = 16$ for each cue combination.

pulcher responses (65.4%) were intermediate relative to the other two predators.

As predicted by its primary foraging strategy, *Crenicichla alta*, a visual ambush predator, was observed in greater numbers when presented with visual cues indicating foraging opportunities of either *P. reticulata* ($F_{1,46} = 24.3$, $P < 0.0001$; Figure 1(a)) or *R. hartii* ($F_{1,46} = 8.78$, $P = 0.0057$; Figure 1(b)); although its response to *P. reticulata* visual cues appears to be greater than to those of *R. hartii*, the difference

is nonsignificant. *Crenicichla alta* did not demonstrate any attraction to chemical cues from either prey species ($P > 0.05$). This is in keeping with its foraging strategy of ambush hunting, as diffusive chemical cues may not reliably indicate the location of potential prey to visual predators.

Aequidens pulcher responded to both *P. reticulata* ($F_{1,46} = 6.52$, $P = 0.014$; Figure 1(c)) and *R. hartii* ($F_{1,46} = 6.25$, $P = 0.016$; Figure 1(d)) chemical cues but not to the visual cues of either prey species ($P > 0.05$). A significant response to

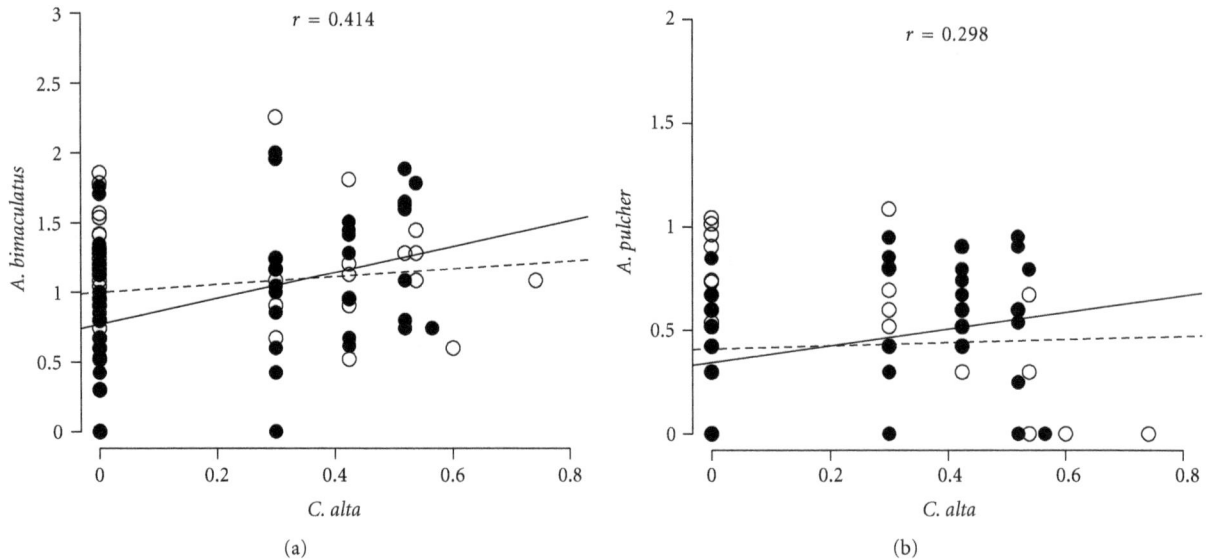

FIGURE 2: Pearson's correlation analyses of square-root transformed mean counts of *Astyanax bimaculatus* (a) and *Aequidens pulcher* (b) observed within 1 m radii of the prey cue introduction sites in the presence of the top predator, *Crenicichla alta*. *Poecilia reticulata* cues, closed points; *Rivulus hartii* cues, open points. Significant linear relationships ($P < 0.05$) are indicated by solid lines; nonsignificant relationships are indicated by dashed lines for illustrative purposes.

the chemical but not the visual cues of both prey species is in keeping with the prediction that the importance of chemical cues is greater than visual cues for a bottom-feeding detritivore.

Astyanax bimaculatus demonstrated a response to *P. reticulata* visual cues ($F_{1,46} = 11.7$, $P = 0.0013$; Figure 1(e)) but not to *R. hartii* visual cues ($P > 0.05$) and responded to the chemical cues of both prey species (*P. reticulata*: $F_{1,46} = 21.3$, $P < 0.0001$; *R. hartii*: $F_{1,46} = 16.9$, $P = 0.0002$; Figure 1(f)). Additionally, the response by *A. bimaculatus* to complementary chemical and visual cues of both prey species appears to be approximately additive (Figures 1(e) and 1(f)). This social foraging species was alone in this study in demonstrating complementary responses to both types of sensory cues indicating the presence of potential prey.

3.2. Predator Interactions. During trials involving *P. reticulata* cues, the mean counts of *C. alta* were positively correlated with the counts of both *A. bimaculatus* ($P = 0.0007$; Figure 2(a)) and *A. pulcher* ($P = 0.017$; Figure 2(b)). Conversely, there were no relationships between the species-pair counts in trials involving the cues of the less locally abundant *R. hartii* (Figure 2).

4. Discussion

These results demonstrate the importance of foraging mode in determining the relative influence of different types of prey cues on predator behaviour and habitat use. Introducing chemical and visual prey cues resulted in increased local abundances of predators, with the demonstrated responses of each species to the cue combinations varying with the primary foraging mode of a predator. Predators involved in this study appear to respond more strongly to cues of the

more locally abundant prey species, *P. reticulata*, particularly in the case of the top predator *C. alta*.

The importance of foraging mode in determining the relevance of prey cue type on predator behaviour in the present study may provide an explanation for earlier findings that damage-released chemical cues did not function as predator attractants. Specifically, Cashner [34] found that juvenile spotted bass *Micropterus punctulatus* did not demonstrate an attraction to the chemical cues of a suite of sympatric prey species. However, *M. punctulatus* may rely more on visual cues than chemical ones, as this species tends to be actively foraging, solitary predators. Although Chivers et al. [35] demonstrated attraction of the visually foraging, ambush predator northern pike *Esox lucius* to the chemical cues of fathead minnows *Pimephales promelas*; their experiment involved releasing chemical cues over a 30-minute period. Similar experiments have been conducted over even longer timescales (hours, [19]). The present study involves five-minute observations, which may be a more ecologically relevant timeframe due to the mechanism of release and intransigence of damage-released chemical cues. Interspecific trophic differences may also be insufficient to predict responses to heterospecific chemical cues. In addition to intraspecific differences in predator behaviour and prey size preference [36], recent findings [2] have established a relative size threshold between antipredator and foraging responses to damage-released chemical cues (predator $L_S >$ 150% prey L_S; [17]) as well as the ability of predators to determine prey quality and condition from information conveyed by these cues. Earlier studies (e.g., [34]) may have failed to include predators above such a relative size threshold or chemical cue donors of ideal size or condition and consequently were not able to elicit foraging responses in heterospecific receivers.

As the top fish predator in this section of the Aripo River [25], *C. alta* preys upon smaller individuals of both of the other predatory species involved in this study and is likely to compete for forage opportunities with larger heterospecific size classes. Likely as a result of these predatory and trophic interactions, *A. bimaculatus* and *A. pulcher* are rarely observed in close proximity to *C. alta* (personal observations), whose presence may indicate relatively high levels of risk. The presence of significant linear relationships between predator species-pair counts in response to *P. reticulata* cues and insignificant relationships in response to *R. hartii* cues is consistent with the notion that predators generally display greater attraction responses to the cues of more abundant or familiar prey species. This observation may imply that *A. bimaculatus* and *A. pulcher* adopt riskier foraging strategies and enter potentially more dangerous areas when presented with more familiar foraging opportunities, increasing the likelihood of encountering *C. alta* and potentially incurring the risks of interspecific competition and/or predation. An adaptationist hypothesis for the evolution of this damage-released chemical signalling system is that, in addition to the survival benefits accrued to conspecific chemical cue receivers through antipredator behavioural responses, chemical signalling may be advantageous to the sender by attracting secondary predators [15]. The differences in localized species-pair abundances in response to less familiar prey cues described above lend some support to this predator attraction/interference hypothesis, in that predators of lower trophic levels appear to avoid predators or competitors of higher trophic levels, sacrificing foraging opportunities in the process.

Prey fishes may experience increased mortality under conditions which eliminate sources of information on the level of predation risk (e.g., damage-released chemical cues lose functionality at pH < 6.6; [37]). The attraction of predators to heterospecific chemical cues demonstrated in the current study suggests that predators may be similarly deprived of information on the presence and quality of foraging opportunities under certain environmental conditions. Predators similarly deprived of sensory information may consequently experience negative fitness consequences. The differences in response in predator species and species-pairs to cues from different prey suggest directions for further research into both the fitness benefits accrued from the use of information on prey availability as well as the interactions between predator species in the context of predator interference.

Acknowledgments

The authors wish to thank the Director of Fisheries in the Trinidadian Ministry of Agriculture, Land and Marine Resources for permission to collect fish for use in this study. They also thank J.-G. J. Godin and I. W. Ramnarine for discussions on experimental design and potential locations for the present study and C. J. Macnaughton, P. H. Malka, and two anonymous reviewers for comments on an earlier version of this paper. Financial support was provided by Concordia University, le Fonds québécois de la Recherché sur la Nature et les Technologies (FQRNT) to C. K. Elvidge and the Natural Sciences and Engineering Research Council of Canada (NSERC) to G. E. Brown. All work reported herein was conducted in accordance with the guidelines of the Canadian Council on Animal Care and the laws of Canada and was approved by the Concordia University Animal Research Ethics Committee (Protocol no. AREC-2008-BROW).

References

[1] E. Danchin, L. A. Giraldeau, T. J. Valone, and R. H. Wagner, "Public information: from nosy neighbors to cultural evolution," *Science*, vol. 305, no. 5683, pp. 487–491, 2004.

[2] C. K. Elvidge, I. W. Ramnarine, J. G. J. Godin, and G. E. Brown, "Size-mediated response to public cues of predation risk in a tropical stream fish," *Journal of Fish Biology*, vol. 77, no. 7, pp. 1632–1644, 2010.

[3] T. W. Cronin, "The visual ecology of predator-prey interactions," in *Behavioural Ecology of Teleost Fishes*, J. G. J. Godin, Ed., pp. 105–138, Oxford University Press, Oxford, UK, 1997.

[4] J. W. Kim, G. E. Brown, I. J. Dolinsek, N. N. Brodeur, A. O. H. C. Leduc, and J. W. A. Grant, "Combined effects of chemical and visual information in eliciting antipredator behaviour in juvenile Atlantic salmon *Salmo salar*," *Journal of Fish Biology*, vol. 74, no. 6, pp. 1280–1290, 2009.

[5] B. A. Hazlett and C. McLay, "Responses to predation risk: alternative strategies in the crab *Heterozius rotundifrons*," *Animal Behaviour*, vol. 69, no. 4, pp. 967–972, 2005.

[6] J. M. Hemmi, "Predator avoidance in fiddler crabs: 2. The visual cues," *Animal Behaviour*, vol. 69, no. 3, pp. 615–625, 2005.

[7] B. D. Wisenden, J. Pogatshnik, D. Gibson, L. Bonacci, A. Schumacher, and A. Willett, "Sound the alarm: learned association of predation risk with novel auditory stimuli by fathead minnows (*Pimephales promelas*) and glowlight tetras (*Hemigrammus erythrozonus*) after single simultaneous pairings with conspecific chemical alarm cues," *Environmental Biology of Fishes*, vol. 81, no. 2, pp. 141–147, 2008.

[8] M. C. O. Ferrari, B. D. Wisenden, and D. P. Chivers, "Chemical ecology of predator-prey interactions in aquatic ecosystems: a review and prospectus," *Canadian Journal of Zoology*, vol. 88, no. 7, pp. 698–724, 2010.

[9] R. J. F. Smith, "Avoiding and deterring predators," in *Behavioural Ecology of Teleost Fishes*, J. G. J. Godin, Ed., pp. 163–190, Oxford University Press, Oxford, UK, 1997.

[10] J. L. Golub, V. Vermette, and G. E. Brown, "Response to conspecific and heterospecific alarm cues by pumpkinseeds in simple and complex habitats: field verification of an ontogenetic shift," *Journal of Fish Biology*, vol. 66, no. 4, pp. 1073–1081, 2005.

[11] G. E. Brown and R. J. F. Smith, "Conspecific skin extracts elicit antipredator responses in juvenile rainbow trout (*Oncorhynchus mykiss*)," *Canadian Journal of Zoology*, vol. 75, no. 11, pp. 1916–1922, 1997.

[12] G. E. Brown, D. P. Chivers, and R. J. Smith, "Localized defecation by pike: a response to labelling by cyprinid alarm pheromone?" *Behavioral Ecology and Sociobiology*, vol. 36, no. 2, pp. 105–110, 1995.

[13] B. D. Wisenden, J. Karst, J. Miller, S. Miller, and L. Fuselier, "Anti-predator behaviour in response to conspecific chemical alarm cues in an esociform fish, *Umbra limi* (Kirtland 1840)," *Environmental Biology of Fishes*, vol. 82, no. 1, pp. 85–92, 2008.

[14] G. E. Brown and J. G. J. Godin, "Chemical alarm signals in wild Trinidadian guppies (*Poecilia reticulata*)," *Canadian Journal of Zoology*, vol. 77, no. 4, pp. 562–570, 1999.

[15] R. J. F. Smith, "Alarm signals in fishes," *Reviews in Fish Biology and Fisheries*, vol. 2, no. 1, pp. 33–63, 1992.

[16] M. C. Harvey and G. E. Brown, "Dine or dash?: ontogenetic shift in the response of yellow perch to conspecific alarm cues," *Environmental Biology of Fishes*, vol. 70, no. 4, pp. 345–352, 2004.

[17] O. A. Popova, "The role of predaceous fish in ecosystems," in *Ecology of Freshwater Fish Production*, S. D. Gerking, Ed., pp. 215–249, Blackwell Scientific, Oxford, UK, 1978.

[18] G. E. Brown, V. J. Leblanc, and L. E. Porter, "Ontogenetic changes in the response of largemouth bass (*Micropterus salmoides*, Centrarchidae, Perciformes) to heterospecific alarm pheromones," *Ethology*, vol. 107, no. 5, pp. 401–414, 2001.

[19] B. D. Wisenden and T. A. Thiel, "Field verification of predator attraction to minnow alarm substance," *Journal of Chemical Ecology*, vol. 28, no. 2, pp. 433–438, 2002.

[20] S. L. Lima and L. M. Dill, "Behavioral decisions made under the risk of predation: a review and prospectus," *Canadian Journal of Zoology*, vol. 68, no. 4, pp. 619–640, 1990.

[21] A. Mathis, D. P. Chivers, and R. J. Smith, "Chemicl alarm signals: predator deterrents or predator attractants?" *American Naturalist*, vol. 145, no. 6, pp. 994–1005, 1995.

[22] O. M. Lonnstedt, M. I. McCormick, and D. P. Chivers, "Well-informed foraging: damage-released chemical cues of injured prey signal quality and size to predators," *Oecologia*, vol. 168, no. 3, pp. 651–658, 2012.

[23] G. E. Brown, C. K. Elvidge, C. J. Macnaughton, I. Ramnarine, and J. G. J. Godin, "Cross-population responses to conspecific chemical alarm cues in wild Trinidadian guppies, *Poecilia reticulata*: evidence for local conservation of cue production," *Canadian Journal of Zoology*, vol. 88, no. 2, pp. 139–147, 2010.

[24] D. P. Croft, L. J. Morrell, A. S. Wade et al., "Predation risk as a driving force for sexual segregation: a cross-population comparison," *American Naturalist*, vol. 167, no. 6, pp. 867–878, 2006.

[25] A. E. Magurran, *Evolutionary Ecology: The Trinidadian Guppy*, Oxford University Press, Oxford, UK, 2005.

[26] R. Froese and D. Pauly, "*FishBase. 2011*," World Wide Web electronic publication, (01/2010)http://www.fishbase.org/.

[27] J. Krause and J. G. J. Godin, "Predator preferences for attacking particular prey group sizes: consequences for predator hunting success and prey predation risk," *Animal Behaviour*, vol. 50, no. 2, pp. 465–473, 1995.

[28] K. E. Esteves, "Feeding ecology of three Astyanax species (Characidae, Tetragonopterinae) from a floodplain lake of Mogi-Guacu River, Parana River Basin, Brazil," *Environmental Biology of Fishes*, vol. 46, no. 1, pp. 83–101, 1996.

[29] B. H. Seghers, *An analysis of geographic variation in the anti-predator adaptations of the guppy, Poecilia reticulata [Ph.D. thesis]*, Department of Zoology, University of British Columbia, Vancouver BC, Canada, 1973.

[30] G. E. Brown, C. J. MacNaughton, C. K. Elvidge, I. Ramnarine, and J. G. J. Godin, "Provenance and threat-sensitive predator avoidance patterns in wild-caught Trinidadian guppies," *Behavioral Ecology and Sociobiology*, vol. 63, no. 5, pp. 699–706, 2009.

[31] J. Pinheiro, D. Bates, S. DebRoy, D. Sarkar, and R Development Core Team, "*nlme: Linear and Nonlinear Mixed Effects Models*," R package version 3.1-97, 2010.

[32] R Development Core Team, *R: A Language and Environment for Statistical Computing*, R Foundation for Statistical Computing, Vienna, Austria, 2010.

[33] E. Paradis, J. Claude, and K. Strimmer, "APE: analyses of phylogenetics and evolution in R language," *Bioinformatics*, vol. 20, no. 2, pp. 289–290, 2004.

[34] M. F. Cashner, "Are spotted bass (*Micropterus punctulatus*) attracted to Schreckstoff? A test of the predator attraction hypothesis," *Copeia*, no. 3, pp. 592–598, 2004.

[35] D. P. Chivers, G. E. Brown, and R. J. F. Smith, "The evolution of chemical alarm signals: attracting predators benefits alarm signal senders," *American Naturalist*, vol. 148, no. 4, pp. 649–659, 1996.

[36] P. A. Nilsson and C. Bronmark, "Prey vulnerability to a gape-size limited predator: behavioural and morphological impacts on Northern pike piscivory," *Oikos*, vol. 88, no. 3, pp. 539–546, 2000.

[37] A. O. H. C. Leduc, J. M. Kelly, and G. E. Brown, "Detection of conspecific alarm cues by juvenile salmonids under neutral and weakly acidic conditions: laboratory and field tests," *Oecologia*, vol. 139, no. 2, pp. 318–324, 2004.

21

Sensory Systems and Environmental Change on Behavior during Social Interactions

S. M. Bierbower,[1] J. Nadolski,[2] and R. L. Cooper[1]

[1] *Department of Biology & Center for Muscle Biology, University of Kentucky, Lexington, KY 40506-0225, USA*
[2] *Department of Mathematical and Computational Sciences, Benedictine University, Lisle, IL 60532, USA*

Correspondence should be addressed to R. L. Cooper; rlcoop1@email.uky.edu

Academic Editor: Randy J. Nelson

The impact of environmental conditions for transmitting sensory cues and the ability of crayfish to utilize olfaction and vision were examined in regards to social interactive behavior. The duration and intensity of interactions were examined for conspecific crayfish with different sensory abilities. Normally, vision and chemosensory have roles in agonistic communication of *Procambarus clarkii*; however, for the blind cave crayfish (*Orconectes australis packardi*), that lack visual capabilities, olfaction is assumed to be the primary sensory modality. To test this, we paired conspecifics in water and out of water in the presence and absence of white light to examine interactive behaviors when these various sensory modalities are altered. For sighted crayfish, in white light, interactions occurred and escalated; however, when the water was removed, interactions and aggressiveness decreased, but, there was an increase in visual displays out of the water. The loss of olfaction abilities for blind cave and sighted crayfish produced fewer social interactions. The importance of environmental conditions is illustrated for social interactions among sighted and blind crayfish. Importantly, this study shows the relevance in the ecological arena in nature for species survival and how environmental changes disrupt innate behaviors.

1. Introduction

Social relationships may take many forms when organisms live in a group, and often times, the individuals must determine their status within a social structure [1–3]. Social dominance is a form of a social relationship in which individuals aggressively interact repeatedly. The interaction between individuals is a well-studied sequential series of interactions, with each individual having the option of terminating or continuing the interaction/contest at any time. The consequence of these interactions most likely results in a dominant individual who repeatedly wins encounters against a subordinate [3]. Therefore, agonistic encounters will generally establish social hierarchies between individuals in a population [4–9]. Dominance hierarchies are known to decrease aggressive interactions between individuals based upon social status, therefore stabilizing the population over time [10, 11].

Smith [12] suggests that rank may be a strategy individuals adopt to maximize fitness in the population based upon the role of other individuals. This correlates with the established Barnard and Sibly [13] producer-scrounger game in which mixes of strategies work better than all one or the other of a specific strategy. There are obvious ecological benefits for being the dominant individual and little point in interacting if there is an absence of benefits with aggressive interactions. Thus, the benefit of interactions must account not only for the resource, but also the cost in obtaining the resource. The dominate individuals often have increased access to resources such as mates, food, and shelters [14, 15]. However, this may not always be the case since many other factors play a role such as the value of the resource [16], the inability to monopolize a resource [17], and the loss of resources' due to stealing of stores/caches by other individuals [18]. Furthermore, females with young often rise in the social ranks to better provide for their young [19], as well as hungry subordinate individuals often win encounters against dominants for access to food [20, 21].

There are many factors involved in the establishment of social dominance, and it is well documented that environmental cues play a major role in the outcome of social interactions whether through chemosensory (odors, [22, 23]), visual (opponent posturing, [24, 25]), and/or tactile cues (physical combat, [4, 5, 7, 8]).

An ideal model system to study social interactions is with crayfish since typical interactions have been well documented for decades. Crayfish are known to form social hierarchies after very aggressive interactions [4–9, 26, 27]. Typically, the encounters escalate from visual threats of defense posturing to actual physical confrontations that include cheliped grasping and more aggressive behaviors where one will try to dismember or even kill another individual.

Currently, most studies observing social interactions occur in a natural field site or a location that mimics the typical environment. While this gives insights into typical behaviors, little is known of the interaction dynamics when natural changes occur such as when an organism leaves the aquatic environment or when sensory systems are diminished. Crayfish leave the water for various reasons such as to find food, mates, or when excessive competition may drive them to look for other niches. The environmental change with the absence of water would eliminate the typical escape response (i.e., tail flip) which allows for a fast retreat from conspecifics. In addition, the absence of water results in other factors influencing social interactions, such as a higher energy demand for movement mainly due to the lack of buoyancy, a greater probability of injury due to a slower response in movement, as well as retreat and also the lack of assessment through chemical cues of not only conspecifics, but also the environment in general. Thus, an organism is at greater risk since they lack the ability to assess their opponent and/or locate a safe place for retreat. This is especially true for a species evolutionarily lacking a sensory modality. Hence, it is of interest to examine the effect of diminished visual and olfactory/chemosensory sensory system. This is possible through studies in the absence of white light and the removal of the primary chemosensory appendage (i.e., antennules) in surface species, as well as studies in an evolutionarily distinct species of crayfish which lack the visual sensory modality.

Although vision and tactile have been suggested to be very important in social interactions for mediating the transfer of information, the full understanding on the ways these two sensory cues are used in agonistic communication remains unclear. It has been well studied and shown that vision is important for agonistic communication in other decapod species, such as fiddler crabs [28–30], hermit crabs [31–35], lobsters [36], and mantis shrimp [37]. Due to this obvious factor in so many other decapod crustaceans, we assume that the visual sensory cue would also be very important for information exchange among crayfish. We chose to separate the roles of vision and chemosensory in the agonistic communication of *P. clarkii* by conducting experiments in red light (not visible to *P. clarkii*) as well as removing the antennules, both independently and additively to a red light environment. The study of vision in this species is particularly appropriate, given that *P. clarkii* are normally active under a wide range of environmental light levels, and we mimic periods of dusk and dawn which are known to be particularly active time points [38].

Blind cave crayfish (*Orconectes australis packardi*) lack visual capabilities; therefore, they provide the opportunity to examine whether behavioral, morphological, and/or physiological evolutionary adaptations may have evolved uniquely to their species based upon the cave environment. Since cave crayfish have a reduced optic system and have more olfactory projection neurons than surface sighted crayfish, it was suggested they have more neural processing related to olfaction [39]. Cave crayfish appear to rely primarily on olfactory and tactile modalities, while surface crayfish rely primarily on visual and olfactory to assess and monitor their surroundings. Since these cave crayfish do have caudal photoreceptors in their 6th abdominal ganglion and respond to white light, studies were performed in white and red lights. The caudal photoreceptors are not sensitive to the red light used, as assayed in behavioral studies [40]. In accordance with the above information, it is logical that cave crayfish do not show the typical postural behaviors (visual display) identified in social encounters within their natural cave environment. We hypothesized that such displays would not be beneficial since conspecifics are not able to observe the visual display. While studies have addressed the neural structure of the optic systems [39] and the effects of light on social interactions [40], the typical behaviors of cave crayfish have not been as thoroughly studied as with surface crayfish. Currently, little is known of interaction dynamics in the absence of water which eliminates the typical escape response (i.e., tail flip) and/or with diminished chemosensory systems in either species of crayfish.

Chemical signals are also important sources of information in aquatic environments where visibility maybe limited when compared to terrestrial open environments [40]. Crayfish are known to have a [41, 42] well-developed olfactory system, and studies have shown that chemical signals play an important role in many aspects of their life [23, 43, 44]. Specifically in agonistic encounters, chemical signals appear to be more important than other offensive displays and signals for settling a fight [45]. Interestingly, research has shown that some species are able to recognize individuals that they have encountered in the recent past such as two species of hermit crabs [46–48], crab [49], mantis shrimp *Gonodactylus festae* [50, 51], lobsters *Homarus americanus* [52], and crayfish [53]. It has been shown that the individual recognition is based upon chemical signals that are emitted during social interactions [54–56] in crustaceans [57]. The chemical signals are important in maintaining the stable dominance hierarchies.

Chemical cues are known to be involved in the establishment of social hierarchies and are known to impact behavior [58, 59]. Bovbjerg [5] first suggested that both vision and tactile are involved in the establishment of social hierarchies, and he also demonstrated that antennae are important for tactile orientation. The antennule is considered the organ most specialized for chemosensory detection and plays a leading role in tracking odor plumes [60] and individual recognition [61]. One way to address the influence of sensory cues is to remove important sensory systems individually and simultaneously. Specifically, by removing the antennules and vision through the absence of white light (provide red light), one can understand the reliance on sensory cues.

It is apparent that many environmental cues determine the outcome of social interactions. With the assumption that all group members begin with equal fighting abilities, environmental effects or diminished sensory cues will most likely disrupt the typical intrinsic behavior. Furthermore, when multiple cues are diminished, the influence may be additive or behave synergistically in altering a behavior. Thus, by examining reliance on single sensory systems on well-defined social behavior, we can begin to understand environmental impacts on populations/species. We compared social interactions in white light to experiments in red light to understand the photoreceptors influence on social interactions.

Past studies have examined many extrinsic factors that influence intraspecific aggression, such as shelter acquisition [19, 62, 63], chemical communication [5, 23, 64], mating [65], food preferences [66], and starvation [67, 68]. An area not yet addressed is the extrinsic factor of "out of water" for crayfish social interactions, and it is unclear whether the hydrodynamics of natural habitats allow for the successful use of chemical signals and typical behavior during social interactions in nature. Thus, the purpose of this study is to present quantitative analysis of environmental influence on social interactions in two species of crayfish with special reference to reliance of different primary sensory modalities.

Intrinsic and extrinsic factors affect intraspecific aggression in many ways, and both should be examined for the impact on agonistic behavior. Herein, a simple additive model for this integration of multiple sensory systems as well as multiple environmental factors in an individual's expected fighting ability determined the impact of additive effects. Examination of environmental influence on behavior was through the measure of fighting strategy and intensity of interaction in two species (*Procambarus clarkii*, sighted surface crayfish and *Orconectes australis packardi*, blind cave crayfish).

Due to distinct behavioral, anatomical, biochemical, morphological, and/or physiological adaptations of cave organisms, there is a fascination and interest in understanding how they are able to adapt and survive in extreme environments. Cave crayfish show the general characteristics of anatomical and morphological adaptations of most cave organisms. Specifically when compared to surface crayfish, cave crayfish are smaller in size, have longer/thinner appendages, possess highly developed nonvisual sensory capabilities, and lack pigment and eyes [69]. In addition, behavioral, physiological, and biochemical adaptations have been identified in cave crayfish such as a decrease in locomotion and oxygen consumption, as well as a decrease in metabolic rates [70]. These are related to a reduction in energy from limited food sources and/or oxygen availability in cave systems [71–73]. Thus, these distinct evolutionary adaptations allows for studies discerning behavioral differences in two species of crayfish. Another goal of this study was to identify species-specific behaviors through comparison of cave and a surface species, as well as determining the environmental and olfactory influence on intrinsic behaviors.

FIGURE 1: Blind cave crayfish, *Orconectes packardi australis*, engaged in an agonistic encounter.

2. Methods

2.1. Animals. Crayfish, *Procambarus clarkii* (sighted), measuring 5.0–6.25 cm in body length were obtained commercially (Atchafalaya Biological Supply Co., Raceland, LA, USA). Crayfish, *Orconectes australis packardi* (Rhoades) (the blind crayfish), measuring 4.6–6.4 cm, were obtained from the Sloan's Valley Cave System near Somerset, KY, USA (state collecting permits were obtained for this study; Figure 1). A total of 25 sighted and 15 blind crayfish were used in the study. Crayfish pairs were randomly chosen from the naïve population stock. The order in which the trials occurred was random. No two crayfish were paired together more than once; thus, all encounters were with conspecifics not previously known to each other. Only male crayfish were used in this study. Animals were housed individually in rectangular plastic containers and cared for in the same manner, except for *O. a. packardi* that were covered with black plastic to omit light in an aquatic facility within our regulated-temperature laboratory (17–20°C). *P. clarkii* were on a 12-hour period light-dark cycle. All crayfish were fed dried fish pellets weekly before and throughout the experiments. Crayfish handling was conducted by using a glass beaker to transfer crayfish from one container to the another. Due to housed containers being cleaned weekly, crayfish were handled often; the limited handling during experimentation is assumed to have little to no effect on the internal status of the crayfish. Only crayfish in their intermolt stage, possessing all walking legs and both chelipeds were used in these studies.

2.2. Social Interactions. Initial experiments (i.e., low light) were focused on characterizing the general behavioral interactions for both species of crayfish. Crayfish were randomly distributed into fourteen different conditions as discussed below. Social interactions were staged in size-matched males. An interaction behavioral scoring index was developed (Table 1(a)) for species comparison of *P. clarkii* and *O. australis packardi*. Observational preexperimental trials identified typical crayfish behavior to establish a quantifiable scale for interactions based on both aggressiveness, as well as intensity (time duration of the interaction, Table 1(b)). Crayfish male pairs of approximate equal size were staged in a glass aquarium and videotaped for one hour, allowing for interaction without outside interference. The crayfish were monitored indirectly with a TV monitor. Trials conducted in low light in the water served as controls for the sighted

TABLE 1: Social interaction scoring bioindex. (a) Indicates the behavioral scoring bioindex used to quantify behavior during each trial in the experimental conditions. (b) Indicates the intensity scale based upon time duration in which the pairs were engaged in a specific behavior.

(a)

0	No interaction
1.	Territory invasion
2.	Intentional touching
3.	Acknowledgment
4.	Threat display
5.	Chase
6.	Grasp/strike
7.	Dismemberment

(b)

0.1	1–15 seconds
0.2	16–30 seconds
0.3	31–45 seconds
0.4	>45 seconds

crayfish, while trials conducted in red light in water act as controls for blind crayfish. The index was then used to quantify each of the trials across conditions and species comparison. Behavioral scores were assigned to pairs of crayfish (not individual scoring) for every interaction that occurred during the 60-minute time period.

2.3. Behavioral Analysis. Previous research and prior observation of aggressive interactions between individuals indicate that the behavior could be classified into several rather distinct categories. These categories represent behavior patterns in what are relatively stereotyped and which are known to be typical behaviors of sighted crayfish. Briefly the behavioral acts established are as follows (also see Table 1(a)).

> 0-*No interaction*: no encounter without any evidence of awareness of other individual.
>
> 1-*Territory invasion/approach/retreat*: deliberate movement towards other individual and a direct, initiation into conspecifics space and/or movement away.
>
> 2-*Intentional touching*: a short rapid movement forward directed at individual.
>
> 3-*Acknowledgment/standoff*: facing one another without visual threat display.
>
> 4-*Meral spread/threat display*: outward raising and spreading of the chelipeds.
>
> 5-*Chase*: pursuit after the individual.
>
> 6-*Grasp/strike*: a blow to or seizing of other individual.
>
> 7-*Dismemberment*: very aggressive action to individual in which dismemberment or likelihood of killing is apparent.

Most of the general characteristics are previously described in Dingle and Caldwell [74]. Interactions typically began with an invasion of territory or an acknowledgment/standoff. Termination of the interaction occurred when the observer determines that individuals no longer appear to be directing behavior at each other. Communication may be occurring, but since the purpose of this study was to concentrate on aggressive interactions, no attempt was made to analyze other possible communicative behaviors. Quantification of behavior was based upon total number of interactions as well as the length of each individual interaction.

Each trial was critically analyzed to categorize crayfish behavior, as well as identify general behavioral trends within and across species. For each environmental condition (i.e., low light, red light, and no antennules), five trials ($N = 5$) were run in the water and five trials ($N = 5$) were run out of the water. All trials were digitally recorded and analyzed through video analysis to record behavioral scores and intensity. To understand behavioral trends, 3D graphs combined all trials together for comparison of the type of behavior as well and intensity of each encounter. The duration of an interaction was used as a measure of interaction intensity. Since interactions are known to be relatively short, a time scale was used (Table 1(b)).

2.4. Environmental Conditions. The various environmental conditions that were used are listed in Table 2. Social interactions were examined in and out of water in low light (25 lux). "In water" studies used a glass aquarium (20 cm × 10 cm × 12 cm) 4 cm filled from the top with aerated water. "Out of water" studies were conducted using the same aquarium but without water and still providing wet sand for the animals to walk on. "In water" studies (control) for both sighted and blind crayfish were compared to other environmental conditions to determine changes in intrinsic behaviors. This part of the study examined: (1) sighted "in water," (2) sighted "out of water," (3) blind "in water," and (4) blind "out of water".

Social interactions were also observed in red light. Red light conditions used a filtered red light (2.5 Lux) to remove the visual sensory stimulation for the sighted crayfish and the stimulation of the caudal photoreceptors in the cave crayfish. The red light (Kodak Adjustable Safeway Lamp, 15 watts), was previously noted to be a wavelength not detected by crayfish [9, 40] thus providing no visual sensory stimulation. The purpose is to examine the reliance of visual cues for sighted crayfish out of water when chemosensory cues are diminished. Furthermore, using blind crayfish in red light allowed us to help determine if low light induces a stress response that influences social behavior. We examined these conditions in this part of the study: (1) sighted/in water/red light, (2) sighted/out of water/red light, (3) blind/in water/red light, and (4) blind/out of water/red light (Table 2).

The removal of olfactory cues was conducted by removing the antennules (primarily sensory system for chemical detection) with sharp scissors at the base of the antennules by the first annuli. There was no death associated with antennulectomy as this is not that invasive of a surgery for crayfish. In

TABLE 2: Social interaction conditions for both species of crayfish. Social interactions were observed both in and out of the water for sighted and blind crayfish. Assessment of different sensory modalities impact on intrinsic behavior was examined through methodical removal of one or many sensory cues.

Sighted		Blind	
In water	Out of water	In water	Out of water
Low light	Low light	Low light	Low light
Red light	Red light	Red light	Red light
Low light/no antennules	Low light/no antennules	—	—
Red light/no antennules	Red light/no antennules	Red light/no antennules	Red light/no antennules

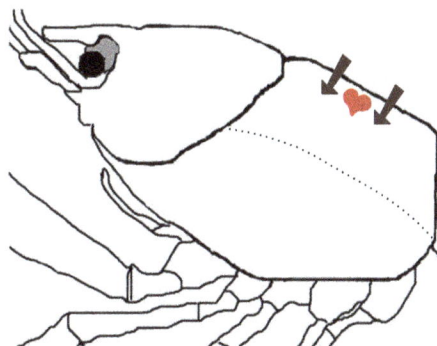

FIGURE 2: Schematic representation for the placement of the recording wires for monitoring the heart from a crayfish (*Procambarus clarkii*.). On the dorsal carapace, large arrows represent the two wires which spanned the rostral-caudal axis of the heart to monitor any change in the dynamic resistance, which is used as a measure of heart rate.

fact, there is little blood loss as well since this is not that wide of a region as compared to the very base of the antennules next to the cephalothorax. The animals were held for 3 days for recovery after antennulectomy. "In water" and "out of water" studies were again conducted for both species of crayfish. The purpose of removing the antennules was to further understand the reliance of visual cues for sighted crayfish and to understand impacts on social behavior for blind crayfish if there was a lack of environmental olfactory cues. These set of conditions compared (1) sighted/in water/no antennules, (2) sighted/out of water/no antennules, (3) blind/in water/no antennules, and (4) blind/out of water/no antennules.

To determine the reliance on environmental cues during social interactions for sighted crayfish, both chemosensory and visual cues were removed. Social interactions were examined for sighted crayfish only in red light and with the removal of antennules in order to compare these two conditions: (1) sighted/in water/red light/no antennules and (2) sighted/out of water/red light/no antennules.

2.5. Recording ECGs. An autonomic response was examined when sighted crayfish ($N = 5$) were placed into the experimental aquarium for interactions. Crayfish pairs were randomly chosen from the naïve population stock for "in water" and "out of water" trials. Order of which trials occurred first was random. There were multiple days between starting the trials to ensure that social recognition was unlikely. Crayfish were wired to record electrocardiograms (ECGs) for heart rate (HR) [75–77]. In brief, two insulated stainless steel wires (diameter 0.005 inches and with the coating 0.008 inches; A-M Systems, Carlsborg, WA, USA) were placed under the dorsal carapace directly over the heart 3 days prior to experimentation. Wires were inserted through holes drilled in the carapace and cemented in place with instant adhesive (Eastman, 5-min drying epoxy). These two wires were placed to span the heart in a rostral-caudal arrangement to insure an accurate impedance measure during each heart contraction as shown in Figure 2. A lid was used to prevent the crayfish from exiting the chamber but left a small section uncovered for the wires to exit the chamber and did not prohibit the crayfish from moving freely. All physiological measures were recorded though an impedance detector which measured dynamic resistance between the

stainless steel wires and recorded on-line to a PowerLab via a PowerLab/4SP interface (AD Instruments). All events were measured and calibrated with the PowerLab Chart software version 5.5.6 (AD Instruments, Australia). Previous studies showed that 3 days was enough time for the animals to return to physiological measurements similar to levels prior to handling [78]. Cave crayfish typically have a thinner, more brittle exoskeleton resulting in more delicate handling during wiring.

Analysis of the response consisted of heart rate in beats per minute (BPM). HR was monitored in and out of water under control conditions to determine physiological responses during social interactions. This provided an internal measure to external cues. The experimental procedure consisted HR recordings before, during, and after social interactions. HR was analyzed to provide a BPM to note changes in the internal response based upon interactions, as well as environmental conditions.

2.6. Statistical Analysis. Parametric tests (ANOVA and *t*-test) were used when comparing differing levels of behavior and levels of intensity. When sufficient evidence for normality was violated, Mann Whitney Rank Sum was used to compare different experimental conditions on the same species. All graphs and statistical tests were performed in SigmaPlot Version 11.0 and R 2.15.0 (Systat Software Inc., San Jose, CA, USA). Additional variables were created such as maximum behavior over the 60-minute trial, time to first maximum behavior observed, the intensity of the first maximum, the number of encounters, number of encounters at maximum behavior, and total intensity of all maximum behavior encounters.

3. Results

3.1. Social Behavior in Low White Light. Five pairs of sighted and five pairs of blind crayfish were allowed to separately interact for 60 minutes in low white light (25 lux) to determine typical behavioral interactions. For sighted crayfish

TABLE 3: Total number of social interaction across study conditions for sighted crayfish. Social interactions were observed for both "in water" and "out of water." Each row corresponds to the total number of interactions for a given behavioral score. Each column corresponds to an environmental condition. In this and succeeding tables, the numbers in brackets are the total "out of water" interactions.

Behavior	Low light	Red light	No antennules	Red light/no antennules
Invasion (1)	$183 (120)^{***}$	$101 (77)^{***}$	$47 (54)^{*}$	$54 (58)^{NS}$
Touching (2)	$160 (97)^{***}$	$107 (120)^{**}$	$98 (135)^{***}$	$70 (99)^{***}$
Acknowledgment (3)	$117 (44)^{***}$	$41 (33)^{*}$	$43 (20)^{**}$	$25 (10)^{**}$
Threat display (4)	$108 (38)^{***}$	$33 (22)^{*}$	$39 (26)^{*}$	$43 (5)^{***}$
Chase (5)	$68 (18)^{***}$	$51 (10)^{***}$	$34 (6)^{***}$	$43 (2)^{***}$
Grasp (6)	$49 (2)^{***}$	$63 (5)^{***}$	$31 (2)^{***}$	$13 (2)^{***}$
Dismemberment (7)	$27 (0)^{***}$	$20 (2)^{***}$	$8 (0)^{***}$	$4 (1)^{***}$

TABLE 4: Total number of social interaction across study conditions for blind crayfish. Social interactions were observed both "in water" and "out of water." Each row corresponds to the total number of interactions for a given behavioral score. Each column corresponds to an environmental condition.

Behavior	Low light	Red light	Red light/no antennules
Invasion (1)	$113 (34)^{***}$	$127 (59)^{***}$	$135 (75)^{***}$
Touching (2)	$87 (27)^{***}$	$160 (89)^{**}$	$104 (97)^{*}$
Acknowledgment (3)	$60 (5)^{***}$	$65 (6)^{***}$	$93 (22)^{***}$
Threat display (4)	$36 (3)^{***}$	$37 (5)^{***}$	$55 (0)^{***}$
Chase (5)	$31 (3)^{***}$	$73 (0)^{***}$	$38 (1)^{***}$
Grasp (6)	$22 (2)^{**}$	$29 (0)^{***}$	$4 (0)^{NS}$
Dismemberment (7)	$3 (1)^{NS}$	$8 (0)^{*}$	$0 (0)^{NS}$

in water, they were shown to interact regularly within the time period, as well as escalate in interactions to high levels of aggression indicated by the total interactions for the behavioral scores (i.e., 6 and 7; Figure 3(a)). Interactions of sighted crayfish out of the water were shown to occur less often, and the interactions were shown to be less aggressive due to the few high scores (Figure 3(b)). There are few interactions overall, for cave crayfish. The cave crayfish show the same trend in decreasing their interactions out of water (Figures 4(a) and 4(b)). Both species exhibited significant differences in maximum behavior with crayfish in water being higher than out of water in low light (cave: $P = 0.031$ and sighted: $P < 0.001$). A similar outcome was found for the total number of interactions with in water being dominant over out of water (cave: $P = 0.018$ and sighted: $P = 0.022$). On average, sighted crayfish had significantly more total interactions compared to blind crayfish in low light ($P < 0.001$) (Figure 5). For either species, there was not a significant difference as to when the maximum behavior occurred. There was also no significant difference found in the duration of the maximal event or the total duration of all maximal events.

3.2. Analysis of Varying Environmental Conditions. Analysis of in water and out of water treatment groups showed significant changes in fighting strategy due to environmental effects. Specifically, out of water results alone or in combination with other conditions reveal that both species do

not tail flip and show less intrusion into the conspecifics territory when compared to social interactions in the water (Figure 5). Blind crayfish were less responsive to the presence of conspecifics (fewer interactions), while surface crayfish showed an increase in visual displays (possible bluffing mechanism) when interacting out of the water, but failed to escalate the interaction when compared to interaction conducted in the water. Thus, for both species, out of the water has the most significant impact on intrinsic behavior and social interactions (Tables 3 and 4).

ANOVA statistical analysis for each environmental condition shows a significant difference between in water and out of water conditions as indicated in the summary tables (Tables 3 and 4). ANOVA values are as follows: sighted in red light ($F_{13,69} = 33.7$, $P < 0.001$), blind in red light ($F_{13,69} = 17.0$, $P < 0.001$), sighted in white light with no antennules ($F_{13,69} = 17.8$, $P < 0.001$), sighted in red light with no antennules ($F_{13,69} = 7.588$, $P < 0.001$), and blind in red light with no antennules ($F_{13,69} = 19.3$, $P < 0.001$). Therefore, interactions occurring out of water showed that both species of crayfish were less likely to interact and more likely to explore their environment.

3.3. White versus Red Light. A significant difference was found in the number of interactions in blind cave crayfish ($P = 0.029$). In general, there were more interactions in red light than white light for the blind cave crayfish. There was no significant difference observed for the surface crayfish in any type of light. Both species did not exhibit any significant difference in the maximum behavior between white and red lights. Also, there was no significant difference detected in duration of interactions (Figures 6 and 7).

In fact, for the red light condition only, there was a marginal difference in the maximum behavior ($P = 0.056$) for sighted crayfish "in water" having a median value of 7 versus 5 "out of water." As for the cave crayfish, there was also a significant difference in median maximum behavior (6 in and 3 out) ($P = 0.008$) and a significant difference in the median number of interactions in red light (in 104 and out 33) ($P = 0.0008$).

3.4. With versus Without Antennules. The cave crayfish had more maximum behavioral encounters without antennules than with antennules ($P = 0.012$). Surface crayfish had more

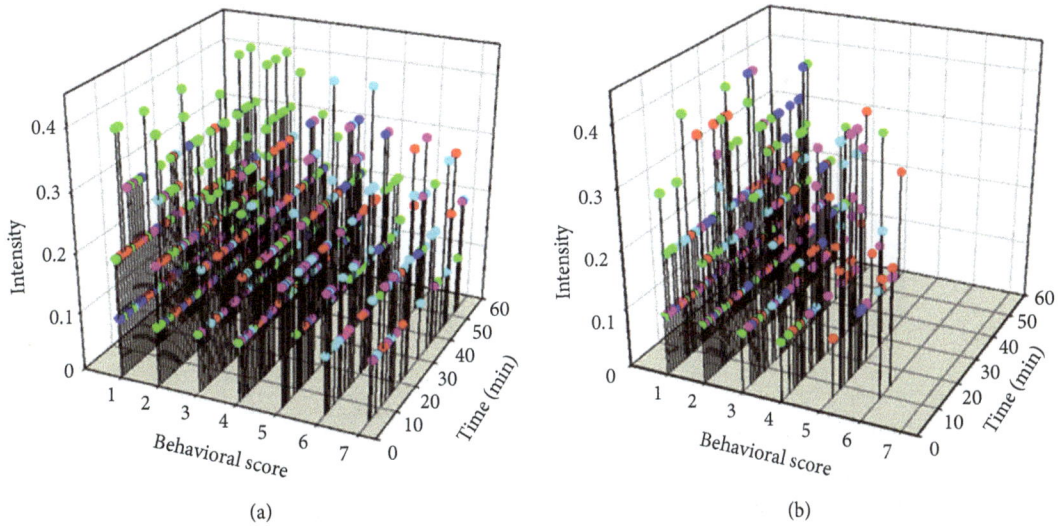

FIGURE 3: Comprehensive representation of social interactions for sighted crayfish in low white light (25 lux). (a) In water. (b) Out of water. A single vertical line indicates a given behavior at a specific point in time as well as the intensity of the behavior. The different colored points represent individual pairs in the trials ($N = 5$).

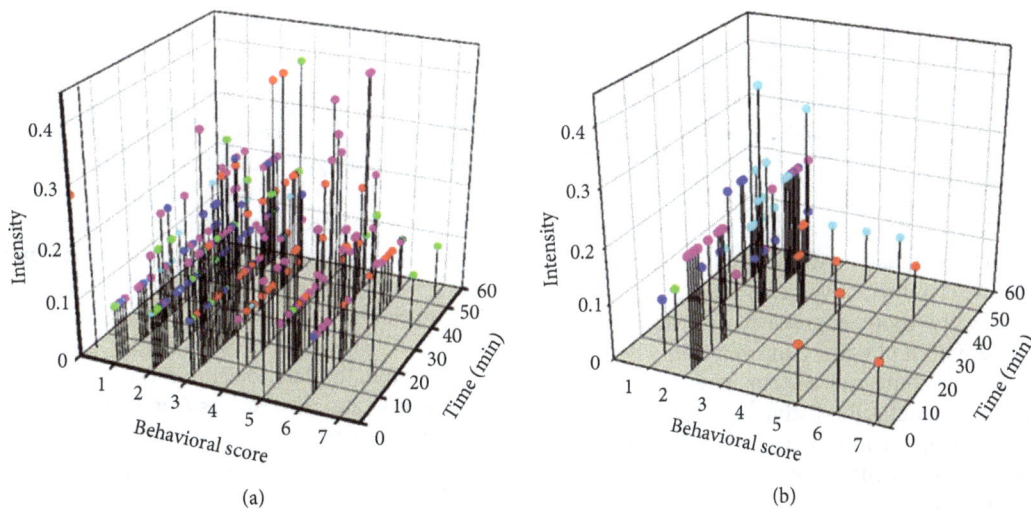

FIGURE 4: Comprehensive representation of social interactions for cave crayfish in low white light (25 lux). (a) In water. (b) Out of water. A single vertical line indicates a given behavior at a specific point in time as well as the intensity of the behavior. The different colored points represent individual pairs in the trials ($N = 5$).

general interactions with antennules rather than without antennules ($P < 0.001$). Again, there was no species identified significant difference in maximum behavior over the 60 minutes with or without antennules.

In red light, the cave crayfish did not show any significant difference over the 60 minutes in the variables measured. However, surface crayfish with antennules had more interactions over the time period than those without ($P = 0.039$). In low light, the sighted crayfish did show significant difference in the average time to the first maximum behavior (in 9 minutes versus out 19.8 minutes, $P = 0.012$).

3.5. *Recording ECGs.* The physiological response of crayfish was recorded to characterize the autonomic response during social interactions as well as for environmental change.

Heart rate (HR) was recorded before, during, and after confrontation, plotted for each crayfish during the entire duration of the trial. A frequency plot of the raw traces shows dramatic changes in HR during interactions when comparing "in water" to "out of water" conditions (Figure 8). Specifically, there is a greater fluctuation for one individual (most likely the subordinate) during and after interactions. As consistent with previously described experiments, it is also shown that the "out of water" condition has fewer interactions. The raw traces show a rapid response during interactions, especially for one individual within a pair, as well as the continued response after the interaction is over. This suggests that "out of water" conditions have a greater effect on intrinsic factors, such as HR, for the individuals. This is most apparent for the individual most likely to become the subordinate since retreat

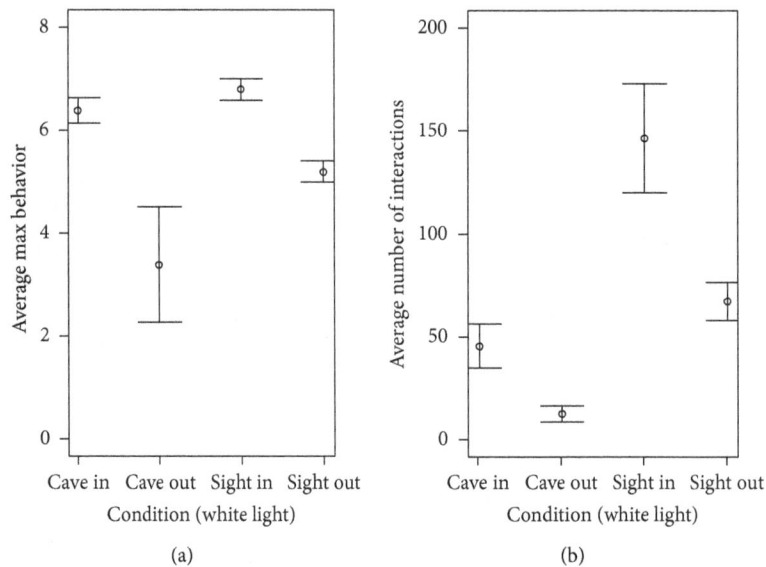

(a)

(b)

FIGURE 5: Comparison of cave and sighted crayfish in and out of water in low/white light. (a) The mean number of maximum behavior (±SEM) is plotted for the four conditions. There is a significant difference between in and out of water for both species. (b) The mean number of total interactions (±SEM) for the four conditions. There is also a significant decrease in both species between in and out of water in low light.

away from the conspecifics is not as feasible out of water and thus a greater chance of being attacked is likely to happen.

4. Discussion

This study demonstrated that environmental factors directly influence crayfish social interactive behavior. Here, we show that interactions were more aggressive and intense and more likely to end with a physical confrontation when they took place "in water" compared to "out of water" for two morphologically and genetically distinct species of crayfish. It is shown that altering environmental conditions induced crayfish to change their intrinsic behavior which resulted in modified social interactions and fighting strategy. For both species in low white light and in water, there was a high value of interactions, and those interactions were likely to escalate to higher levels of aggression (behavioural score of 5, 6, or 7). The duration of interaction was consistently longer in time (intensity of 0.3 or 0.4) when in water. Interestingly, when water was removed from the environment, the total number of interactions, as well as the aggression level and duration of each interaction, dramatically decreased for both species. Across all environmental conditions and exclusion of sensory systems (i.e., vision and chemosensory), removal of water produced the greatest and most consistent change in social interactions. For "out of water" trials, both species were shown not to tail flip (typical escape response), and they showed less intrusion into the conspecific's territory as well as being less likely to engage in social interactions. Importantly, while sighted crayfish did show an increase in visual displays out of the water, a possible bluffing mechanism [79], they failed to escalate in social interactions.

Interactions in red light for sighted crayfish did not appear to decrease aggression levels. This is most likely due to the chemical cues providing enough information about the environment and the conspecific. The removal of antennules along with red light showed a reduction in the number of interactions but did not diminish the aggression levels since many of the interactions escalated to a behavioral score of 5 (chase) and 6 (grasp/strike). When these crayfish were taken out of water in combination with the diminished sensory cues, there was a dramatic decrease in aggression of social interaction. This pattern was similar for blind crayfish in red light and the lack of antennules. There were very few interactions, and the aggression levels were dramatically decreased. Furthermore, heart rate measures during social interactions for a single pair of crayfish showed that "out of water" interactions have a large effect on the organism. It is likely that the dramatic effect on one of the individuals in the pair (most likely the subordinate) is due to an increased probability of injury which could occur in the absence of water. Although heart rate remained relatively unchanged when the crayfish were placed into the chamber, heart rate was shown to immediately decrease for one individual upon interaction with the conspecific.

Agonistic behavior is a fundamental factor of ecological nature, and aggression has been studied extensively in many invertebrate species such as bees [80, 81], ants [82–84], termites [85], wasps [86], lobsters [87–90], crabs [91], and crayfish [92–94]. Ritualized displays and cues that are predicative of agonistic success enable the assessment a rivals' relative fighting ability [95]. Fights occurring in nature are known to be shorter, less intense, and more likely to end with a tail flip, but the animals do show the fundamental fight dynamics as seen in laboratory studies [96]. Fighting is potentially costly to each contestant for a variety of factors including time and energy [97–101] and physical injuring [101–106]. A limited number of studies integrate multiple

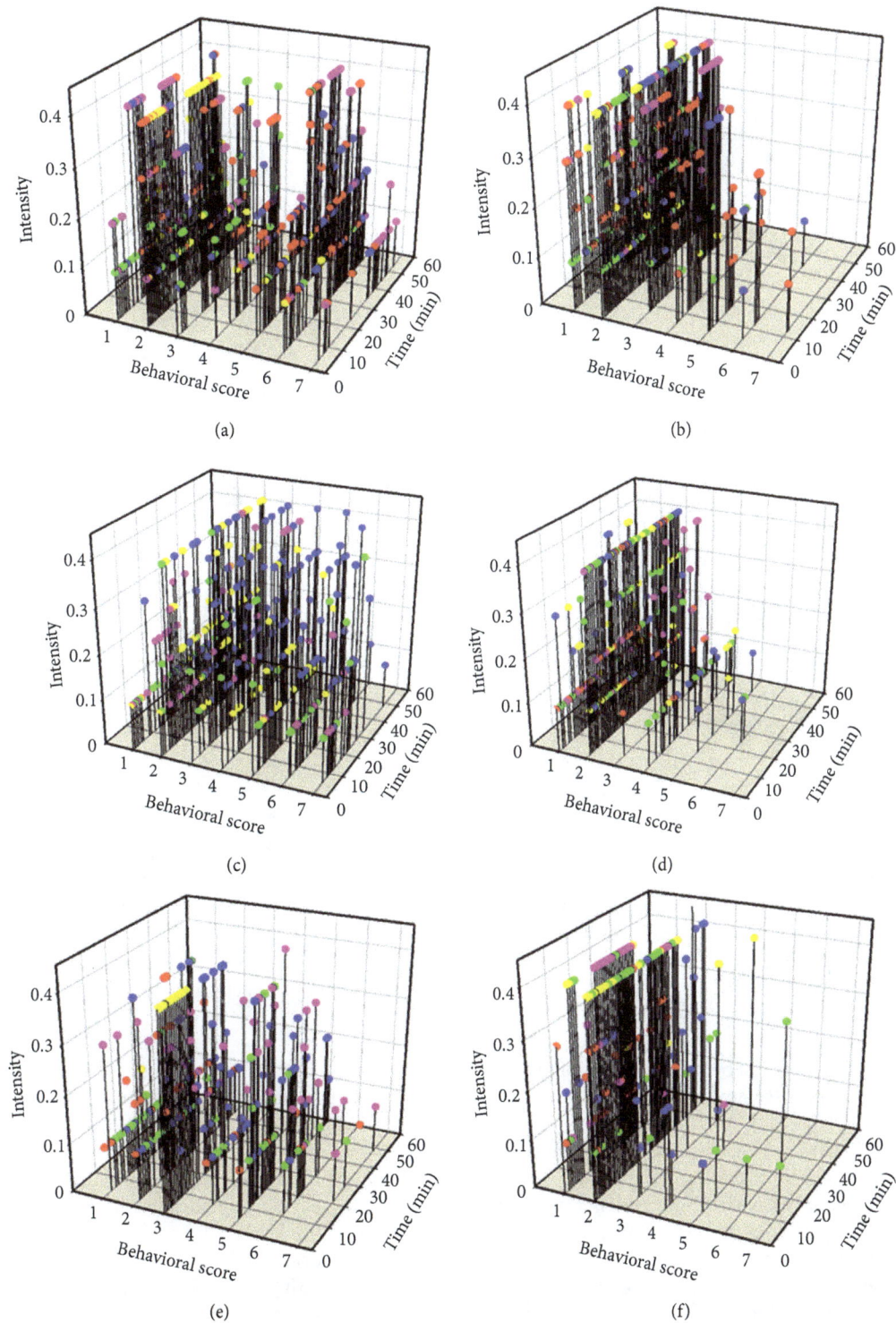

FIGURE 6: Comprehensive representation of social interactions for sighted crayfish in varying environmental conditions. (a) Red light and in water. (b) Red light and out of water. (c) Low white light, no antennules, and in water. (d) Low white light, no antennules and, out of water. (e) Red light, no antennules and, in water. (f) Red light, no antennules and, out of water. A single vertical line indicates a given behavior at a specific point in time as well as the intensity of the behavior. The different colored points represent individual pairs in the trials ($N = 5$).

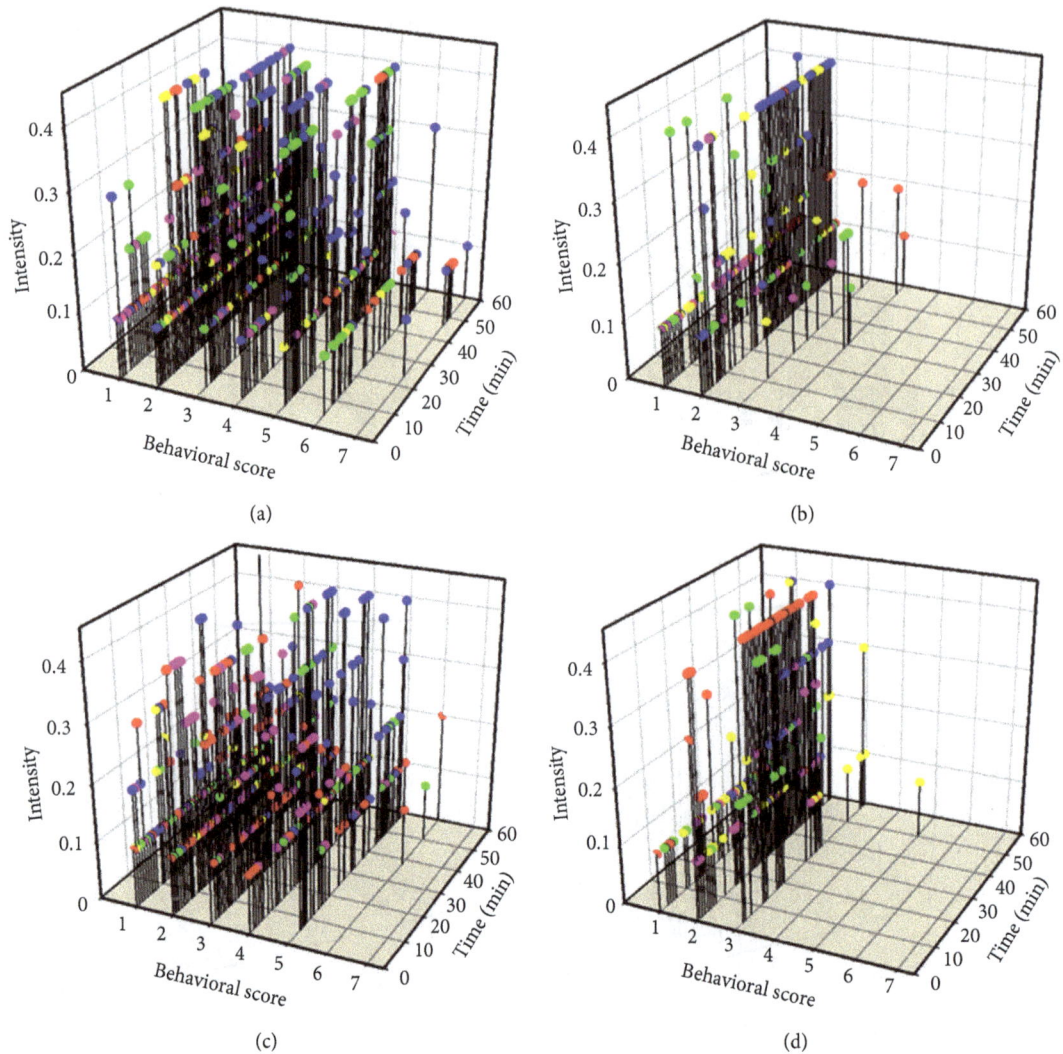

FIGURE 7: Comprehensive representation of social interactions for cave crayfish in varying environmental conditions (a) Red light and in water. (b) Red light and out of water. (c) Red light, no antennules and, in water. (d) Red light, no antennules and, out of water. A single vertical line indicates a given behavior at a specific point in time as well as the intensity of the behavior. The different colored points represent a total of each individual pairs of crayfish in the trials ($N = 5$).

factors that can influence contest behavior. Details of multiple factor sensory integration for any one species are virtually unknown.

The types of behavioral repertories we described are similar to those indexed by Huber and Kravitz [107] in the American lobster *Homarus americanus* and Bergman and Moore [96] in two species of crayfish *Orconectes rusticus* and *Orconectes virilis*. However, we used a scale of 0 to 7, while Bergman and Moore used from (−2) to 5 scale. While the general descriptions were similar for each behavioral level, there were modified classifications in areas described in holding an opponent as a "do-see-do," which relates to a dance term, where we considered this behavior as a dismemberment grasp since they would try to twist the others cheliped off. We also indexed the time of interaction along with the aggression score and duration so that we could assess over time, the complexity of the repetitive interactions.

As expected, behavioral scores incrementally decreased with increasing the aggression levels and duration of interaction, as the hierarchy is likely established. Observational data from video as well as graph summaries document that the interactions do occur throughout the entire hour of the observation period. Specifically, interactions are just as likely to occur in the last ten minutes as they are in the first ten minutes. So even though a social status is being determined within the early interactions, there are continuous bouts to confirm or test the opponent within this initial hour of being introduced. Previous work on the crayfish *Astacus astacus* showed that the number of agonistic challenges, mean duration, and maximum intensity of encounters, were also initially high but then decreased steadily as the hierarchy developed [8]. Thus, the fact that interactions are still common after 50 minutes suggest that development of dominance relationships is incomplete. However, it should be noted

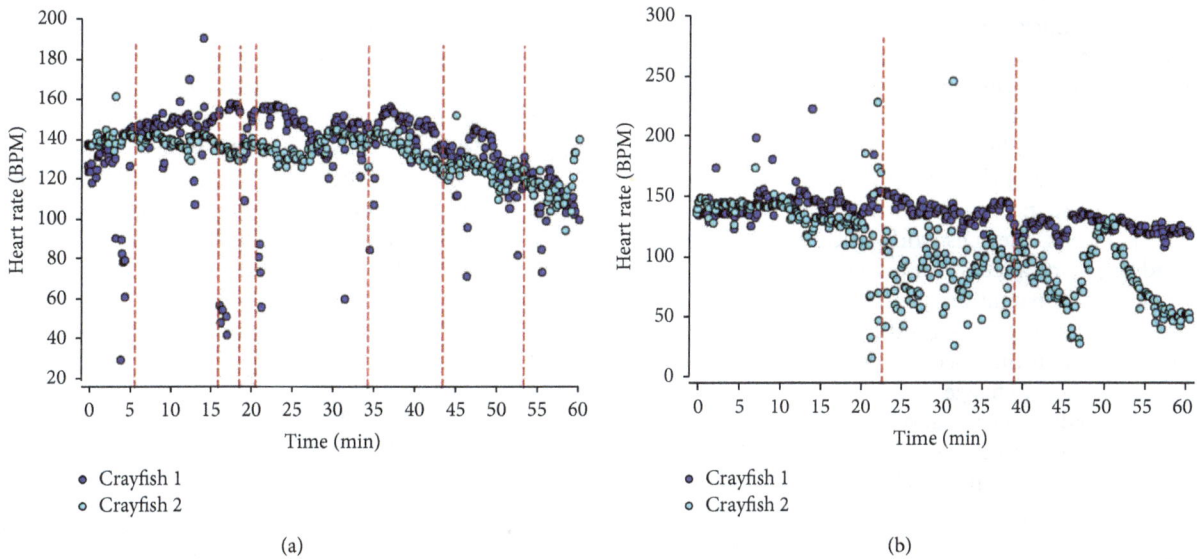

FIGURE 8: Physiological response of a single pair of crayfish. (a) "In water". (b) "Out of water". The dark blue line indicates crayfish one and light blue indicates crayfish two. Each point represents direct counts of each beat over 10-s intervals and then converted to beats per minute (BPM). The red dotted vertical lines indicate a physical interaction. The same pair was used in both conditions with multiple days in between each trial.

that a limitation to laboratory studies is the restriction of escape from an opponent. This would be less of an issue in natural ecosystems; however, small interaction arenas in the laboratory may lead to more aggressive interactions [95, 107].

If one were to document the sensory cues necessary for social dominance and maintenance of social hierarchy, a more in-depth study is required. In this study, the type of interactions and the effect of environment on these general levels of interactions were the focus. Many observations of crayfish behavior have been made to examine specific factors influencing intraspecific aggression such as in shelter acquisition [19, 62, 63], chemical communication [5, 23, 64], mating [65], food preferences [66], and starvation [67, 68]. These studies provide valuable information to determine intrinsic and extrinsic factors that affect agonistic interactions.

There are other extrinsic factors that influence intraspecific interactions such as previous history in agonistic encounters [96, 108, 109], different fighting strategies [110], and prior residence [63, 111]. These can all significantly impact the outcome of social interactions. While we cannot control all these factors due to these organisms not being raised exclusively in the lab, we can use crayfish that have never been before placed together into a new environment that is not previously occupied by either in the past. Crayfish housed individually have been shown to be more aggressive [112] and that previous agonistic encounters with the same individuals can change the outcome of encounters [108, 113]. Since we did house the crayfish as individuals this might have raised their aggressiveness upon interacting.

While the use of a new environment will eliminate a prior residence variable, it does still pose other variables that need to be considered. The use of the new environment introduces the problem of the animals wanting to explore the new surroundings and thus could take away interest in the opponent. Searching/exploring behavior for both species

of crayfish is likely a major drive. A previous study of cave crayfish showed this was especially true [114, 115]. Therefore, animals might be in an anxious state in the conditions of pairing in this study (new environment), and upon meeting an opponent, they could be hesitant to interact as compared to an intruder invading one's space when an opponent is introduced to a resident's tank.

Studies examining short-term changes in behavior, specifically social interaction outcomes, have shown that physiological changes occur in both learning and the neuroendocrine system. The changes in either of these are associated with effects of experience on the neuroendocrine system of the individuals. Encounter behavior is modified as a result of learning [116–118]. Learning itself is a physiological change in synaptic transmission in specific neuronal pathways. Whether the changes are pre- or postsynaptic is not the issue, but only that physiological changes occur through experience [119]. Neuroendocrine changes such as in corticosteroids and androgens as a relation to fighting strategy has been well studied in vertebrates [120–126]. The relationship between dominance status and corticosteroid levels is less clear since in many cases the hormone levels can correlate positively, negatively, or not all with social rank as there appears to more of a species specific response [126–129]. Serotonin (5-HT) has been associated with aggressive behavior [88, 130–133]. In invertebrates, increased serotonin shows an increase in aggression [134] since infusion of 5-HT in the hemocoel cavity of the crayfish Astacus astacus caused the animal to fight longer in an encounter [6, 135]. It is most likely that after aggressive interactions, further physiological changes are associated with energy metabolism in modifying the neuroendocrine system due to energy depletion and hormonal actions which may even alter synaptic communication [135, 136].

An intrinsic index is more reliable than a visual assessment of the animal's responsiveness and basal status. The autonomic control of the cardiovascular and respiratory systems can regulate the availability of oxygen and other nutrients needed for a behavioral response without causing any outward behavioral change [76]. Due to this fact, observational data alone incorrectly assess environmental factors effecting organisms. Previous work on crayfish showed that visual and/or chemical cues from other crayfish altered HR without any real apparent behavioral changes [9, 75]. Thus, HR has been shown to be a good index in crayfish to use of environmental disturbances to fully understand if the animals can detect a change. Schapker et al. [76] showed that crayfish rapidly alter HR and ventilatory rate (VR) with changes in the environment and that HR and VR indicators were far more sensitive than behavioral data alone [137]. In addition, HR has been used to assess physiological state during copulation in crayfish which gave surprising results with a slowing of HR response in females but not males during copulation [138].

Our assessment of HR showed rapid and prolonged response to social interactions when out of water. This seems to be more prominent for one individual in the dyad since one will become the subordinate. This may also account for a stress response when one individual cannot retreat. Out of water conditions remove typical rapid escape responses; thus, individuals are more likely to sustain injuries during intense interactions. The disruption of social behavior "out of water" was consistently demonstrated with each environmental or physiological modification. There are many distinct experimental advantages to the use of crayfish in behavioral and physiological studies. In particular, the present experiments have shown that crayfish are suited to bring invertebrate studies of environmental effects on behavior and physiology to a level of more complex behavioral phenomena. This also provides a rich and vast foundation to study much broader evolutionary representations among taxa. The present study is built upon a wealth of existing research that has explored the social hierarchies in crayfish.

Acknowledgments

The authors are especially grateful to the undergraduates, Easter Bocook, Becca Liberty, Mary-Catherine Wright, Jessica Simpson, and Jessica McQuerry for their work on this project. Each of them worked very hard and conducted many of the behavioral interactions as well as watching videos to score behavior. Support was provided by a G. Ribble Fellowship for undergraduate studies in the Department of Biology at the University of Kentucky (EB, BL, and JM) and personal funds (RLC). There is no financial gain by the authors in relation to publishing this paper.

References

[1] W. C. Allee and R. H. Masure, "A comparison of maze behavior in paired and isolated shellparrakeets (*Melopsittacus undulatus* Shaw) in a two-alley problem box," *Journal of Comparative Psychology*, vol. 22, no. 1, pp. 131–155, 1936.

[2] T. E. Rowell, "The concept of social dominance," *Behavioral Biology*, vol. 11, no. 2, pp. 131–154, 1974.

[3] C. Drews, "The concept and definition of dominance in animal behavior," *Behavior*, vol. 125, pp. 283–313, 1993.

[4] R. V. Bovbjerg, "Dominance order in the crayfish *Orconectes virilis* (Hagen)," *Physiological Zoology*, vol. 26, pp. 173–178, 1953.

[5] R. V. Bovbjerg, "Some factors affecting aggressive behavior in crayfish," *Physiological Zoology*, vol. 29, pp. 127–136, 1956.

[6] R. Huber, K. Smith, A. Delago, K. Isaksson, and E. A. Kravitz, "Serotonin and aggressive motivation in crustaceans: altering the decision to retreat," *Proceedings of the National Academy of Sciences of the United States of America*, vol. 94, no. 11, pp. 5939–5942, 1997.

[7] F. A. Issa, D. J. Adamson, and D. H. Edwards, "Dominance hierarchy formation in juvenile crayfish *Procambarus clarkii*," *Journal of Experimental Biology*, vol. 202, no. 24, pp. 3497–3506, 1999.

[8] C. Goessmann, C. Hemelrijk, and R. Huber, "The formation and maintenance of crayfish hierarchies: behavioral and self-structuring properties," *Behavioral Ecology and Sociobiology*, vol. 48, no. 6, pp. 418–428, 2000.

[9] H. Li, L. R. Listeman, D. Doshi, and R. L. Cooper, "Heart rate measures in blind cave crayfish during environmental disturbances and social interactions," *Comparative Biochemistry and Physiology*, vol. 127, no. 1, pp. 55–70, 2000.

[10] A. Lomnicki, *Population Ecology of Individuals*, Princeton University Press, Princeton, NJ, USA, 1988.

[11] N. Jiménez-Morales, A. R. Espinoza, K. Mendoza-Angeles, G. Roldán, and J. Hernandez-Falcon, "Memory and social interactions in crayfish," in *Neuroscience Meeting Planner*, Society for Neuroscience, New Orleans, La, USA, 2012.

[12] J. M. Smith, *Evolution and the Theory of Games*, Cambridge University Press, Cambridge, UK, 1982.

[13] C. J. Barnard and R. M. Sibly, "Producers and scroungers: a general model and its application to captive flocks of house sparrows," *Animal Behaviour*, vol. 29, no. 2, pp. 543–550, 1981.

[14] N. E. Collias, "Statistical analysis of factors which make for success in initial encounters between hens," *American Naturalist*, vol. 77, pp. 519–538, 1943.

[15] C. Barrette and D. Vandal, "Social rank, dominance, antler size, and access to food in snow-bound wild woodland caribou," *Behaviour*, vol. 97, no. 1-2, pp. 118–146, 1986.

[16] S. A. Frank, "Hierarchical selection theory and sex ratios I. General solutions for structured populations," *Theoretical Population Biology*, vol. 29, no. 3, pp. 312–342, 1986.

[17] J. Stahl, P. H. Tolsma, M. J. J. E. Loonen, and R. H. Drent, "Subordinates explore but dominants profit: resource competition in high arctic barnacle goose flocks," *Animal Behaviour*, vol. 61, no. 1, pp. 257–264, 2001.

[18] B. Sklepkovych, "The influence of kinship on foraging competition in Siberian jays," *Behavioral Ecology and Sociobiology*, vol. 40, no. 5, pp. 287–296, 1997.

[19] M. H. Figler, H. M. Cheverton, and G. S. Blank, "Shelter competition in juvenile red swamp crayfish (*Procambarus clarkii*): the influences of sex differences, relative size, and prior residence," *Aquaculture*, vol. 178, no. 1-2, pp. 63–75, 1999.

[20] S. Rohwer and P. W. Ewald, "The cost of dominance and advantage of subordination in a badge signaling system," *Evolution*, vol. 35, pp. 441–454, 1981.

[21] A. J. Hansen, "Fighting behavior in bald eagles: a test of game theory," *Ecology*, vol. 67, no. 3, pp. 787–797, 1986.

[22] P. L. Rutherford, D. W. Dunham, and V. Alllison, "Antennule use and agonistic success in the crayfish *Orconectes rusticus* (Girard, 1852) (Decapoda, Cambaridae)," *Crustaceana*, vol. 69, pp. 117–122, 1996.

[23] R. A. Zulandt Schneider, R. W. S. Schneider, and P. A. Moore, "Recognition of dominance status by chemoreception in the red swamp crayfish, *Procambarus clarkii*," *Journal of Chemical Ecology*, vol. 25, no. 4, pp. 781–794, 1999.

[24] C. A. Bruski and D. W. Dunham, "The importance of vision in agonistic communication of the crayfish *Orconectes rusticus*. I. An analysis of bout dynamics," *Behaviour*, vol. 103, no. 1–3, pp. 83–107, 1987.

[25] M. R. Smith and D. W. Dunham, "Chela posture and vision: compensation for sensory deficit in the crayfish *Orconectes propinquus* (Girard) (Decapoda, Cambaridae)," *Crustaceana*, vol. 59, pp. 309–313, 1990.

[26] R. V. Bovbjerg, "Ecological isolation and competitive exclusion in two crayfish (*Orconectes virilis* and *Orconectes immunis*)," *Ecology*, vol. 51, pp. 225–236, 1970.

[27] E. A. Kravitz, S. Glusman, R. M. Harris-Warrick, M. S. Livingstone, T. Schwarz, and M. F. Goy, "Amines and a peptide as neurohormones in lobsters: actions on neuromuscular preparations and preliminary behavioural studies," *Journal of Experimental Biology*, vol. 89, pp. 159–175, 1980.

[28] J. Crane, "Combat, display and ritualization in fiddler crabs (Ocypodidae, Genus Uca)," *Philosophical Transactions of the Royal Society B*, vol. 251, no. 772, pp. 459–472, 1966.

[29] H. O. Wright, "Visual displays in brachyuran crabs: field and laboratory studies," *Integrative and Comparative Biology*, vol. 8, no. 3, pp. 655–665, 1968.

[30] G. W. Hyatt and M. Salmon, "Comparative statistical and information analysis of combat in the fiddler crabs, *Uca pugilator* and *U. pugnax*," *Behavior*, vol. 68, pp. 1–23, 1979.

[31] B. A. Hazlett and W. H. Bossert, "A statistical analysis of the aggressive communications systems of some hermit crabs," *Animal Behaviour*, vol. 13, no. 2–3, pp. 357–373, 1965.

[32] B. A. Hazlett, "Shell fighting and sexual behaviour in the hermit crab genera *Paguristes* and *Calcinus* with comments on *Pagurus*," *Bulletin of Marine Science*, vol. 22, pp. 806–823, 1972.

[33] P. J. Dunham, "Effect of chela white on agonistic success in a Diogenid hermit crab (Calcinus laevimanus)," *Marine Behavior and Physiology*, vol. 5, pp. 137–144, 1978.

[34] P. J. Dunham, "Sex pheromones in Crustacea," *Biological Review*, vol. 53, pp. 555–583, 1978.

[35] D. W. Dunham and A. J. Tierney, "The communicative cost of crypsis in a hermit crab *Pagurus marshi*," *Animal Behaviour*, vol. 31, no. 3, pp. 783–785, 1983.

[36] J. C. E. Scrivener, "Agonistic behavior of the American lobster, *Homarus americanus* (Mime-Edwards)," Fish Research Board Canada Techical Report 235, 1971.

[37] H. Dingle, "A statistical and information analysis of aggressive communication in the mantis shrimp Gonodactylus bredini Manning," *Animal Behaviour*, vol. 17, no. 3, pp. 561–575, 1969.

[38] A. R. Tilden, R. Brauch, R. Ball et al., "Modulatory effects of melatonin on behavior, hemolymph metabolites, and neurotransmitter release in crayfish," *Brain Research*, vol. 992, no. 2, pp. 252–262, 2003.

[39] R. L. Cooper, H. Li, L. Y. Long, J. L. Cole, and H. L. Hopper, "Anatomical comparisons of neural systems in sighted epigean and troglobitic crayfish species," *Journal of Crustacean Biology*, vol. 21, no. 2, pp. 360–374, 2001.

[40] H. Li and R. L. Cooper, "The effect of ambient light on blind cave crayfish: social interactions," *Journal of Crustacean Biology*, vol. 22, no. 2, pp. 449–458, 2002.

[41] T. Eisner and J. Meinwald, *Chemical Ecology*, National Academic Press, Washington, DC, USA, 1995.

[42] J. W. Bradbury and S. L. Vehrencamp, "Animal communication," in *Encyclopædia Britannica*, 2009, http://www.britannica.com/EBchecked/topic/25653/animal-communication.

[43] A. J. Tierney and D. W. Dunham, "Chemical communication in the reproductive isolation of the crayfishes *Orconectes propinquus* and *Orconectes virilis* (Decapoda, Cambaridae)," *Journal of Crustacean Biology*, vol. 2, pp. 544–548, 1982.

[44] R. A. Zulandt Schneider and P. A. Moore, "Urine as a source of conspecific disturbance signals in the crayfish *Procambarus clarkii*," *Journal of Experimental Biology*, vol. 203, no. 4, pp. 765–771, 2000.

[45] T. Breithaupt and J. Petra, "Evidence for the use of urine signals in agonistic interactions of the American lobster," *Biological Bulletin*, vol. 185, pp. 318–323, 2003.

[46] B. A. Hazlett, "'individual' recognition and agonistic behaviour in pagurus bernhardus," *Nature*, vol. 222, no. 5190, pp. 268–269, 1969.

[47] F. Gherardi and J. Atema, "Memory of social partners in hermit crab dominance," *Ethology*, vol. 111, no. 3, pp. 271–285, 2005.

[48] F. Gherardi and J. Tiedemann, "Binary individual recognition in hermit crabs," *Behavioral Ecology and Sociobiology*, vol. 55, no. 6, pp. 524–530, 2004.

[49] M. Vannini and F. Gherardi, "Dominance and individual recognition in *Potamon fluviatile* (Decapoda, Brachyura) possible role of visual cues," *Marine Behavior and Physiology*, vol. 8, pp. 13–20, 1981.

[50] R. L. Caldwell, "Cavity occupation and defensive behaviour in the stomatopod *Gonodactylus festai*: evidence for chemically mediated individual recognition," *Animal Behaviour*, vol. 27, no. 1, pp. 194–201, 1979.

[51] R. L. Caldwell, "A test of individual recognition in the stomatopod *Gonodactylus festate*," *Animal Behaviour*, vol. 33, no. 1, pp. 101–106, 1985.

[52] C. Karavanich and J. Atema, "Individual recognition and memory in lobster dominance," *Animal Behaviour*, vol. 56, no. 6, pp. 1553–1560, 1998.

[53] M. Lowe, "Dominance-subordinance relationships in the crawfish *Cambarellus shufeldtii*," *Tulane Studies Zoology*, vol. 4, pp. 139–170, 1956.

[54] J. L. Hurst, "Urine marking in populations of wild house mice *Mus domesticus* rutty. II. Communication between females," *Animal Behaviour*, vol. 40, no. 2, pp. 223–232, 1990.

[55] J. L. Hurst, "Urine marking in populations of wild house mice *Mus domesticus* rutty. I. Communication between males," *Animal Behaviour*, vol. 40, no. 2, pp. 209–222, 1990.

[56] J. L. Hurst, "Urine marking in populations of wild house mice *Mus domesticus* Rutty. III. Communication between the sexes," *Animal Behaviour*, vol. 40, no. 2, pp. 233–243, 1990.

[57] J. Atema and M. A. Steinbach, "Chemical communication and social behavior of the lobster, *Homarus americanus*, and other Decapod Crustacea," in *Evolutionary Ecology of Social and Sexual Systems: Crustaceans as Model Organisms*, J. E. Duffy and M. Thiel, Eds., pp. 115–144, Oxford University Press, New York, NY, USA, 2007.

[58] C. Ameyaw-Akumfi and B. A. Hazlett, "Sex recognition in the crayfish *Procambarus clarkii*," *Science*, vol. 190, no. 4220, pp. 1225–1226, 1975.

[59] J. H. Thorp and K. S. Ammerman, "Chemical communication and agonism in the crayfish *Procambarus acutus acutus*," *American Midland Naturlist*, vol. 100, pp. 471–474, 1978.

[60] D. V. Devine and J. Atema, "Function of chemoreceptor organs in spatial orientation of the lobster, *Homarus americanus*: differences and overlap," *Biological Bulletin*, vol. 163, no. 1, pp. 144–153, 1982.

[61] C. Karavanich and J. Atema, "Olfactory recognition of urine signals in dominance fights between male lobster, *Homarus americanus*," *Behaviour*, vol. 135, no. 6, pp. 719–730, 1998.

[62] G. M. Capelli and P. A. Hamilton, "Effects of food and shelter on aggressive activity in the crayfish *Orconectes rusticus* (Girard)," *Journal of Crustacean Biology*, vol. 4, pp. 252–260, 1984.

[63] H. V. S. Peeke, J. Sippel, and M. H. Figler, "Prior residence effects in shelter defense in adult signal crayfish (*Pacifastacus leniusculus* (Dana)): results in same- and mixed-sex dyads," *Crustaceana*, vol. 68, no. 7, pp. 873–881, 1995.

[64] R. A. Zulandt Schneider, R. Huber, and P. A. Moore, "Individual and status recognition in the crayfish, *Orconectes rusticus*: the effects of urine release on fight dynamics," *Behaviour*, vol. 138, no. 2, pp. 137–153, 2001.

[65] A. M. Hill and D. M. Lodge, "Replacement of resident crayfishes by an exotic crayfish: the roles of competition and predation," *Ecological Applications*, vol. 9, no. 2, pp. 678–690, 1999.

[66] G. M. Capelli and B. L. Munjal, "Aggressive interactions and resource competition in relation to species displacement among crayfish of the genus orconectes," *Journal of Crustacean Biology*, vol. 2, no. 4, pp. 486–492, 1982.

[67] B. D. Hazlett, D. Rubenstein, and D. Rittschof, "Starvation, energy reserves, and aggression in the crayfish," *Orconectes Virilis (Hagen), Crustaceana*, vol. 28, pp. 11–16, 1975.

[68] A. M. Stocker and R. Huber, "Fighting strategies in crayfish *Orconectes rusticus* (Decapoda, Cambaridae) differ with hunger state and the presence of food cues," *Ethology*, vol. 107, no. 8, pp. 727–736, 2001.

[69] A. T. Gannon, V. G. Demarco, T. Morris, M. G. Wheatly, and Y. H. Kao, "Oxygen uptake, critical oxygen tension, and available oxygen for three species of cave crayfishes," *Journal of Crustacean Biology*, vol. 19, no. 2, pp. 235–243, 1999.

[70] E. A. Caine, "A comparative ecology of epigean and hypogean crayfish (Crustacea: Cambaridae) from northwestern Florida," *American Midland Naturalist*, vol. 99, pp. 315–329, 1978.

[71] P. W. Hochachka, *Living without Oxygen: Closed and Open Systems in Hypoxia Tolerance*, Harvard University Press, Cambridge, Mass, USA, 1980.

[72] K. Hüppop, "The role of metabolism in the evolution of cave animals," *The National Speleological Society Bulletin*, vol. 47, pp. 136–146, 1985.

[73] D. C. Culver, *Cave Life. Evolution and Ecology*, Harvard University Press, Cambridge, Mass, USA, 1982.

[74] H. Dingle and R. L. Caldwell, "The aggressive and territorial behaviour of the mantis shrimp Gonodactylus bredini manning (crustacea: stomatopoda)," *Behaviour*, vol. 33, no. 1, pp. 115–136, 1969.

[75] L. R. Listerman, J. Deskins, H. Bradacs, and R. L. Cooper, "Heart rate within male crayfish: social interactions and effects of 5-HT," *Comparative Biochemistry and Physiology*, vol. 125, no. 2, pp. 251–263, 2000.

[76] H. Schapker, T. Breithaupt, Z. Shuranova, Y. Burmistrov, and R. L. Cooper, "Heart and ventilatory measures in crayfish during environmental disturbances and social interactions,"

Comparative Biochemistry and Physiology, vol. 131, no. 2, pp. 397–407, 2002.

[77] S. M. Bierbower and R. L. Cooper, "Measures of heart and ventilatory rates in freely moving crayfish," *Journal of Visualized Experiments*, vol. 32, article e1594, 2009.

[78] J. L. Wilkens, A. J. Mercier, and J. Evans, "Cardiac and ventilatory responses to stress and to neurohormonal modulators by the shore crab, Carcinus maenas," *Comparative Biochemistry and Physiology C*, vol. 82, no. 2, pp. 337–343, 1985.

[79] C. Ameyaw-Akumfi, "Appeasement displays in cambarid crayfish (Decapoda, Astacoidea)," *Crustaceana*, vol. 5, pp. 135–141, 1979.

[80] R. J. Paxton, P. F. Kukuk, and J. Tengö, "Effects of familiarity and nestmate number on social interactions in two communal bees, *Andrena scotica* and *Panurgus calcaratus* (Hymenoptera, Andrenidae)," *Insectes Sociaux*, vol. 46, no. 2, pp. 109–118, 1999.

[81] L. A. Halling, B. P. Oldroyd, W. Wattanachaiyingcharoen, A. B. Barron, P. Nanork, and S. Wongsiri, "Worker policing in the bee Apis florea," *Behavioral Ecology and Sociobiology*, vol. 49, no. 6, pp. 509–513, 2001.

[82] M. Beye, P. Neumann, M. Chapuisat, P. Pamilo, and R. F. A. Moritz, "Nestmate recognition and the genetic relatedness of nests in the ant Formica pratensis," *Behavioral Ecology and Sociobiology*, vol. 43, no. 1, pp. 67–72, 1998.

[83] E. Nowbahari, R. Feneron, and M. C. Malherbe, "Effect of body size on aggression in the ant, *Cataglyphis niger* (hymenoptera, formicidae)," *Aggressive Behavior*, vol. 25, pp. 369–379, 1999.

[84] W. D. Brown, C. Liautard, and L. Keller, "Sex-ratio dependent execution of queens in polygynous colonies of the ant *Formica exsecta*," *Oecologia*, vol. 134, no. 1, pp. 12–17, 2003.

[85] J. M. Polizzi and B. T. Forschler, "Factors that affect aggression among the worker caste of *Reticulitermes spp.* Subterranean termites (Isoptera: Rhinotermitidae)," *Journal of Insect Behavior*, vol. 12, no. 2, pp. 133–146, 1999.

[86] J. Ruther, S. Sieben, and B. Schricker, "Nestmate recognition in social wasps: manipulation of hydrocarbon profiles induces aggression in the European hornet," *Naturwissenschaften*, vol. 89, no. 3, pp. 111–114, 2002.

[87] B. L. Antonsen and D. H. Paul, "Serotonin and octopamine elicit stereotypical agonistic behaviors in the squat lobster *Munida quadrispina* (Anomura, Galatheidae)," *Journal of Comparative Physiology A*, vol. 181, no. 5, pp. 501–510, 1997.

[88] M. S. Livingstone, R. M. Harris-Warrick, and E. A. Kravitz, "Serotonin and octopamine produce opposite postures in lobsters," *Science*, vol. 208, no. 4439, pp. 76–79, 1980.

[89] H. V. S. Peeke, G. S. Blank, M. H. Figler, and E. S. Chang, "Effects of exogenous serotonin on a motor behavior and shelter competition in juvenile lobsters (*Homarus americanus*)," *Journal of Comparative Physiology A*, vol. 186, no. 6, pp. 575–582, 2000.

[90] S. B. Doernberg, S. I. Cromarty, R. Heinrich, B. S. Beltz, and E. A. Kravitz, "Agonistic behavior in naïve juvenile lobsters depleted of serotonin by 5,7-dihydroxytryptamine," *Journal of Comparative Physiology A*, vol. 187, no. 2, pp. 91–103, 2001.

[91] L. U. Sneddon, A. C. Taylor, F. A. Huntingford, and D. G. Watson, "Agonistic behaviour and biogenic amines in shore crabs Carcinus maenas," *Journal of Experimental Biology*, vol. 203, no. 3, pp. 537–545, 2000.

[92] R. Huber, A. G. Daws, S. B. Tuttle, and J. B. Panksepp, "Quantitative techniques for the study of crustacean aggression," in *The Crustacean Nervous System*, K. Wiese, Ed., pp. 186–203, Springer, Berlin, Germany, 2001.

[93] L. Schroeder and R. Huber, "Fight strategies differ with size and allometric growth of claws in crayfish, *Orconectes rusticus*," *Behaviour*, vol. 138, no. 11-12, pp. 1437–1449, 2001.

[94] J. B. Panksepp and R. Huber, "Chronic alterations in serotonin function: dynamic neurochemical properties in agonistic behavior of the crayfish, *Orconectes rusticus*," *Journal of Neurobiology*, vol. 50, no. 4, pp. 276–290, 2002.

[95] E. A. Kravitz and R. Huber, "Aggression in invertebrates," *Current Opinion in Neurobiology*, vol. 13, no. 6, pp. 736–743, 2003.

[96] D. A. Bergman and P. A. Moore, "Field observations of intraspecific agonistic behavior of two crayfish species, *Orconectes rusticus* and *Orconectes virilis*, in different habitats," *Biological Bulletin*, vol. 205, no. 1, pp. 26–35, 2003.

[97] J. Haller and C. Wittenberger, "Biochemical energetics of hierarchy formation in Betta splendens," *Physiology and Behavior*, vol. 43, no. 4, pp. 447–450, 1988.

[98] J. Haller, "Muscle metabolic changes during the first six hours of cohabitation in pairs of male Betta splendens," *Physiology and Behavior*, vol. 49, no. 6, pp. 1301–1303, 1991.

[99] K. E. Thorpe, A. C. Taylor, and F. A. Huntingford, "How costly is fighting? Physiological effects of sustained exercise and fighting in swimming crabs, Necora puber (L.) (Brachyura, Portunidae)," *Animal Behaviour*, vol. 50, no. 6, pp. 1657–1666, 1995.

[100] J. R. P. Halperin, T. Giri, J. Elliott, and D. W. Dunham, "Consequences of hyper-aggressiveness in Siamese fighting fish: cheaters seldom prospered," *Animal Behaviour*, vol. 55, no. 1, pp. 87–96, 1998.

[101] F. C. Neat, A. C. Taylor, and F. A. Huntinford, "Proximate costs of fighting in male cichlid fish: the role of injuries and energy metabolism," *Animal Behavior*, vol. 55, pp. 875–882, 1998.

[102] S. N. Austad, "A game theoretical interpretation of male combat in the bowl and doily spider (Frontinella pyramitela)," *Animal Behaviour*, vol. 31, no. 1, pp. 59–73, 1983.

[103] B. Gottfried, K. Andrews, and M. Haug, "Breeding robins and nest predators: effect of predator type and defense strategy on initial vocalization patterns," *Wilson Bulletin*, vol. 97, pp. 183–190, 1985.

[104] J. G. M. Robertson, "Male territoriality, fighting and assessment of fighting ability in the Australian frog Uperoleia rugosa," *Animal Behaviour*, vol. 34, no. 3, pp. 763–772, 1986.

[105] M. A. McPeek and P. H. Crowley, "The effects of density and relative size on the aggressive behaviour, movement and feeding of damselfly larvae (Odonata: Coenagrionidae)," *Animal Behaviour*, vol. 35, no. 4, pp. 1051–1061, 1987.

[106] P. H. Crowley, S. Gillett, and J. H. Lawton, "Contests between larval damselflies: empirical steps toward a better ESS model," *Animal Behaviour*, vol. 36, no. 5, pp. 1496–1510, 1988.

[107] R. Huber and E. A. Kravitz, "A quantitative analysis of agonistic behavior in juvenile American lobsters (*Homarus americanus* L.)," *Brain, Behavior and Evolution*, vol. 46, no. 2, pp. 72–83, 1995.

[108] D. I. Rubenstein and B. A. Hazlett, "Examination of the agonistic behaviour of the crayfish Orconectes virilis by character analysis," *Behavior*, vol. 50, no. 3-4, pp. 193–216, 1974.

[109] A. G. Daws, J. Grills, K. Konzen, and P. A. Moore, "Previous experiences alter the outcome of aggressive interactions between males in the crayfish, *Procambarus clarkii*," *Marine and Freshwater Behaviour and Physiology*, vol. 35, no. 3, pp. 139–148, 2002.

[110] R. C. Guiaşu and D. W. Dunham, "Initiation and outcome of agonistic contests in male form I *Cambarus robustus* girard, 1852 crayfish (Decapoda, Cambaridae)," *Crustaceana*, vol. 70, no. 4, pp. 480–496, 1997.

[111] H. V. S. Peeke, M. H. Figler, and E. S. Chang, "Sex differences and prior residence effects in shelter competition in juvenile lobsters, *Homarus americanus* Milne-Edwards," *Journal of Experimental Marine Biology and Ecology*, vol. 229, no. 1, pp. 149–156, 1998.

[112] P. J. Dunham, "Some effects of group housing upon the aggressive behavior of the lobster *Homarus americanus*," *Journal of Fisheries Research Board Canada*, vol. 29, pp. 598–601, 1972.

[113] T. Burk, *An analysis of the social behaviour of crickets [Ph.D. thesis]*, Oxford University, 1979.

[114] S. Kellie, J. Greer, and R. L. Cooper, "Alterations in habituation of the tail flip response in epigean and troglobitic crayfish," *Journal of Experimental Zoology*, vol. 290, no. 2, pp. 163–176, 2001.

[115] Z. Shuranova, Y. Burmistrov, and C. I. Abramson, "Habituation to a novel environment in the crayfish Procambarus cubensis," *Journal of Crustacean Biology*, vol. 25, no. 3, pp. 488–494, 2005.

[116] R. F. Oliveira, P. K. McGregor, and C. Latruffe, "Know thine enemy: fighting fish gather information from observing conspecific interactions," *Proceedings of the Royal Society B*, vol. 265, no. 1401, pp. 1045–1049, 1998.

[117] L. A. Dugatkin, "Bystander effects and the structure of dominance hierarchies," *Behavior Ecology*, vol. 12, pp. 348–352, 2001.

[118] S. M. Bierbower, Z. P. Shuranova, K. Viele, and R. L. Cooper, "Comparative study of environmental factors influencing motor task learning and memory retention in sighted and blind crayfish," *Brain and Behavior*, vol. 3, no. 1, pp. 4–13, 2013.

[119] Y. Hsu, R. L. Earley, and L. L. Wolf, "Modulation of aggressive behaviour by fighting experience: mechanisms and contest outcomes," *Biological Reviews of the Cambridge Philosophical Society*, vol. 81, no. 1, pp. 33–74, 2006.

[120] R. P. Hannes, D. Franck, and F. Liemann, "Effects of rank order fights on whole-body and blood concentrations of androgens and corticosteroids in the male swordtail (Xiphophorus helleri)," *Zeitschrift fur Tierpsychologie*, vol. 65, pp. 53–65, 1984.

[121] K. L. Huhman, T. O. Moore, C. F. Ferris, E. H. Mougey, and J. L. Meyerhoff, "Acute and repeated exposure to social conflict in male golden hamsters: increases in plasma POMC-peptides and cortisol and decreases in plasma testosterone," *Hormones and Behavior*, vol. 25, no. 2, pp. 206–216, 1991.

[122] K. L. Huhman, T. O. Moore, E. H. Mougey, and J. L. Meyerhoff, "Hormonal responses to fighting in hamsters: separation of physical and psychological causes," *Physiology and Behavior*, vol. 51, no. 5, pp. 1083–1086, 1992.

[123] G. W. Schuett, H. J. Harlow, J. D. Rose, E. A. Van Kirk, and W. J. Murdoch, "Levels of plasma corticosterone and testosterone in male copperheads (Agkistrodon contortrix) following staged fights," *Hormones and Behavior*, vol. 30, no. 1, pp. 60–68, 1996.

[124] Y. Sakakura, M. Tagawa, and K. Tsukamoto, "Whole-body cortisol concentrations and ontogeny of aggressive behavior in yellowtail (Seriola quinqueradiata Temminck and Schlegel; Carangidae)," *General and Comparative Endocrinology*, vol. 109, no. 2, pp. 286–292, 1998.

[125] G. W. Schuett and M. S. Grober, "Post-fight levels of plasma lactate and corticosterone in male copperheads, Agkistrodon contortrix (Serpentes, Viperidae): differences between winners and losers," *Physiology and Behavior*, vol. 71, no. 3-4, pp. 335–341, 2000.

[126] O. Øverli, W. J. Korzan, E. Höglund et al., "Stress coping style predicts aggression and social dominance in rainbow trout," *Hormones and Behavior*, vol. 45, no. 4, pp. 235–241, 2004.

[127] K. A. Sloman, K. M. Gilmour, A. C. Taylor, and N. B. Metcalfe, "Physiological effects of dominance hierarchies within groups of brown trout, Salmo trutta, held under simulated natural conditions," *Fish Physiology and Biochemistry*, vol. 22, no. 1, pp. 11–20, 2000.

[128] M. N. Muller and R. W. Wrangham, "Dominance, cortisol and stress in wild chimpanzees (Pan troglodytes schweinfurthii)," *Behavioral Ecology and Sociobiology*, vol. 55, no. 4, pp. 332–340, 2004.

[129] J. Sands and S. Creel, "Social dominance, aggression and faecal glucocorticoid levels in a wild population of wolves, Canis lupus," *Animal Behaviour*, vol. 67, no. 3, pp. 387–396, 2004.

[130] F. Saudou, D. A. Amara, A. Dierich et al., "Enhanced aggressive behavior in mice lacking 5-HT(1B) receptor," *Science*, vol. 265, no. 5180, pp. 1875–1878, 1994.

[131] O. Cases, I. Self, J. Grimsby et al., "Aggressive behavior and altered amounts of brain serotonin and norepinephrine in mice lacking MAOA," *Science*, vol. 268, no. 5218, pp. 1763–1766, 1995.

[132] D. H. Edwards and E. A. Kravitz, "Serotonin, social status and aggression," *Current Opinion in Neurobiology*, vol. 7, no. 6, pp. 812–819, 1997.

[133] W. A. Weiger, "Serotonergic modulation of behaviour: a phylogenetic review," *Biological Reviews*, vol. 72, pp. 61–95, 1997.

[134] R. Huber and A. Delago, "Serotonin alters decisions to withdraw in fighting crayfish, Astacus astacus: the motivational concept revisited," *Journal of Comparative Physiology A*, vol. 182, no. 5, pp. 573–583, 1998.

[135] W. H. Wu and R. L. Cooper, "The regulation and packaging of synaptic vesicles as related to recruitment within glutamatergic synapses," *Neuroscience*, vol. 225, pp. 185–198, 2012.

[136] W. H. Wu and R. L. Cooper, "Role of serotonin in the regulation of synaptic transmission in invertebrate NMJs," *Experimental Neurobiology*, vol. 21, no. 3, pp. 101–112, 2012.

[137] Y.-S. Chung, R. M. Cooper, J. Graff, and R. L. Cooper, "The acute and chronic effect of low temperature on survival, heart rate and neural function in crayfish (*Procambarus clarkii*) and prawn (*Macrobrachium rosenbergii*) species," *Open Journal of Molecular and Integrative Physiology*, vol. 2, pp. 75–86, 2012.

[138] R. M. Cooper, H. Schapker-Finucane, H. Adami, and R. L. Cooper, "Heart and ventilatory measures in crayfish during copulation," *Open Journal of Molecular and Integrative Physiology*, vol. 1, no. 3, pp. 36–42, 2011.

22

Butterfly Species Richness in Selected West Albertine Rift Forests

Patrice Kasangaki,[1] Anne M. Akol,[2] and Gilbert Isabirye Basuta[2]

[1] *National Livestock Resources Research Institute (NaLIRRI), P.O. Box 96, Tororo, Uganda*
[2] *Department of Biological Sciences, Makerere University, P.O. Box 7062, Kampala, Uganda*

Correspondence should be addressed to Patrice Kasangaki, pkasangaki2005@yahoo.com

Academic Editor: Alan Hodgson

The butterfly species richness of 17 forests located in the western arm of the Albertine Rift in Uganda was compared using cluster analysis and principal components analysis (PCA) to assess similarities among the forests. The objective was to compare the butterfly species richness of the forests. A total of 630 butterfly species were collected in 5 main families. The different species fell into 7 ecological groupings with the closed forest group having the most species and the swamp/wetland group with the fewest number of species. Three clusters were obtained. The first cluster had forests characterized by relatively high altitude and low species richness despite the big area in the case of Rwenzori and being close to the supposed Pleistocene refugium. The second cluster had forests far away from the supposed refugium except Kisangi and moderate species richness with small areas, whereas the third cluster had those forests that were more disturbed, high species richness, and low altitudinal levels with big areas.

1. Introduction

Butterflies populate the entire land area of the earth except for the polar regions and the most arid deserts [1]. Each species occupies a definable geographical area, which is known as its area of distribution or, more simply, its range. Some species have ranges that cover very small areas while others have large ranges.

Butterflies also occur as distinct communities, which may be specific not only to geographical subregions but also to disparate ecological conditions [2, 3]. Butterflies are known to respond to environmental changes and there have been considerable amounts of data collected on how particular species contend with alteration in land-use [3, 4]. Because of their sensitivity to environmental conditions, butterflies have also been classified into ecological/functional groups that correspond more accurately to specific habitat conditions. The explicit environmental requirements of many species mean that they can have considerable value as indicators of community or habitat health [3] and may also play a valuable role in ecological monitoring [5].

In Uganda, about 1245 butterfly species have been recorded [6] from a variety of habitats and it is thus feasible to evaluate the butterfly fauna of the region as well as deriving reasonably accurate comparisons of sites and subsequently identify conservation requirements. The forests of the western arm of the Albertine Rift within Uganda are remnants of a once widespread forest ecosystem that has since become highly fragmented. Inspite of the fragmentation, the forests are still significant ecologically with respect to hydrological cycles and species conservation. The forests are under pressure from logging/deforestation and land-use change arising from increasing human populations and other development concerns. The impact of these pressures on these forests needs to be understood so that appropriate conservation requirements can be made.

The main objective of this study was to compare the butterfly species richness in selected forests of the West Albertine Rift within Uganda.

2. Materials and Methods

This study was based on data collected by the former Forest Department in Uganda, now National Forestry Authority (NFA) over a period of three years from January 1993 to December 1995 as part of a National Forestry Biodiversity in seventeen forests (Figure 1).

Twelve fine-mesh cylindrical traps (approximately 70 × 40 cm diameter) were set at a range of heights from 1–10 m above ground level for the duration of the survey. A variety of baits, namely, fermenting banana, dog feces, chicken offals, urine and locally distilled alcohol were used in the traps. Traps were checked regularly and representative specimens of each species collected. For those species not usually attracted to traps, sweep netting was carried out daily in a range of habitat types within the forests (Table 1). The average sampling efforts for each forest were measured in terms of man days. All the collected specimens were put in papers with their wing folded on the back and later identified.

3. Data Analysis

The butterfly species were assessed based on presence or absence of species for the different study forests.

(1) Cluster Analysis was used to determine the levels of similarities among the forests based on the presence or absence of butterfly species. This is a technique that sorts objects (such as sampling units) into groups or clusters based upon their overall resemblance to one another [7]. To establish the similarity among the forests, species presence (=1) or absence (=0) in the 17 forests was scored. These scores provided the basis for cluster analysis. To determine similarity of sites, total species richness × number of forests (17) array was used to calculate percent similarity indices [7]. The percent similarity ranged from near 0 (for a site pair highly dissimilar with respect to butterfly species) to near 1 (for a site pair very similar). An agglomerative clustering technique (weighted centroid) provided in the Multivariate Statistical Package [8] was used to produce a dendrogram containing all 17 forests. A minimum similarity index of 0.0 was used for defining clusters.

(2) Principal component analysis (PCA) was used to relate butterfly species distributions in the respective forests and to enhance the results of the cluster analysis. The data was centred, and two axes were extracted at the "low" ($1E - 4$) level of accuracy. This is an ordination technique [9] which breaks down or partitions a resemblance matrix (variance-covariance or correlation) into a set of orthogonal (perpendicular) axes or PCA "components" [7]. The first few PCA components explain the largest percentage of variation in the data set [10] and ordinations of sampling units on these axes provide information about the ecological relationship between them.

4. Results

A total of 630 different butterfly species belonging to 5 families were recorded for all the 17 forests. All the five major families of butterflies were recorded in all the forests

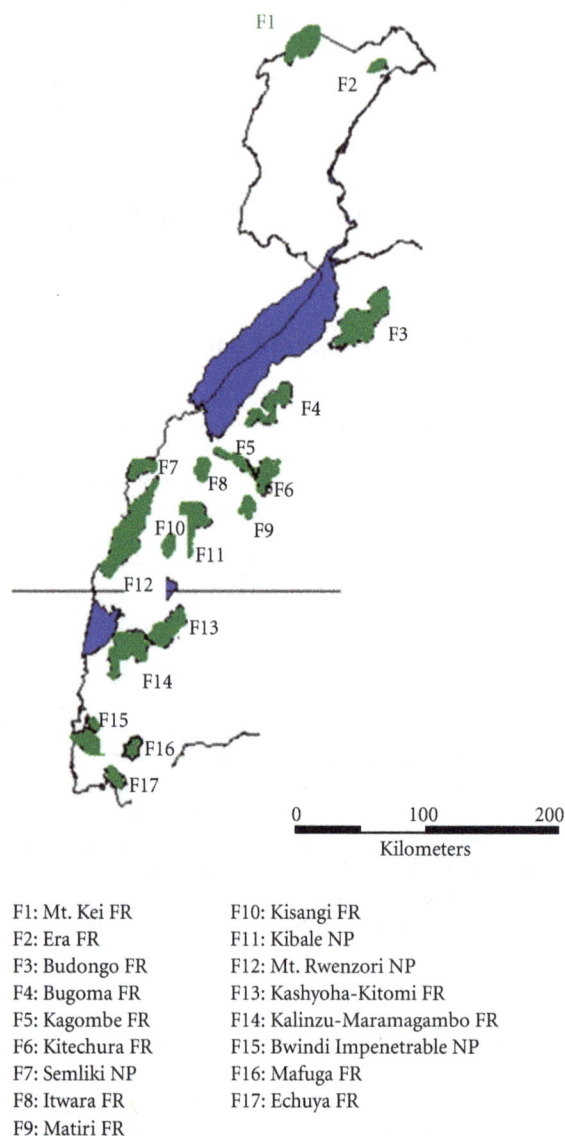

F1: Mt. Kei FR	F10: Kisangi FR
F2: Era FR	F11: Kibale NP
F3: Budongo FR	F12: Mt. Rwenzori NP
F4: Bugoma FR	F13: Kashyoha-Kitomi FR
F5: Kagombe FR	F14: Kalinzu-Maramagambo FR
F6: Kitechura FR	F15: Bwindi Impenetrable NP
F7: Semliki NP	F16: Mafuga FR
F8: Itwara FR	F17: Echuya FR
F9: Matiri FR	

FIGURE 1: Location of the selected study forests in the Western Albertine Rift, Uganda. FR: Forest Reserve; NP: National Park.

except Mafuga in which only four were recorded. The family Nymphalidae had the highest species richness in all the forests followed by Lycaenidae and Hesperiidae. The number of species and subfamilies varied in all the forests. Papilionidae, a small family, had the lowest number of species recorded in a single forest with none in Mafuga (Figure 2).

The highest number of species was recorded in Semliki NP while the lowest number was in Mafuga FR (Table 2). Most forests had few open habitat species except Budongo, Era, and Mt. Kei Forest reserves (Table 2).

High species richness was recorded in forests within which the sampling effort was higher (Figure 3), suggesting that species richness in some forests could have been underestimated due to low sampling effort. For example, Semliki NP was sampled for more days than all the other forests.

TABLE 1: Key characteristics of the study forests.

Forest	Size (km^2)	Altitude (masl)	Location (latitude/longitude)	When sampled	Sampling intensity (man days)	Average sampling effort (man days/km^2)
Kalinzu-Maramagambo (Kal)	854	915−1845	0°17′−0°36′N and 29°47′−30°10′E	November and December 1993; August, October, and December 1994	44	0.08
Itwara (Itw)	87	1220−1510	0°45′−0°52′N and 30°25′−30°32′E	July 1992 and January and February 1993	25	0.29
Bugoma (Bug)	401	990−1295	1°07′−1°25′N and 30°48′−31°07′E	March to April 1993 and July to August 1994	59	0.15
Kisangi (Kis)	54	914−1100	0°17′−0°20′N and 30°14′−30°18′E	July 1993	6	0.11
Budongo (Bud)	793	700−1270	1°37′−2°03′N and 31°22′−31°46′E	August to September 1993 and September to October 1994	53	0.06
Rwenzori (Rwe)	996	1700−5109	0°06′−0°46′N and 29°47′−30°11′E	February, November, and December 1994	30	0.03
Echuya (Ech)	34	2270−2570	1°14′−1°21′S and 29°47′−29°52′E	August 1993, July, November, and December 1994	8	0.24
Mafuga (Maf)	34	2270−2570	1°00′−1°05′S and 29°51′−29°55′E	August 1993, July, November, and December 1994	6	0.16
Kagombe (Kag)	113	1112−1372	0°34′−0°54′N and 30°32′−30°58′E	April and May 1993	22	1.9
Matiri (Mat)	54	1112−1372	0°34′−0°54′N and 30°32′−30°58′E	April and May 1993	12	0.22
Kitechura (Kit)	53	1189−1372	0°34′−0°54′N and 30°32′−30°58′E	April and May 1993	14	0.26
Kasyoha-Kitomi (Kas)	399	975−2136	0°05′−0°25′S and 30°05′−30°20′E	May and June 1993 and September 1994	45	0.12
Semliki (Sem)	219	670−760	0°44′−0°53′N and 29°57′−30°11′E	January to April 1993 and December 1994	98	0.45
Kibale (Kib)	679	1110−1590	0°12′−0°40′N and 30°20′−30°35′E	May and June 1993 and September 1994	35	0.06
Bwindi (Bwi)	231	1190−2607	0°53′−1°08′S and 29°35′−29°50′E	February and September 1994	24	0.07
Mt. Kei (Kei)	384	915−1332	03°34′−03°48′N and 31°00′−31°16′E	July and August 1993; June 1994 and September 1995	53	0.14
Era (Era)	74	850−1040	03°29′−03°36′N and 31°36′−31°46′E	July and August 1993, April and May 1994, and then February 1995	17	0.24

There were 49 widely occurring butterfly species (recorded in at least 10 or more of the 17 forests) with *Danaus chrysippus* being the most common (found in 16 out of 17 forests), *Gnophodes betsimena* and *Ypthima albida* (recorded in 14 out of 17 forests), and *Bicyclus jefferyi*, *Charaxes tiridates*, and *Neptidopsis ophione* (recorded in 13 out of 17 forests). On the other hand, there were 394 rare butterfly species with members recorded in less than 5 forests each. 150 of these were recorded in only one forest each.

There were more closed forest species recorded in forests such as Kalinzu-Maramagambo, Bugoma, Budongo, Kasyoha-Kitomi, Semliki, and Kibale compared to the others

with Semliki having the largest number of the closed forest species and Era, Mafuga and Echuya FRs having the least (Table 2). Edge species were also relatively abundant in all forests and so were the nonspecific habitat species. Swamp/wetland species were very few in all the forests as these were not their characteristic habitats.

Kibale and Bugoma forests had the highest percent similarity index of 63.4 (Figure 4), this was followed by Kasyoha-Kitomi and Kalinzu-Maramagambo and, Semliki and Budongo pairs both with 60.7, while other sites clustered at lower values. Using a minimum index of 0.00 for defining clusters (Figure 4, dashed line), the analysis produced three

TABLE 2: Species richness in the 17 forests represented as ecological groupings.

| Number | Forest | Species richness per habitat | | | | | | | Total |
		F	FH	FL	f	O	S	Ns	
1	Kalinzu-Maramagambo	109	4	11	50	6	3	44	227
2	Itwara	51	1	6	38	3	1	19	119
3	Bugoma	141	2	13	59	9	1	53	278
4	Kisangi	10	0	1	8	2	0	20	41
5	Budongo	109	2	11	45	21	3	63	254
6	Rwenzori	20	14	2	24	3	0	15	78
7	Echuya	6	11	0	15	6	0	16	54
8	Mafuga	4	7	0	8	3	2	8	32
9	Kagombe	86	0	7	41	6	1	49	190
10	Matiri	46	0	4	18	1	2	27	98
11	Kitechura	45	0	5	31	2	2	29	114
12	Kasyoha-Kitomi	119	2	11	47	7	1	48	235
13	Semliki	166	2	17	49	12	1	62	309
14	Kibale	105	1	10	55	3	3	43	220
15	Bwindi	61	19	3	39	6	0	34	162
16	Mt. Kei	12	0	2	19	30	3	60	126
17	Era	3	0	1	2	19	0	31	56

F: Closed forest, f: forest edge/woodland, FH: closed highland forest, O: open habitat, FL: closed lowland forest, S: swamp/wetland species, Ns: nonspecific habitat.

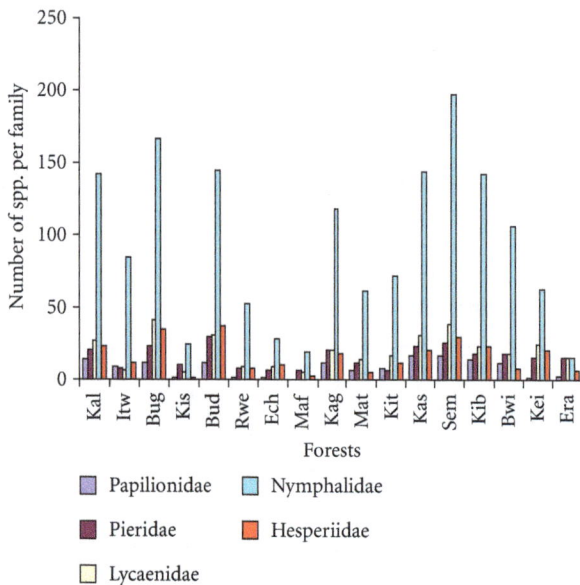

FIGURE 2: Number of butterfly species per family in the 17 West Albertine Rift Forests.Era: Era; Kei: Mt. Kei; Bwi: Bwindi; Kib: Kibale;Sem: Semliki; Kas: Kasyoha-Kitomi; Kit: Kitechura; Mat: Matiri;Kag: Kagombe; Maf: Mafuga;Ech: Echuya; Rwe: Rwenzori; Bud: Budongo; Kis: Kisangi;Bug: Bugoma; Itw: Itwara;Kal: Kalinzu-Maramagambo.

variance in the binary species data. A scatterplot of sites on PC1 and PC2 (Figure 5) suggested results similar to the cluster analysis shown in Figure 4. All forests with relatively high loadings on PC1 and having high scores, such as the Kallinzu-Maramagambo, Itwara, Bugoma, Kagombe, Matiri, Kasyoha-Kitomi, and Kibale are grouped in cluster C in the cluster analysis (Figure 4). On the other hand, forests with high loadings on PC2 and having high scores, such as Rwenzori, Echuya, and Mafuga, are members of cluster A (Figure 4).

5. Discussion

The observed variation in butterfly species richness among the forests (Figure 2) can be attributed to the sampling effort and the physical and environmental factors (forest size, altitude, number of plant species, forest disturbance, rainfall, temperature, and distance from the supposed Pleistocene refugium). This is supported by the works of Wood and Gillman [11], Cleary and Mooers [12], Posa and Sodhi [13] and Clark et al. [14] who separately studied species richness including that of butterflies in forests in relation to plant species richness and disturbance. They found some correlation between the number of butterfly species and forest area. Baz and Garcin-Boyero [15], on the other hand, found that there was no correlation between butterfly species richness and forest area. Our study shows that most of the bigger forests (cluster C forests) have higher species richness compared to those of clusters A and B except for Mt. Rwenzori which is a very high altitude forest.

While the similarity of Kasyoha-Kitomi and Kalinzu-Maramagambo is easy to explain (contiguous and therefore

distinct groups of sites A, B, and C. Observed species richness recorded for the different forest clusters was 104 species in A, 177 species in B, and 561 species in C.

Using principal components analysis, the first two principal components (PC1 and PC2) accounted for 35.2% of

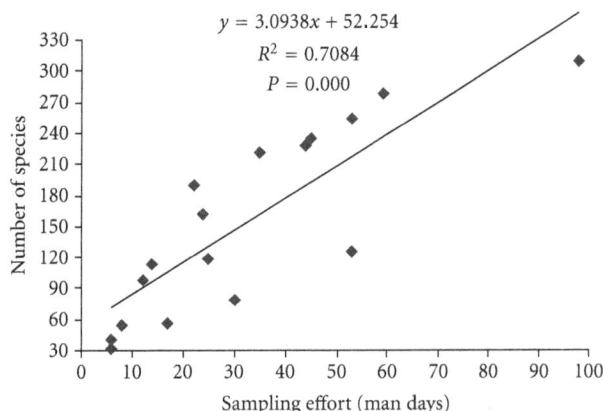

$$y = 3.0938x + 52.254$$
$$R^2 = 0.7084$$
$$P = 0.000$$

FIGURE 3: Sampling effort and number of species.

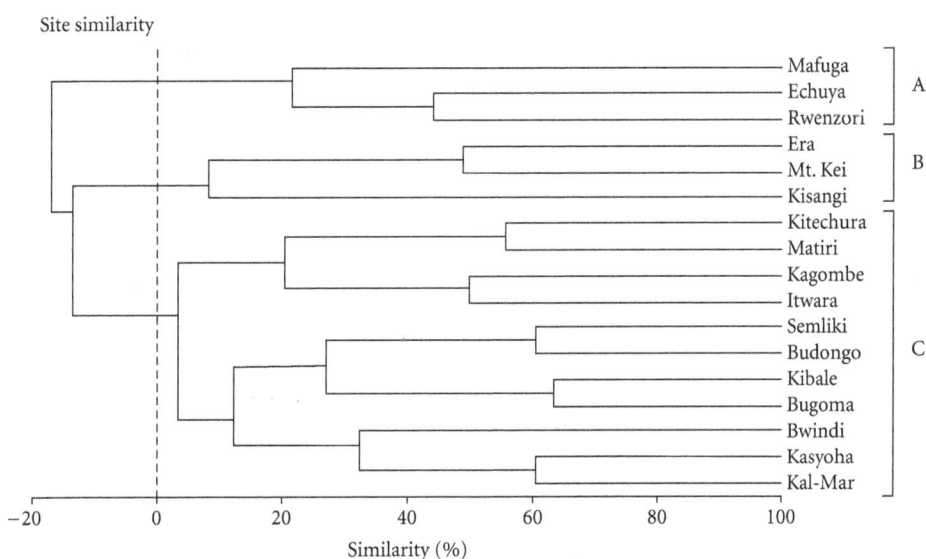

FIGURE 4: Clustering of the study forests based on the presence or absence of butterfly species.

are expected to have similar butterfly fauna), others are difficult to explain. It is possible that Bugoma and Kibale were contiguous in the far past but Semliki is very far from Budongo. Other factors such as isolation could have also influenced the butterfly species richness in the forests [12].

The butterfly species richness per forest indicated that forests (reserves) in the same cluster had comparable species richness (Figures 4 and 5). According to Diamond [16], reserves or habitat patches are considered to be "islands". Islands as ecological systems have such salient features as simple biotas and variability in isolation, shape, and size [17]. According to the theory of island biogeography [18], islands which are close to each other tend to have similar species compared to the isolated ones (this is also true for any other habitat). For example, Kitechura forest reserve which is a small forest ($53 \, km^2$) and is contiguous with Kagombe forest reserve ($113 \, km^2$) had a slightly higher butterfly species richness compared to Mt. Rwenzori NP forest reserve ($996 \, km^2$) which is separated from other forests (Table 1 and Figure 2) and had low butterfly species richness (78 species). Kagombe which is contiguous with Kitechura and Matiri

forest reserves had higher species richness than the other two. This may be attributed to the influence of a big forest nearby (Bugoma) which can also be explained by the MacArthur and Wilson's theory of island biogeography. In addition to the contiguity of forests as a variable accounting for the variation in species richness among the WARF's, the cluster analysis suggests that each of the three groups of forests may have similar physical and environmental attributes (Table 1).

Although Semliki NP appears as a small forest in Uganda, it is part of a very large forest in the eastern DRC, and this reason probably has been responsible for the high number of species in this forest. Also, Semliki NP was sampled more intensely than the other forests (an average sampling effort of 0.45 man days/km^2) as shown in Figure 3. Sampling effort showed a positive correlation with species richness. For example, only 162 butterfly species were obtained from Bwindi NP which was sampled for only 24 days during the survey compared to 181 species obtained from the same forest in 1991 by Omoding [19] who did intense sampling for about six months. On the other hand, Mafuga which is a young secondary forest and Echuya which is a bamboo forest

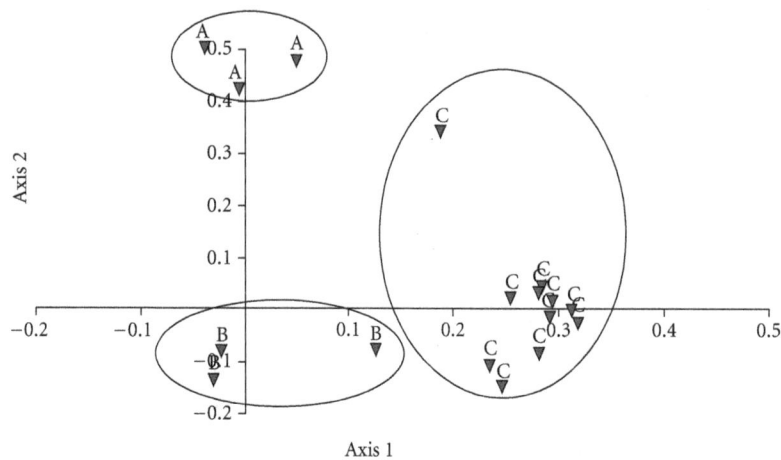

FIGURE 5: A scatter plot showing 17 forests plotted on principal component axes 1 and 2 (PC1 and PC2). Axes are linear combinations of butterfly species.

[20] all had lower species richness. These factors together with the high altitudes of the forests may have contributed to the small number of species obtained despite their close proximity to the supposed refugium. Mt. Rwenzori NP which is the largest forest among the study forests had a small number of species recorded. This could also be attributed to the small sampling effort (average of 0.03 man days/km^2) and then its location at a very high altitude and also the high rainfall associated with this area [2] which may not be favourable to butterflies.

The butterfly species which were most common to all the forests were from the family Nymphalidae. This is expected since the family constitutes a very diverse group [1] and occupies a wide range of habitats with about 440 species recorded in Uganda [4]. Species such as *Danaus chrysippus* and *Gnophodes betsimena* were recorded in almost all the 17 forests.

The butterfly fauna of the West Albertine Rift Forests was dominated by closed forest species (with Semliki recording as high as 166 species) [21]. This is in agreement with a study by Hill et al. [22] on tropical butterfly communities which found that these were diverse communities with many endemic species dependent on closed-canopy forests. This is because closed-canopy forests create microhabitats which are suitable for the butterflies. The present study also revealed that forest edge species were relatively abundant in all the forests in the West Albertine Rift. Waltert et al. [23] in a study on effects of land-use on bird species richness in Indonesia found that the forest edge could play an important role in the conservation of many species, but, although suitable for colonization, its potential to sustain populations over the long term is unknown. This is in agreement with this study which showed that forest edge species were relatively abundant in all the forests in the West Albertine Rift.

Wide range species (nonspecific habitat species) also had many representatives in all the forests. These are mainly generalist species that utilize a wide range of habitats in order to look for food and habitats for laying eggs. Kunte [24] observed that some butterfly species (e.g., *Ypthima* spp.,

family Nymphalidae) showed interesting trends with larval stage being grass feeders and adults feeding on a variety of fruits and nectar. These species are bound to occupy a wide range of habitats. Closed highland forest, closed lowland forest, swamp/wetland species, and open habitat species were generally few in all the forests implying that they are occupied by mainly specialist groups of butterflies.

6. Conclusions

There are more closed forest butterfly species in the WARFs than the swamp/wetland ecological group. This suggests that the WARF's have not changed very much from earlier times except for Era and Mt. Kei forest reserves which are very far from the supposed refugium and Echuya and Mafuga which are newly established forests.

References

[1] T. B. Larsen, *The Butterflies of Kenya and Their Natural History*, Oxford University Press, Oxford, UK, 1991.

[2] P. C. Howard, T. R. B. Davenport, and M. Baltzer, Eds., *Forest Biodiversity Reports*, vol. 1–33, Forest Department, Kampala, Uganda, 1996.

[3] T. R. B. Davenport, *Endemic Butterflies of the Albertine Rift—An Annotated Checklist*, The Wildlife Conservation Society, Mbeya, Tanzania, 2002.

[4] N. Carder and L. Tindimubona, *Butterflies of Uganda: A Field Guide to Butterflies and Silk Moths from the Collection of the Uganda Society*, Uganda Wildlife Society, 2002.

[5] G. C. Daily and P. R. Ehrlich, "Preservation of biodiversity in small rainforest patches: rapid evaluations using butterfly trapping," *Biodiversity and Conservation*, vol. 4, no. 1, pp. 35–55, 1995.

[6] T. R. B. Davenport, *The Butterflies of Uganda—An Annotated Checklist*, Uganda Forest Department, Kampala, Uganda, 2003.

[7] J. A. Ludwig and J. F. Reynolds, *Statistical Ecology: A Primer on Methods and Computing*, John Wiley and Sons, New York, NY, USA, 1988.

[8] W. I. Kovach, *A Multivariate Statistical Package for Windows, Version 3.1*, Kovach Computing Services, Pentraeth, Wales, UK, 1999.

[9] E. C. Pielou, *Interpretation of Ecological Data*, John Wiley and Sons, New York, NY, USA, 1984.

[10] H. G. Gauch, *Multivariate Analyses in Community Ecology*, Cambridge University Press, Cambridge, UK, 1982.

[11] B. Wood and M. P. Gillman, "The effects of disturbance on forest butterflies using two methods of sampling in Trinidad," *Biodiversity and Conservation*, vol. 7, no. 5, pp. 597–616, 1998.

[12] D. F. R. Cleary and A. Mooers, "Butterfly species richness and community composition in forests affected by ENSO-induced burning and habitat isolation in Borneo," *Journal of Tropical Ecology*, vol. 20, no. 4, pp. 359–367, 2004.

[13] M. R. C. Posa and N. S. Sodhi, "Effects of anthropogenic land use on forest birds and butterflies in Subic Bay, Philippines," *Biological Conservation*, vol. 129, no. 2, pp. 256–270, 2006.

[14] P. J. Clark, J. M. Reed, and F. S. Chew, "Effects of urbanization on butterfly species richness, guild structure, and rarity," *Urban Ecosystems*, vol. 10, no. 3, pp. 321–337, 2007.

[15] A. Baz and A. Garcia-Boyero, "The effects of forest fragmentation on butterfly communities in central Spain," *Journal of Biogeography*, vol. 22, no. 1, pp. 129–140, 1995.

[16] J. M. Diamond, "The island dilemma: lessons of modern biogeographic studies for the design of natural reserves," *Biological Conservation*, vol. 7, no. 2, pp. 129–146, 1975.

[17] J. Wu and J. L. Vankat, "Island biogeography: theory and applications," in *Encyclopedia of Environmental Biology*, W. A. Nierenberg, Ed., vol. 2, pp. 371–379, Academic Press, San Diego, Calif, USA, 1995.

[18] R. H. Macarthur and O. E. Wilson, *The Theory of Island Biogeography*, Princenton University Press, Princenton, NY, USA, 1967.

[19] J. Omoding, *Status, distribution and ecology of butterflies in the impenetrable (Bwindi) forest, South-West Uganda [M.S. thesis]*, 1992.

[20] A. C. Hamilton, *Environmental History of East Africa: A Study of the Quaternary*, Academic Press, 1982.

[21] P. Kasangaki, A. M. Akol, and G. Isabirye Basuuta, "Butterfly species list for selected West Albertine Rift Forests," *Dataset Papers in Biology*, vol. 2013, Article ID 451461, 4 pages, 2013.

[22] J. K. Hill, K. C. Hamer, M. M. Dawood, J. Tangah, and V. K. Chey, "Rainfall but not selective logging affect changes in abundance of a tropical forest butterfly in Sabah, Borneo," *Journal of Tropical Ecology*, vol. 19, no. 1, pp. 35–42, 2003.

[23] M. Waltert, A. Mardiastuti, and M. Mühlenberg, "Effects of land use on bird species richness in Sulawesi, Indonesia," *Conservation Biology*, vol. 18, no. 5, pp. 1339–1346, 2004.

[24] K. J. Kunte, "Seasonal patterns in butterfly abundance and species diversity in four tropical habitats in northern Western Ghats," Life Research Foundation, 1997.

Permissions

The contributors of this book come from diverse backgrounds, making this book a truly international effort. This book will bring forth new frontiers with its revolutionizing research information and detailed analysis of the nascent developments around the world.

We would like to thank all the contributing authors for lending their expertise to make the book truly unique. They have played a crucial role in the development of this book. Without their invaluable contributions this book wouldn't have been possible. They have made vital efforts to compile up to date information on the varied aspects of this subject to make this book a valuable addition to the collection of many professionals and students.

This book was conceptualized with the vision of imparting up-to-date information and advanced data in this field. To ensure the same, a matchless editorial board was set up. Every individual on the board went through rigorous rounds of assessment to prove their worth. After which they invested a large part of their time researching and compiling the most relevant data for our readers. Conferences and sessions were held from time to time between the editorial board and the contributing authors to present the data in the most comprehensible form. The editorial team has worked tirelessly to provide valuable and valid information to help people across the globe.

Every chapter published in this book has been scrutinized by our experts. Their significance has been extensively debated. The topics covered herein carry significant findings which will fuel the growth of the discipline. They may even be implemented as practical applications or may be referred to as a beginning point for another development. Chapters in this book were first published by Hindawi Publishing Corporation; hereby published with permission under the Creative Commons Attribution License or equivalent.

The editorial board has been involved in producing this book since its inception. They have spent rigorous hours researching and exploring the diverse topics which have resulted in the successful publishing of this book. They have passed on their knowledge of decades through this book. To expedite this challenging task, the publisher supported the team at every step. A small team of assistant editors was also appointed to further simplify the editing procedure and attain best results for the readers.

Our editorial team has been hand-picked from every corner of the world. Their multi-ethnicity adds dynamic inputs to the discussions which result in innovative outcomes. These outcomes are then further discussed with the researchers and contributors who give their valuable feedback and opinion regarding the same. The feedback is then collaborated with the researches and they are edited in a comprehensive manner to aid the understanding of the subject.

Apart from the editorial board, the designing team has also invested a significant amount of their time in understanding the subject and creating the most relevant covers. They scrutinized every image to scout for the most suitable representation of the subject and create an appropriate cover for the book.

The publishing team has been involved in this book since its early stages. They were actively engaged in every process, be it collecting the data, connecting with the contributors or procuring relevant information. The team has been an ardent support to the editorial, designing and production team. Their endless efforts to recruit the best for this project, has resulted in the accomplishment of this book. They are a veteran in the field of academics and their pool of knowledge is as vast as their experience in printing. Their expertise and guidance has proved useful at every step. Their uncompromising quality standards have made this book an exceptional effort. Their encouragement from time to time has been an inspiration for everyone.

The publisher and the editorial board hope that this book will prove to be a valuable piece of knowledge for researchers, students, practitioners and scholars across the globe.

List of Contributors

M. James C. Crabbe
Institute of Biomedical and Environmental Science Technology and Faculty of Creative Arts, Technologies and Science, University of Bedfordshire, Park Square, Luton LU1 3JU, UK

Andrew K. Davis
Odum School of Ecology, The University of Georgia, Athens, GA 30602, USA

Nathan P. Nibbelink
D.B. Warnell School of Forestry and Natural Resources, The University of Georgia, Athens, GA 30602, USA

Elizabeth Howard
Journey North, 1321 Bragg Hill Road, Norwich, VT 05055, USA

Hrefna Sigurjonsdottir
School of Education, University of Iceland, Stakkahlíð, 105 Reykjavík, Iceland

Anna G. Thorhallsdottir
Bioforsk Ost, Heggenes, 2940 Volbu, Norway
Division of Environmental Sciences, The Agricultural University of Iceland, Hvanneyri, 311 Borgarnes, Iceland

Helga M. Hafthorsdottir
Division of Environmental Sciences, The Agricultural University of Iceland, Hvanneyri, 311 Borgarnes, Iceland

Sandra M. Granquist
Institute of Freshwater Fisheries and The Icelandic Seal Center, Brekkugata 2, 530 Hvammstangi, Iceland

Ken Yoda, Tadashi Tajima and Sachiho Sasaki
Graduate School of Environmental Studies, Nagoya University, Furo-cho, Chikusa-ku, Nagoya 464-8601, Japan

Katsufumi Sato
International Coastal Research Center, Atmosphere and Ocean Research Institute, University of Tokyo, 5-1-5 Kashiwanoha, Kashiwa, Chiba 277-8564, Japan

Yasuaki Niizuma
Faculty of Agriculture, Meijo University, 1-501 Shiogamaguchi, Tenpaku-ku, Nagoya 468-9502, Japan

Paria Parto
Biological Department, Faculty of Science, Razi University, Kermansha 6714967346, Iran
Mina Tadjalli and S. Reza Ghazi
School of Veterinary Medicine, Shiraz University, Shiraz 1731-71345, Iran

Mohammad Ali Salamat
Biological Department, Faculty of Science, Razi University, Kermansha 6714967346, Iran

Jörn Buse
Department of Ecology, Institute of Zoology, Johannes Gutenberg-University Mainz, Becherweg 13, 55099 Mainz, Germany
Ecosystem Analysis, Institute for Environmental Sciences, University of Koblenz-Landau, Fortstrasse 7, 76829 Landau, Germany

Eva Maria Griebeler
Department of Ecology, Institute of Zoology, Johannes Gutenberg-University Mainz, Becherweg 13, 55099 Mainz, Germany

Víctor M. Hernández-Vel ázquez, Laura P. Lina-García and Verónica Obregón-Barboza
Centro de Investigación en Biotecnolog´ıa, Universidad Autónoma del Estado de Morelos, Avenida Universidad No. 1001, Colonia Chamilpa, 62210 Cuernavaca, MOR, Mexico

Adriana G. Trejo-Loyo and Guadalupe Peña-Chora
Centro de Investigaciones Biológicas, Universidad Autónoma del Estado de Morelos, Avenida Universidad No. 1001, Colonia Chamilpa, 62210 Cuernavaca, MOR, Mexico

Jane R. Lloyd
Department of Natural and Social Sciences, University of Gloucestershire, Cheltenham GL50 4AZ, UK

Miguel Á. Maldonado
Centro Ecológico Akumal, Akumal, 77730 Quintana Roo, Mexico

Richard Stafford
Luton Institute of Research in the Applied Natural Sciences, Division of Science, University of Bedfordshire, Luton, LU1 3JU, UK

Heather R. Cunningham and Charles A.Davis
The Natural History Society of Maryland, P.O. Box 18750, Baltimore, MD 21206, USA

Christopher W. Swarth
Jug Bay Wetlands Sanctuary, 1361 Wrighton Road, Lothian, MD 20711, USA

Glenn D. Therres
Wildlife and Heritage Service, Maryland Department of Natural Resources, 580 Taylor Avenue, Annapolis, MD 21401, USA

Christina L. Catlin-Groves
Department of Natural and Social Sciences, University of Gloucestershire, Cheltenham GL50 4AZ, UK

Anna Pérez-Beloborodova and Damir Hernández-Martínez
Conservation Genetic Group, Marine Research Centre, University of Havana, Street 16 No 114, Playa Havana, CP 11300, Cuba

Adriana Artiles-Valor, Lourdes Pérez-Jar and Missael Guerra-Aznay
Molecular Biology Laboratory, Aquaculture Division, Fisheries Research Centre, 5th Avenue and 246, Barlovento, Playa, Havana, CP 19100, Cuba

Georgina Espinosa-López
Department of Biochemistry, Faculty of Biology, Havana University, Street 25 No. 455 between J. and I. Vedado, Havana, Cuba

Ulrika Candolin and Marita Selin
Department of Biosciences, University of Helsinki, P.O. Box 65, 00014 Helsinki, Finland

Silvio F. B. Lima and Martin L. Christoffersen
Departamento de Sistemática e Ecologia, Universidade Federal da Paraíba (UFPB), 58059-900 João Pessoa, PB, Brazil

José C. N. Barros
Laboratório de Malacologia, Departamento de Pesca e Aquicultura, Universidade Federal Rural de Pernambuco (UFRPE), Avenida Dom Manuel de Medeiros S/N, Dois Irm˜aos, 52171-030 Recife, PE, Brazil

Manuella Folly
Departamento de Zoologia, Instituto de Biologia, Centro de Ciências da Sa´ude, Universidade Federal do Rio de Janeiro (UFRJ), Ilha do Fund˜ao, 21941-570 Rio de Janeiro, RJ, Brazil

Monsuru Oladimeji Abioja, Kabir Babatunde Ogundimu, Titilayo Esther Akibo, Kayode Ezekiel Odukoya, Oluwatosin Olawanle Ajiboye, John Adesanya Abiona, Tolulope JuliusWilliams, Emmanuel Oyegunle Oke, and Olusegun Ayodeji Osinowo
Department of Animal Physiology, College of Animal Science and Livestock Production, University of Agriculture, Abeokuta PMB 2240, Nigeria

Takeshi Shirai
Department of Biology, Tokyo Metropolitan University, Minamiosawa 1-1, Hachioji, Tokyo 192-0397, Japan

Louisa K. Higby
School of Ocean Sciences, Bangor University, Menai Bridge, Anglesey LL59 5AB, UK
School of Engineering and Natural Sciences, Faculty of Life and Enviromental Sciences, Elding Whale Watching, Ægisgata 7, 101 Reykjavik, Iceland

Richard Stafford
Division of Science, Institute of Biomedical and Environmental Science and Technology, University of Bedfordshire, Luton LU1 3JU, UK

Chiara G. Bertulli
School of Engineering and Natural Sciences, Faculty of Life and Enviromental Sciences, Elding Whale-Watching, Ægisgata 7, 101 Reykjavik, Iceland
School of Engineering and Natural Sciences, Faculty of Life and Enviromental Sciences, University of Iceland, Sturlugata 7, 101 Reykjavik, Iceland

Claudine Tekounegning Tiogué
The University of Dschang, Faculty of Agronomy and Agricultural Sciences, Laboratory of Applied Ichthyology and Hydrobiology, P.O. Box 222, Dschang, Cameroon

Minette Tabi Eyango Tomedi
The University of Douala, Institute of Fisheries and Aquatic Sciences of Yabassi, P.O.Box 2701, Douala, Cameroon

Joseph Tchoumboué
The University of Mountains, P.O. Box 208, Bangant´e, Cameroon

Marco Ferretti, Gisella Paci and Marco Bagliacca
Department of Animal Production, University of Pisa, Viale delle Piagge 2, 56100 Pisa, Italy

Marco Foi
Department of the Earth Science, University of Milan, Via Mangiagalli 34, 20133 Milan, Italy

Walter Tosi
Geographic Information System Office, Province of Pistoia, Corso Gramsci 110, 51100 Pistoia, Italy

Gerd Mayer, Andreas Maas, and Dieter Waloszek
Workgroup Biosystematic Documentation, University of Ulm, Helmholtzstrasse 20, 89081 Ulm, Germany

Chris K. Elvidge and Grant E. Brown
Department of Biology, Concordia University, 7141 Sherbrooke St. West, Montreal, QC, Canada H4B 1R6

S. M. Bierbower and R. L. Cooper
Department of Biology & Center for Muscle Biology, University of Kentucky, Lexington, KY 40506-0225, USA

J. Nadolski
Department of Mathematical and Computational Sciences, Benedictine University, Lisle, IL 60532, USA

Patrice Kasangaki
National Livestock Resources Research Institute (NaLIRRI), P.O. Box 96, Tororo, Uganda

Anne M. Akol and Gilbert Isabirye Basuta
Department of Biological Sciences, Makerere University, P.O. Box 7062, Kampala, Uganda

www.ingramcontent.com/pod-product-compliance
Lightning Source LLC
Chambersburg PA
CBHW070154240326
41458CB00126B/4839